AREA ARRAY PACKAGING PROCESSES

For BGA, Flip Chip, and CSP

Ken Gilleo, Ph.D. Editor

McGRAW-HILL

New York Chicago San Francisco Lisbon London
Madrid Mexico City Milan New Delhi San Juan
Seoul Singapore Sydney Toronto

The McGraw-Hill Companies

Cataloging-in-Publication Data is on file with the Library of Congress

1 2 3 4 5 6 7 8 9 0 DOC/DOC 0 9 8 7 6 5 4 3

ISBN 0-07-142829-1

The sponsoring editor for this book was Stephen S. Chapman, the editing supervisor was Stephen M. Smith, and the production supervisor was Sherri Souffrance. It was set in Times Roman by Wayne A. Palmer of McGraw-Hill Professional's Hightstown, N.J., composition unit. The art director for the cover was Margaret Webster-Shapiro.

Printed and bound by RR Donnelley.

Portions of this book were previously published in *Area Array Packaging Handbook,* © 2002.

This book is printed on recycled, acid-free paper containing a minimum of 50% recycled, de-inked fiber.

CONTENTS

Preface ix

Chapter 1. Next-Generation Flip Chip Materials and Processing 1

1.0 INTRODUCTION / 1
1.1 CONCEPT / 1
1.1.1 Limits of Conventional FC Processing / 2
1.1.2 Next-Generation FC Technology Overview / 2
1.1.3 Primary Requirements for IC to Package/Substrate Assembly / 5
1.2 COST ANALYSIS / 7
1.2.1 Conventional FC Processing Using Capillary Flow Underfills / 7
1.2.2 Low-Cost, High-Throughput FC Processing Using No-Flow Underfills / 7
1.2.3 Cost Implications of Throughput, Chip Size, and Number of Chips per Panel / 10
1.3 HIGH-THROUGHPUT FC PROCESSING USING NO-FLOW UNDERFILLS / 11
1.3.1 Process Yield Analysis / 14
1.3.2 Yield Analysis Results and Discussion / 18
1.3.3 Placement Voids / 22
1.3.4 Summary of Process Yield Analysis / 22
1.3.5 Reliability Analysis / 23
1.3.6 No-Flow Underfill Reliability Results / 26
1.3.7 Reliability Data Discussion / 33
1.3.8 Summary of No-Flow Underfill Reliability / 35
1.3.9 Failure-Mode Analysis / 35
1.3.10 Summary of Failure-Mode Analysis / 42
1.4 WAFER-LEVEL FC PROCESSING / 42
1.4.1 Desired Coating Characteristics / 47
1.4.2 Coating Methods and Materials / 49
1.4.3 Chip Imaging and Placement / 53
1.4.4 Automated Vision Analysis of Underfill-Coated Chips / 55
1.4.5 Analysis of Underfill-Coated Chip Placement Accuracy / 59
1.4.6 Wafer-Level FC Processing Summary / 61
1.5 ANALYSIS OF NEXT-GENERATION FC PROCESSES / 62
1.5.1 Potential Void Formation / 62
1.5.2 Design Guidelines / 64
1.5.3 Placement Force Characterization of Compression Flow Underfill Processing / 65
1.5.4 Simulation Analysis of Compression Flow Underfill Processing / 67
1.6 COMPRESSION FLOW PLACEMENT MODEL AND ANALYSIS / 77
1.7 MODELING AND ANALYSIS OF CHIP FLOATING / 79
1.8 FC INTERCONNECTION YIELD ANALYSIS DURING THE REFLOW PROCESS / 85
1.8.1 FC Interconnection Yield Analysis / 86
1.8.2 Steady State Forces / 87
1.8.3 Solder Force / 88
1.8.4 Underfill Force / 90

1.8.5 Chip Collapse Time / 91
1.8.6 Underfill Viscosity Effect / 91
1.8.7 Processing Windows for Complete Chip Collapse / 92
1.8.8 Interconnection Density Effect / 93
1.8.9 Bumps in Contact Due to Placement / 93
1.9 SUMMARY / 94
1.10 REFERENCES / 96

Chapter 2. Flip Chip Assembly and Underfilling 99

2.0 INTRODUCTION / 99
2.1 PROCESS OVERVIEW / 100
2.1.1 Capillary Flow Underfills / 100
2.1.2 Fluxing Underfills / 100
2.1.3 Wafer-Applied Underfills / 101
2.2 SUBSTRATE DESIGN / 101
2.2.1 Layout / 101
2.2.2 Copper and Solder Mask Requirements / 107
2.3 ASSEMBLY WITH CAPILLARY FLOW UNDERFILL / 109
2.3.1 Die Presentation / 109
2.3.2 Flux and Flux Application / 110
2.3.3 Pick and Place / 113
2.3.4 Reflow / 114
2.3.5 Substrate Dehydration / 116
2.3.6 Underfill Dispense and Cure / 116
2.3.7 Rework / 121
2.4 ASSEMBLY WITH FLUXING UNDERFILLS / 123
2.4.1 Substrate Dehydration / 124
2.4.2 Application of Fluxing Underfill / 125
2.4.3 Die Placement / 125
2.4.4 Reflow / 126
2.4.5 Rework / 127
2.5 WAFER-APPLIED UNDERFILLS / 128
2.5.1 Application to Wafer / 128
2.5.2 Placement / 130
2.5.3 Reflow / 130
2.5.4 Rework / 130
2.6 RELIABILITY / 130
2.6.1 Component-Level Testing / 131
2.6.2 Environmental and Board-Level Testing / 133
2.7 SUMMARY / 140
2.8 REFERENCES / 141

Chapter 3. BGA and CSP Rework: What Is Involved? 143

3.0 INTRODUCTION / 143
3.1 REWORK PROCESS CONSIDERATIONS / 143
3.1.1 The Need for Process Control / 143
3.1.2 What Is the Interconnection Method? / 144
3.1.3 Do I Use Solder Paste or Flux? / 146
3.2 BASIC REWORK STEPS / 148
3.2.1 Prebaking Components and PWBs / 148
3.2.2 Component Removal / 148
3.2.3 Pad Cleaning and Preparation Methods / 149
3.2.4 Material Deposition Techniques / 150
3.2.5 Component Placement / 154
3.2.6 Component Reflow Profiling / 155
3.3 IDENTIFICATION OF PROCESS YIELD PROBLEMS / 156
3.3.1 Balancing Profile Settings / 156

3.3.2 Problems Caused by Warping / 158
3.3.3 Simultaneous Reflow Issues / 159
3.4 COMPONENT REBALLING CONSIDERATIONS / 161
3.4.1 Reballing Using Solder Preforms / 161
3.4.2 Reballing Using Fixtures / 163
3.5 ANALYSIS USING X-RAY INSPECTION / 164
3.5.1 Solder-Joint Size Variability / 165
3.5.2 Solder-Joint Shape / 165
3.5.3 Voids / 165
3.5.4 Solder Shorts / 166
3.5.5 Open Circuits / 166
3.6 PWB DESIGN GUIDELINES FOR REWORK / 167

Chapter 4. BGA Assembly Reliability 171

4.0 INTRODUCTION / 171
4.1 BALL GRID ARRAY (BGA) / 172
4.2 BGA AND QFP COMPARISON / 174
4.3 RELIABILITY OF BGA ASSEMBLY BY JPL-LED BGA CONSORTIUM / 175
4.4 BGA THERMAL CYCLING / 177
4.5 FAILURE CRITERIA / 180
4.6 CBGA FAILURE MECHANISMS / 181
4.6.1 Damage Monitoring / 181
4.6.2 Failure Mechanisms for Ceramic Assemblies / 181
4.7 PBGA FAILURE MECHANISMS / 182
4.8 PBGA THERMAL CYCLING RESULTS / 184
4.9 CBGA THERMAL CYCLING RESULTS / 185
4.9.1 Fatigue Equation / 186
4.9.2 Projections / 187
4.10 SHOCK AND THERMAL CYCLING SYNERGISM EFFECTS ON RELIABILITY OF CBGAs / 188
4.10.1 Literature Review of Dynamic Behavior of BGAs / 189
4.10.2 Vibration / 189
4.10.3 Natural Frequency Measurement/Projection Test Results / 190
4.10.4 Damage Induced by Thermal Cycling Alone / 190
4.10.5 Damage Induced by a Priori Vibration / 191
4.10.6 Synergism of Vibration and Thermal Cycling / 191
4.11 BGA CONCLUSIONS FOR QUALITY AND RELIABILITY ASSURANCE PROGRAMS / 192
4.12 REFERENCES / 193

Chapter 5. Die Attach and Rework 195

5.0 INTRODUCTION / 195
5.1 POLYMER BASICS / 195
5.1.1 Thermoplastic versus Thermoset Polymers / 196
5.2 DIE-ATTACH MATERIALS / 197
5.2.1 Metallurgic; Solders / 197
5.2.2 Glass Die-Attach Materials / 197
5.2.3 Polymer Die-Attach Adhesives / 197
5.3 DIE-ATTACH APPLICATION METHODS / 198
5.3.1 Needle Dispensing / 198
5.3.2 Stenciling / 198
5.3.3 Screen Printing / 199
5.3.4 Fluid Jetting / 199
5.3.5 Spin Coating on Wafer / 199
5.3.6 Film Application / 199
5.4 APPLICATIONS (USES) / 202
5.4.1 BGAs / 202
5.4.2 Lid Seal / 202
5.4.3 Known Good Dies / 203
5.5 PROPERTIES / 203

5.5.1 Bond Strength / 203
5.5.2 Bonding Temperature/Curing / 204
5.5.3 Outgassing / 204
5.5.4 Electrical Conductivity / 204
5.5.5 Coefficient of Thermal Expansion / 204
5.5.6 Glass Transition Temperature T_g / 204
5.5.7 Thermal Conductivity Properties / 205
5.5.8 Thermal Stability / 205
5.5.9 Modulus / 205
5.5.10 Solubility/Chemical Resistance / 205
5.5.11 Debonding Characteristics / 205
5.5.12 Test Methods / 206
5.6 RELIABILITY / 206
5.6.1 Heat Aging / 206
5.6.2 Thermocycling / 206
5.6.3 Temperature and Humidity (T&H) / 206
5.6.4 Failure Modes and Mechanisms / 207
5.7 REWORK / 207
5.7.1 Thermosets / 207
5.7.2 Thermoplastics / 207
5.8 SUMMARY AND CONCLUSIONS / 208
5.9 REFERENCES / 208

Chapter 6. Liquid Encapsulation Equipment and Processes 209

6.0 INTRODUCTION / 209
6.1 MODERN TRENDS IN PACKAGING / 209
6.1.1 COB / 209
6.1.2 BGAs / 210
6.1.3 Flip Chips (FCs) / 210
6.1.4 Chip Scale Packages (CSPs) / 210
6.2 ENCAPSULATION MATERIALS / 210
6.2.1 Basic Chemistry / 211
6.2.2 Functionality / 211
6.2.3 Performance Characteristics / 211
6.2.4 Handling and Storage / 212
6.3 ENCAPSULATION METHODS / 212
6.3.1 Overmolding / 212
6.3.2 Overmolding Equipment / 212
6.3.3 Needle Dispensing / 213
6.3.4 Needle Dispensing Equipment / 214
6.4 PLATFORM DESIGN AND CONSTRUCTION / 214
6.5 *XY* POSITIONING MECHANISMS / 214
6.5.1 Conventional Positioning Assemblies / 214
6.5.2 *z* Positioning (Vertical Control) / 215
6.5.3 Multigantry or Modular-Style Systems / 215
6.6 DISPENSING HEAD / 217
6.7 NEEDLES / 218
6.8 PUMPS / 218
6.8.1 Pneumatic Pumps / 218
6.8.2 Piston-Type Pumps / 218
6.8.3 Rotary Pumps / 220
6.8.4 Auger Pumps / 220
6.8.5 Fluid Jet Pumps / 220
6.8.6 Overall Fluid-Handling System / 221
6.9 SYSTEM CONTROL / 222
6.9.1 Advanced Process/Software Controls / 222
6.9.2 Human Interface / 223

6.9.3 Computers / 223
6.9.4 Software / 223
6.9.5 System Calibration / 224
6.10 THE DISPENSING PROCESS / 224
6.10.1 Material Constraints and Considerations / 224
6.10.2 Substrates / 224
6.10.3 Applications and Processes / 225
6.10.4 Chip-on-Board (COB) / 225
6.10.5 Dam/Fill / 226
6.10.6 Underfill / 227
6.11 ENCAPSULATING WITH VACUUM ASSIST FOR CSPs AND OTHER PACKAGES / 229
6.12 PERFORMANCE / 231
6.12.1 Positioning Accuracy / 231
6.12.2 Shot Accuracy / 232
6.12.3 Application Rate / 232
6.12.4 Selecting the Right Equipment / 232
6.13 SUMMARY AND CONCLUSIONS / 232
6.14 REFERENCES / 233

Chapter 7. Molding for Area Array Packages 235

7.0 INTRODUCTION / 235
7.1 MOLDING MATERIALS / 235
7.2 TRANSFER-MOLDING EQUIPMENT / 237
7.3 MOLDING PROCESSES / 237
7.3.1 Recommended Transfer-Molding Parameters / 240
7.3.2 Area or Flood Molding / 244
7.3.3 Injection Molding / 244
7.3.4 Molding Underfill (MUF) / 245
7.3.5 Preventing Molding Process Problems / 246
7.4 TROUBLESHOOTING GUIDE / 247
7.4.1 Incomplete Fill and Knit Lines / 248
7.4.2 Voids and Blisters / 249
7.4.3 Flashing and Resin Bleed / 249
7.4.4 Mold Staining and High Molding Temperature / 249
7.4.5 Sticking / 250
7.5 TOOLING / 250
7.6 PRODUCTIVITY AND COST / 250
7.6.1 Volume Dependency / 250
7.6.2 Advantages of Area Molding / 250
7.6.3 Liquid versus Solid Encapsulants / 251
7.7 RELIABILITY / 251
7.8 SUMMARY AND CONCLUSIONS / 251
7.9 FUTURE EXPECTATIONS / 252

Index 253

PREFACE

Area array packaging provides the highest density with maximum performance of any concept yet devised, making it the most important technology in advanced packaging. Older and more familiar perimeter connection–style packages that first appeared about 40 years ago still serve many applications, but designs exploiting package area for connections are now essential for modern electronics, especially high-performance systems and portable products. Area array packages range from the simplest, like small flip chips, to very complex multichip categories. These modern designs can therefore require specialized equipment and unusual processes not found in perimeter package products. This book describes all of the key methods, techniques, processes, and equipment used to fabricate and assemble ball grid arrays (BGAs) and chip scale packages (CSPs) as well as the bumping, attachment, and underfilling of flip chips.

Flip chip technology is extensively covered in several chapters that describe how this seemingly simple component requires advanced equipment and carefully controlled processes. Flip chip bumping methods are compared and include solder paste stenciling, electrolytic plating, maskless electroless nickel plating, mechanical stud bumping, and several conductive adhesive methods. Assembly options range from simple reflow of solder bumps to low-temperature conductive adhesive bonding suitable for temperature-sensitive devices and substrates. Underfill application is analyzed from the process alternatives perspective, and all known methods are compared with cost ramifications. The description starts with conventional capillary flow (CUF) as the reference point. Pre-applied, or no flow (NUF), is examined where material begins as a flux and is transformed into underfill by thermal action. Pros and cons of these two main processes are discussed. Molded underfill (MUF) and wafer-level (WUF) are also described, although these concepts are still emerging.

BGA and CSP manufacturing processes for fabricating various package types are covered in detail along with equipment requirements. The entire sequence from die bonding to solder ball attach is included. BGA assembly is considered from the reliability perspective and also the affects of reworking. Area array assembly challenges include the inability to see the solder joints directly. A complete chapter is devoted to reliability and another to rework methods and equipment.

Discussions of die attach compare thermoplastic and thermoset materials and their very different processes. Although thermoplastic films can provide better attributes and performance, the more specialized equipment requirement is a limitation. However, thermoplastics permit die rework that can be especially valuable for multichip packages. Die attachment and rework processes are explained.

Placement of die in package and flip chip on substrate is covered from the view of equipment accuracy and throughput in a comprehensive chapter. Placement strategies are compared with the rationale for each approach. Subsystems are discussed, and the role that each plays in the final result is shown. They include vision, feeds, mechanical accuracy, and fluxing methods versus throughput.

The topic of package protection is well-covered. Processes and the varied equipment needed are described in several chapters. The liquid dispensing chapter describes the equipment and machine details relating to design, subsystems, pump selection, process alternatives, and machine control. The chapter devoted to encapsulation compares common transfer molding to liquid dispenses that can be used for organic, ceramic, and metal cavity substrates. Cost, productivity, performance, and time-to-market considerations are included. Special CSP methods, like vacuum-assisted filling, are also covered. Finally, transfer molding is discussed and compared to competing processes, especially liquid dispensing. A valuable troubleshooting guide for molding is provided.

Process and equipment topics include other methods useful for area array, but also employed in general electronics manufacturing. Screen printing and stenciling, important for package fabrication and board-level assembly, are thoroughly covered by an expert who takes a step-by-step approach. Topics include screen printing equipment, machine setup, and how to develop the right process. Another chapter describes oven technology and considerations for package production and assembly.

The reader should conclude that every important packaging topic has been included and that a convincing case has been made for the continued success and growth of the area array revolution far into the foreseeable future.

Ken Gilleo, Ph.D.

CHAPTER 1
NEXT-GENERATION FLIP CHIP MATERIALS AND PROCESSING

Daniel F. Baldwin, Ph.D.

1.0 INTRODUCTION

One of the significant developments to improve cost, reliability, and productivity in the electronic packaging industry has been flip chip (FC) technology. This technology has been enhanced further to enable direct assembly to printed wiring boards (PWBs) in high-volume surface-mount assembly called FC-on-board technology. The FC process was first introduced for ceramic substrates as the solid logic technology by IBM in 1962. In 1970, IBM introduced the controlled collapse chip connection (C4 technology) for integrated circuits (ICs) in high-volume FC-in-package (FCIP) production. FC technology is an advanced form of surface mount technology (SMT) in which bare semiconductor chips are turned upside down and hence called flip chips (i.e., active face down) and bonded directly to a PWB or chip carrier substrate. Development of solder-bump FC interconnections was initiated in an attempt to eliminate the expense, low reliability, and low productivity of manual wire bonding of the era. In contrast to wire bonding, which is a peripheral and time-consuming bonding technique in which bonds are formed sequentially, the FC allows all input-outputs (I/Os) to be connected simultaneously. Like its predecessors, FC technology was applied initially to peripheral contacts but quickly progressed to area arrays, which allow for high I/O counts at larger pitches and reduced die size because solder bumps can be put over active device areas on ICs, unlike wire bonds.

1.1 CONCEPT

FC interconnection is the connection of an IC chip to a carrier or substrate with the active face of the chip facing toward the substrate. Interconnection between the chip I/O and substrate is achieved using a bump structure on the chip and a bonding material typically on the substrate forming an electrical interconnection between the chip and the substrate. FC bonding typically involves solder interconnections that make the electrical and mechanical connection between the chip and the carrier although alternate material systems such as conductive adhesives are also used. A schematic representation of an FC interconnection configuration is shown in Fig. 1.1 showing a bumped chip interconnected to a substrate with the active face of the IC toward the substrate surface.

As an example, IBM's C4 FC technology uses high lead solder—usually 93 percent Pb, 7 percent Sn—bumps deposited on solder wettable metal terminals on the active surface of the semiconductor chip that connect to matching wettable substrate pads. During chip assembly, the solder bumps are aligned to the corresponding substrate metal pads and then reflowed at high temperature to simultaneously form electrical and mechanical connections. During reflow, the wetting action of the solder driven by surface tension forces will align the chip's bump pattern to the corresponding substrate pad.

To better understand the FC interconnection system, it is helpful to systematically break down the structural elements of the assembly. One such breakdown is shown in Fig. 1.1. The basic structure of a FC consists of an IC or chip, an interconnection system, and a substrate. The ICs can be made of silicon (the most common), gallium arsenide (GaAs), indium phosphide (IP), silicon germanium (SiGe), etc. The substrate materials could be ceramic (used in IBM's original C4 process), epoxy-glass laminate, polymer thin-film buildup, resin-coated copper (RCC) buildup, glass, silicon, dielectric-coated metal, liquid-crystal polymer, metal-matrix composite, low-temperature cofired ceramic (LTCC), ceramic thick-film, multilayer high-temperature cofired ceramic, etc.

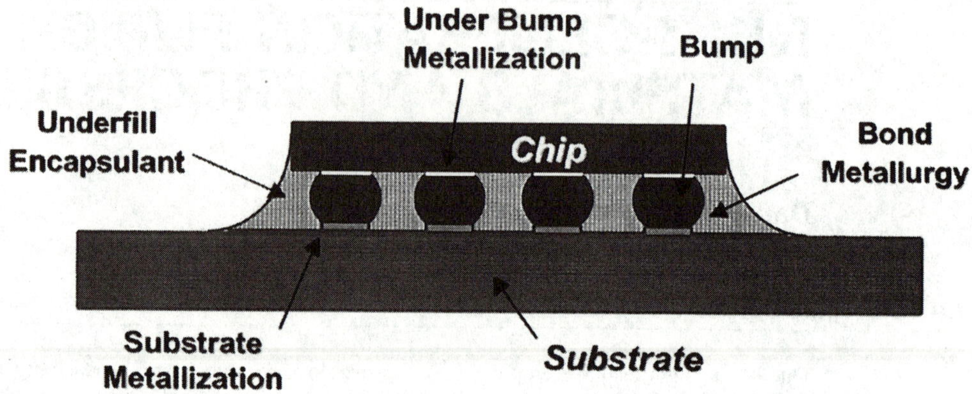

FIGURE 1.1 Schematic of an FC interconnection system.

1.1.1 Limits of Conventional FC Processing

In many cases, the present state of the art FC on-board (FCOB) assembly technology is not capable of achieving the high-throughputs required for integrated high-volume SMT processing and low-cost electronics packaging. Elimination of processing steps enables high-throughput, reduce process complexity, reduce capital equipment requirements, reduce equipment maintenance, and increase process robustness. Processing steps that could be eliminated are flux application, flux residue cleaning, underfill flow, underfill fillet processing, and secondary thermal curing of underfill. Highly populated FC assemblies and large chips compound current FC process limitations. Based on semiconductor and packaging roadmaps, it is clear that trends for larger devices and higher packaging densities will continue, therefore inhibiting cost-effective FC assembly based on conventional FC process techniques. Underfill flow and cure processes dramatically reduce assembly line throughput and therefore exert significant pressure on profit margins. This has been verified by cost modeling and analysis comparing a typical industry FC process, the propposed low-cost, high-throughput process, and surface-mount assembly.[1,16,19,25–30]

The current method for manufacture of FCOB assemblies is time-consuming leading to low throughputs. The processes that increase production time are the flux application process, flux residue cleaning, the underfill dispense process, the underfill fillet dispense process, and underfill cure. The need for underfill derives from the difference in the coefficient of thermal expansion (CTE) between the silicon chip and the organic substrate. The CTE mismatch causes large stresses in the interconnections and eventual failure of the interconnections due to cyclic fatigue induced by thermal cycling. To prevent premature interconnection failure, the space between the chip and the substrate is filled with epoxy, which is then cured, to reduce stress in the interconnection by distributing it over the entire chip area.[5,6,20]

1.1.2 Next-Generation FC Technology Overview

To enable low-cost, high-throughput processing that is compatible with SMT assembly and high-volume electronics packaging, several new processing techniques been developed, including low-cost FC processing based on fast-flow, "snap" cure underfills, high-throughput FC processing using no-flow underfills, and low-cost FC processing based on wafer-applied underfills called wafer-level FC processing.[1,7–11,14,15,19,24–30,36,38–40,42–44]

1.1.2.1 Low-Cost FC Processing Based on Fast-Flow, Snap Cure Underfills. Since underfills are necessary for the reliability of FCOB, key factors to be considered in production are processability (e.g., flow time, cure time, shelf life, flow characteristics) and reliability (e.g., moisture sensitivity, adhesion, thermal mechanical characteristics, corrosion, electromigration). From a processability standpoint, some of the main reasons for incompatibility between FCOB and SMT or FCIP and high-volume electronics packaging are the underfill flow time and the underfill cure time. Conventional underfills can take anywhere from 1 to 10 min to flow under a 5 mm square die while requiring a cure time of 30 to 90 min at temperatures ranging from 125 to 165°C. New commercial fast-flow underfills take between 2 to 30 s to flow under a 5 mm square die. Cure times have also been cut with the introduction of snap-cure underfills. The cure schedules have been reduced to 5 to 10 min at temperatures between 150 and 165°C. A target processing metric for high-throughput FC processing based on fast-flow, snap-cure underfills is a maximum flow time of 5 s on a 5 mm chip and a maximum cure time of 5 min.

The advent of snap-cure underfills has also made possible a new FC assembly process leveraging double-sided board assembly. The new low-cost FC assembly process is shown in Fig. 1.2.[14] The process flow is to print paste for the top-side SMT components, place the SMT components, flux the FC bumps or bond site, place the FCs, reflow the top side forming interconnections, and perform an electrical test. The in-circuit test allows rework of any failed devices prior to dispensing underfill but can be eliminated based on throughput requirements. If there are no defects, the underfill is dispensed on the FC sites, underfill fillets are dispensed if necessary, the board is flipped, the second-side SMT components are processed, and the second-side reflow is used to simultaneously cure the underfill and form the second-side interconnections. The second-side reflow profile is

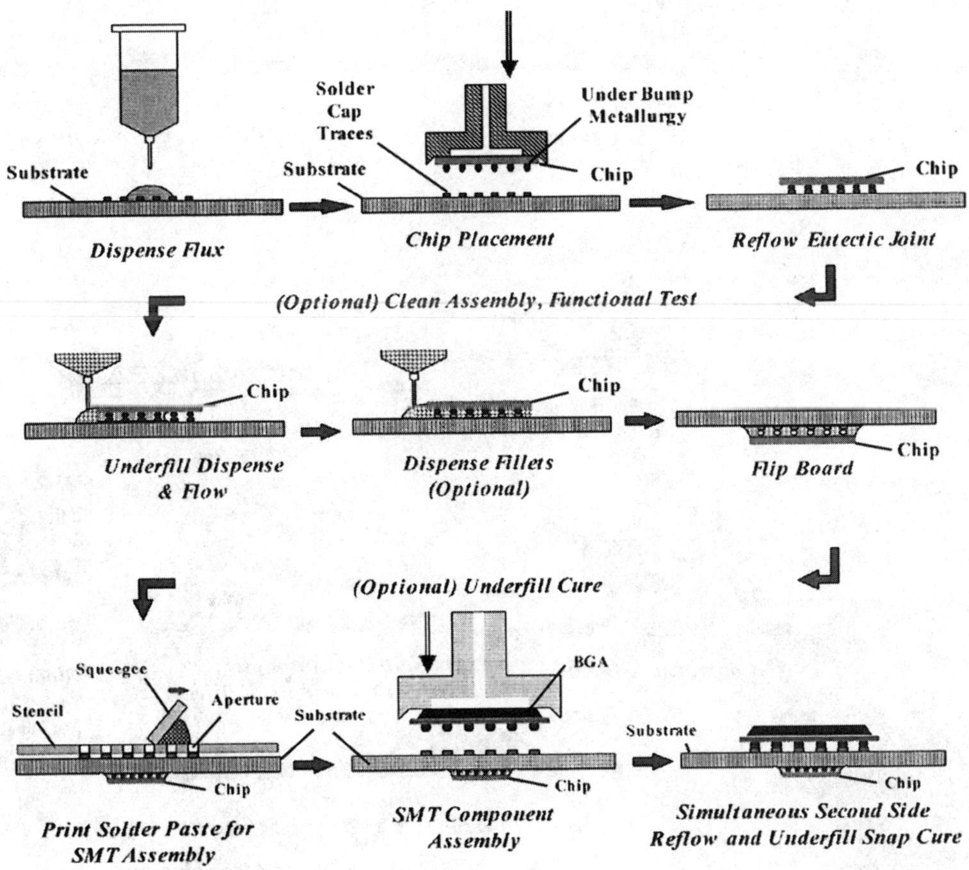

FIGURE 1.2 A fast-flow snap cure underfill process.

slightly modified to accommodate underfill snap-cure in the soak stage of the profile. Typically, this involves increasing the soak time and possibly the soak temperature of a conventional eutectic solder reflow profile. A key element of the process is verifying that the underfills are fully converted during the soak temperature exposure. This is done using differential scanning calorimetry (DSC) analysis to verify underfill conversion. The new FC assembly process is much more conducive to high-throughput, double-sided SMT than conventional FC processes.

1.1.2.2 High-Throughput FC Processing Using No-Flow Underfills. As a concept to achieve high-throughput FCOB and FCIP assembly, a new high-speed FC process has been developed, implementing next-generation FC processing based on large area underfill printing/dispensing, integrated chip-placement and underfill flow, and simultaneous solder interconnection reflow and underfill cure. To enable high-throughput that is compatible with high-speed in-line automated assembly, a new process shown in Fig. 1.3 has been developed.[1,7–10,19,24–30,36,38–40,42–44] Key elements of the

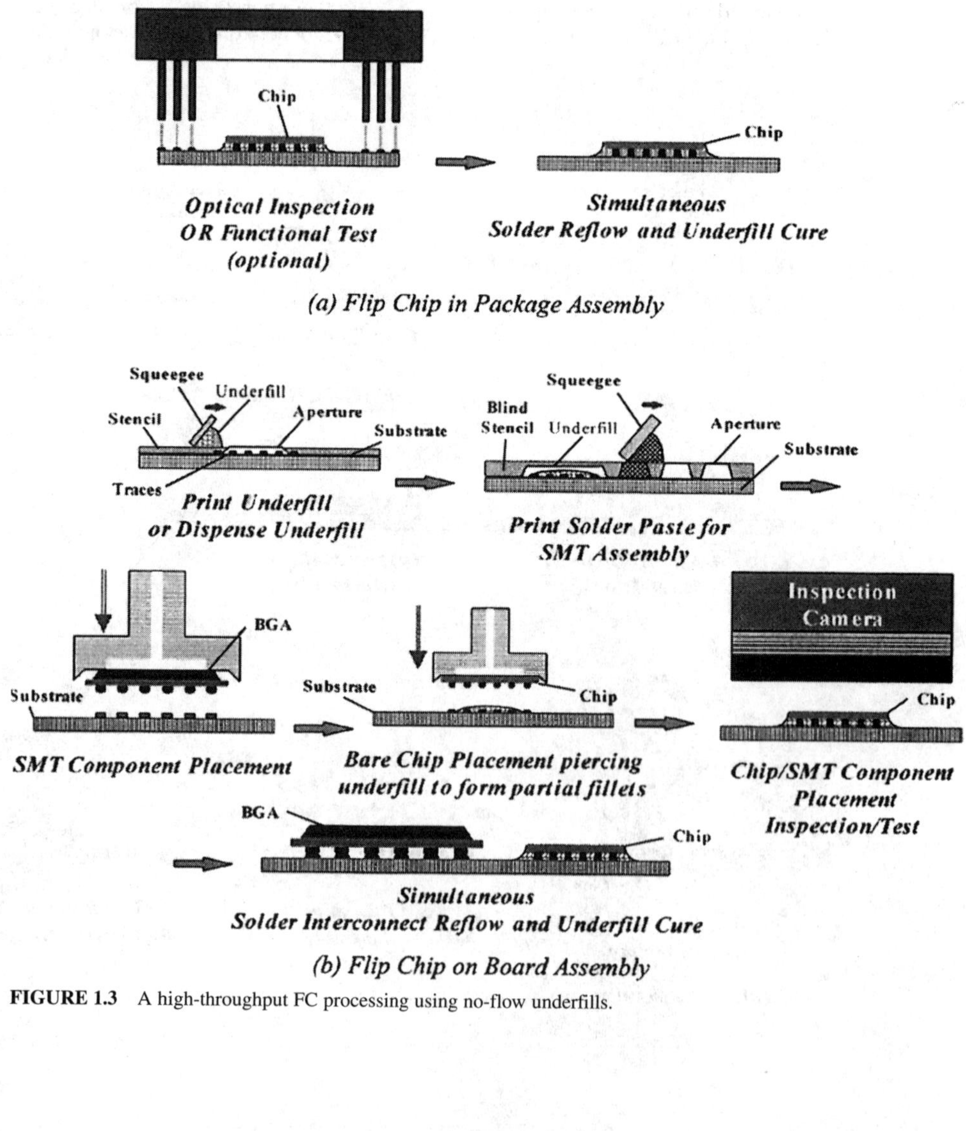

FIGURE 1.3 A high-throughput FC processing using no-flow underfills.

process are elimination of discrete flux application, flux materials management, flux residue cleaning, underfill flow, underfill fillet dispensing, and secondary underfill thermal cure.

The overall process flow begins with a known good substrate and die. A controlled volume of underfill material is stencil printed over the bond pads on the substrate an area-based process compared with conventional point-to-point underfill dispensing. Next, solder paste is printed onto the board using a blind stencil for SMT component assembly. Alternately, the no-flow underfills can be dispensed on the chip site after solder paste printing. Next, the SMT components are placed. The bare chips are then aligned using a vision system to orient the chips relative to bond site fiducials on the substrate. The chips are placed on the substrate, compressing the liquid underfill. Note that the placement forces required are well within those of commercial high-speed FC placement systems and that the process requires no special high-force capability, unlike anisotropic conductive adhesive or compression-bond adhesive processing. A significant advantage here is the elimination of lengthy capillary flow times for underfill processing, particularly with large devices. Finally, the solder interconnections are reflowed simultaneously while curing the polymer encapsulant underfill. Care is taken here to prevent premature gelation of the underfill prior to solder reflow using innovative underfill materials called no-flow underfills that provide latent gelation such that the cure reaction is inhibited until a critical temperature is reached above the solder liquidus temperature. Moreover, it is critical that proper fluxing action be achieved during reflow. The major advantages of the new process are that it increases throughput of FC processing, transforms a chip-based process into an area-based process, reduces the required number of processing steps, and reduces the ratio of wafer to assembly cost by 2 to 5 times over the competing processes.

1.1.2.3 Low-Cost FC Processing Based on Wafer-Applied Underfills: Wafer-Level FC Processing. The key to low-cost and high profit margins in FC assembly is high process throughput. While FC technology has been widely publicized, little attention has been paid to process throughput, whereas the majority of work has concentrated on interconnection technologies. New efforts have focused on the development of process technologies and material systems that could make the FC assembly process transparent to SMT assembly lines and high-volume packaging processes by eliminating FC flux application, underfill application, and underfill cure processes. The innovative process uses reflowable encapsulants similar to no-flow underfills applied to bumped FC wafers forming a wafer scale package and a drop-in solution for low-cost FC assembly.[15]

The new process is unique in that it uses a wafer-level underfill approach, a placement process comparable with chip scale packages (i.e., low placement forces), and underfill systems that enable simultaneous reflow of solder interconnections and cure of the underfill (Fig. 1.4). The basic concept is to apply modified no-flow (reflowable) underfill encapsulants to FC bumped wafers as a form of wafer scale packaging. The wafer-applied underfills dry to a solid film, enabling packing, shipping, handling, and board assembly similar to chip scale packages (CSPs). On reflow, the wafer-applied underfills liquefy, wet the substrate, and provide fluxing action on the solder joints/bond pads, resulting in a high-reliability underfilled FC. The wafer-applied fluxing underfill solid presents several nontrivial challenges, but none of the required properties is mutually exclusive from a materials standpoint.

The new assembly process under development is shown in Fig. 1.5.[11,15] It provides numerous advantages over conventional FC processes because it saves floor space and capital costs by removing underfill dispensers and cure ovens. Moreover, the wafer application of underfill is an enabling technology for fine-pitch FC processing (i.e., $\leq$150 μm). The new process could also increase worldwide FC implementation by reducing the process learning curve and by eliminating special training requirements necessary for current FC dedicated equipment.

1.1.3 Primary Requirements for IC to Package/Substrate Assembly

There are five primary requirements for IC assembly:

1. To provide *acceptable electrical properties,* including capacitance, resistance, and inductance. Each IC assembly has unique electrical characteristics along with mechanical characteristics and manufacturability. For example, wire bonds have long lengths and parallel proximity, resulting in high impedance and longer signal delay times.

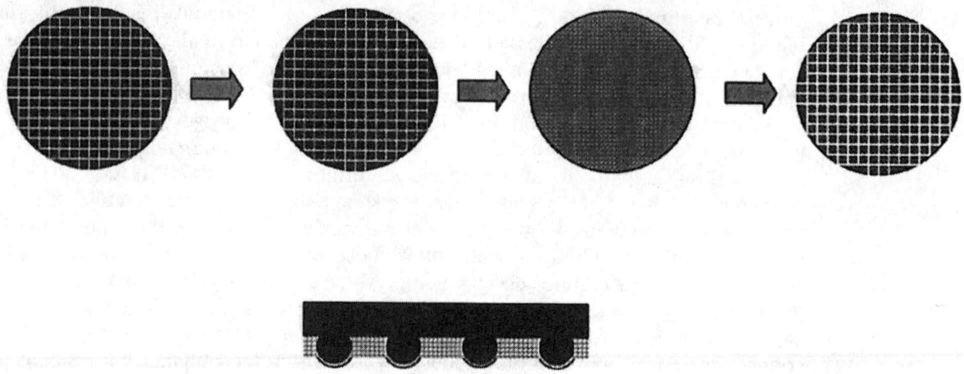

FIGURE 1.4 Wafer-level processing of FC underfills.

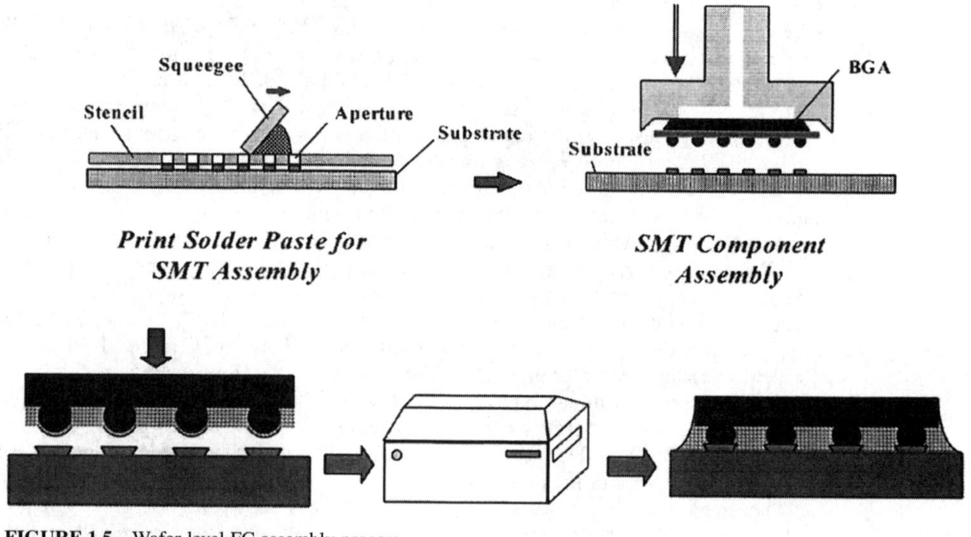

FIGURE 1.5 Wafer-level FC assembly process.

2. IC assembly technologies should provide a *low-cost* solution for providing the electrical interface between the chip and package. Tape automated bonding (TAB) is not typically a low-cost interconnection technique. It requires specialty tooling, equipment, and circuit tapes for production, significantly increasing cost.
3. The third requirement is *high-throughput manufacturing*. For finer pitch packaging applications, TAB provides a low-cycle-time, high-throughput IC assembly approach.
4. The fourth requirement is *high reliability*. FC on-ceramic technology has been a highly reliable interconnection technique. For instance, IBM has used this IC assembly technology for over 40 years with no reported field failures due to thermal cycle fatigue.

5. The final requirement is repairability or replaceability, where the interconnection between the IC and package should provide for removal of a failed IC and replacement with a high-quality new IC or a new part number.

These functional requirements must be met by all IC-to-package/substrate interconnection methods, whether it be wire bonding, TAB, wafer-level packages, or FC interconnection. It is therefore important to consider the degree that each of these functional elements can be met by a particular FC assembly process, particularly in terms of next-generation processes.

1.2 COST ANALYSIS

In order to assess the cost implications of underfill processes and of FC processing in general, activity-based cost modeling has been performed.[1,16,25–27,29,30] Activity-based cost analysis is used to quantify the costs associated with each step of the FC assembly process. Cost model predictions estimate a two to six fold cost reduction over a conventional FC on board assembly process[26,27,29,30] using the new low-cost, high-throughput FC processes of Fig. 1.3. Moreover, the low-cost next-generation FC process also reduces the process cycle time by a factor of two to six. In this case, the conventional process uses capillary flow underfill processing, whereas the low-cost process uses the compression flow underfill process.

1.2.1 Conventional FC Processing Using Capillary Flow Underfills

For the analysis of the conventional FC assembly process, the process flow outlined in Fig. 1.6 is used. A silicon wafer with appropriate solder-wettable metallization is bumped with high-melting-point lead-tin solder, and the bumps are reflowed. The wafer is cleaned and is then diced into individual chips prior to placement. Bond metallurgy made up of a low melting point lead tin solder (typically eutectic) is deposited onto the substrate circuit traces by electroplating or stencil printing. In the case of solder plated substrates, a rosin or no clean flux is deposited onto the chip site, and the chip is placed onto the substrate. The tacky flux holds the aligned chip in position prior to reflow. The assembly is reflowed, and then cleaned to remove any flux residues. Underfill material is dispensed along a single or two adjacent edges of the chip, and is drawn underneath the chip by capillary action. The underfill fillets are applied, and the material is cured in an oven. Modeling of the process was performed for a 5000 panel per day throughput with 30.5 cm (12 in) square substrates populated with 10 1.27 cm (0.5 in) square devices. Cost breakdown results of the baseline process are shown in Fig. 1.7.

Electroplating of the solder (for the interconnection) onto both the wafer (for bumping) and the substrate (for a solder cap) accounts for almost thirty percent of the total assembly cost. These two electroplating steps are expensive due to their limited batch sizes, process time, waste stream environmental clean up, and costly equipment. It was also found that cleaning during FC assembly was expensive due to the numerous cleaning steps involved in the process as well as the limited batch size. The underfilling process accounted for nearly thirty percent of the total assembly cost resulting from the slow processing time, added capital equipment requirements, and material costs.

1.2.2 Low-Cost, High-Throughput FC Processing Using No-Flow Underfills

Cost analysis of the low-cost, high-throughput FC assembly process using the same operating parameters resulted in reduction of the total assembly cost. A breakdown of the costs associated with each process step of Fig. 1.5 is shown in Fig. 1.8. It shows a reduction in process time over previous processes. A cost reduction of approximately 60 percent from the previous benchmark process was achieved as shown in Fig. 1.9. One of the major cost components in the new process was underfill processing, but the overall process cost was reduced. The model showed that the material cost for the underfill step was the major contributor in its expense accounting for approximately ninety percent of the underfill processing costs.

Figure 1.9 shows a comparison of the cost for the conventional FC on board process (specified as the benchmark process), an initial process concept using no-flow underfills, and the low-cost,

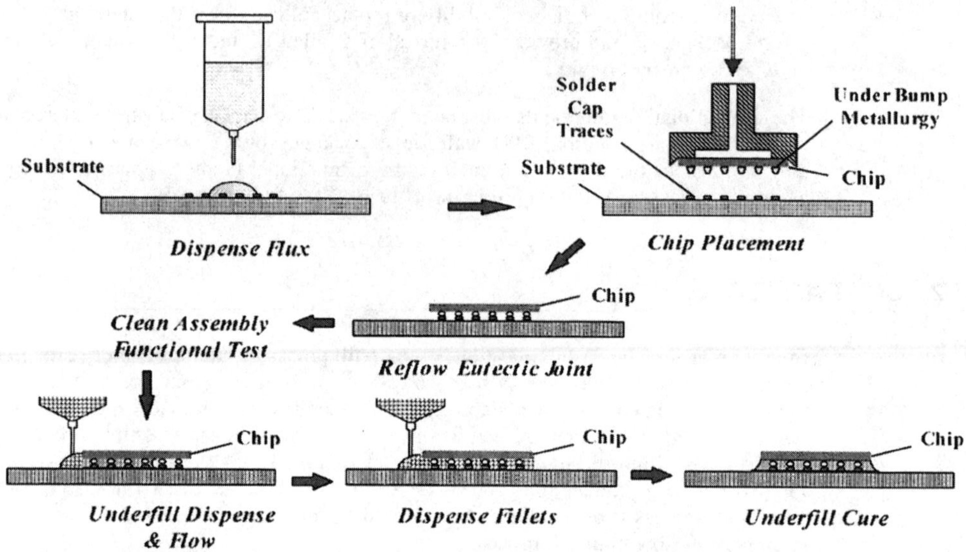

FIGURE 1.6 Conventional FCOB assembly process flow.

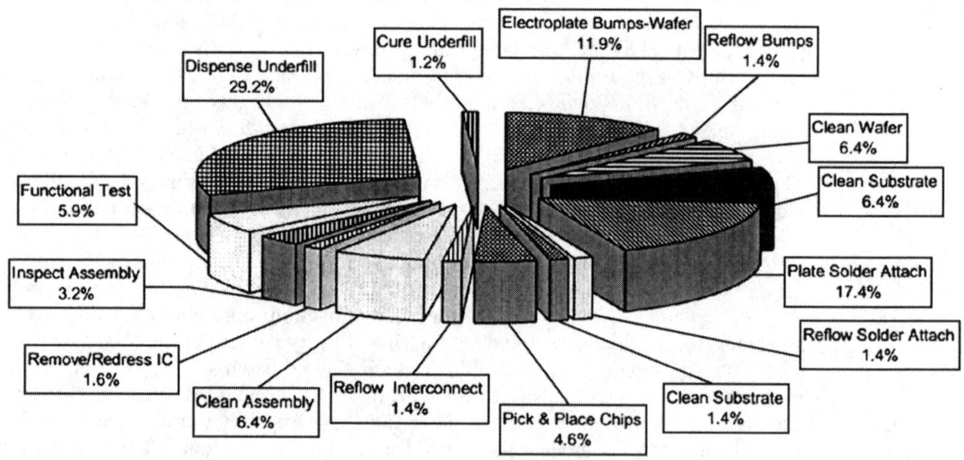

FIGURE 1.7 Breakdown of baseline process costs based on activity based cost model.

high-throughput FC assembly process (specified by the next-generation process). The initial concept process reduced the assembly cost by only twenty percent while the refined process concept reduced cost by approximately sixty percent, representing a significant reduction in assembly process cost. Furthermore, process time has been reduced by over fifty percent as shown in Fig. 1.10. This is mainly due to the reengineering of the process from a point-to-point process to an area-based process. It is important to note that cost and cycle time changed significantly with changes in panel and device sizes.

Cost savings in the low-cost, high-throughput FC assembly process stem from a number of factors. The next-generation process does not require nitrogen inerting during reflow because the encapsulation underfill protects the solder/metallization from oxidation during reflow. Underfill material costs are potentially lower for the high-throughput process in that fillers may not be needed and any

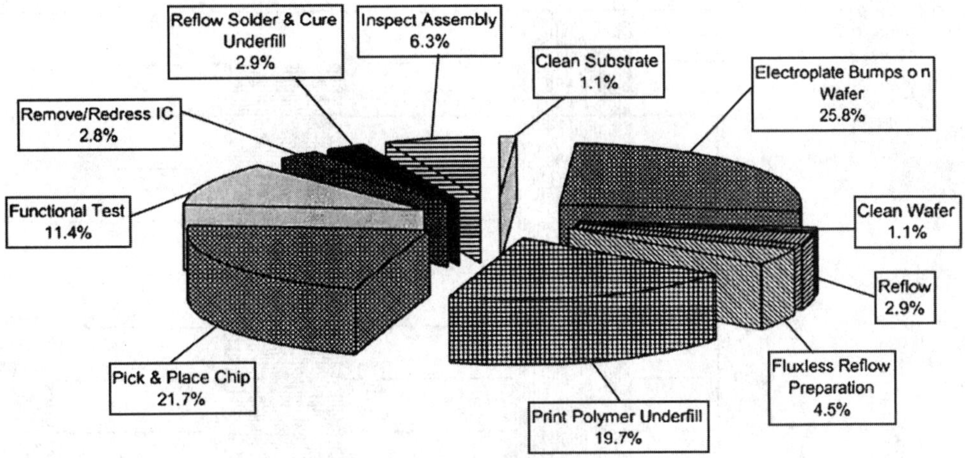

FIGURE 1.8 Breakdown of low-cost high thoughput FC assembly process cost, based on activity-based cost model.

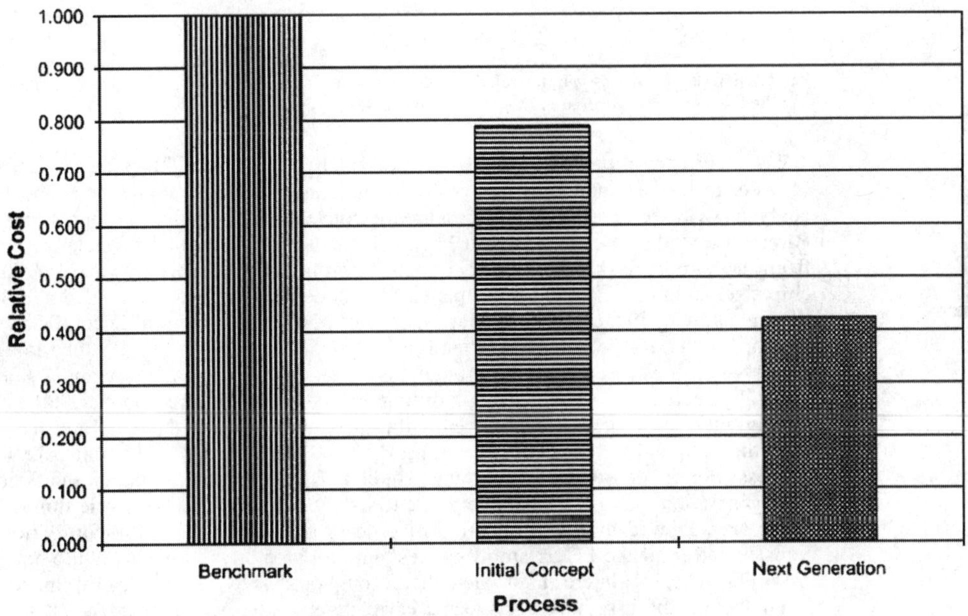

FIGURE 1.9 Comparison of the relative wafer-to-assembly cost for the baseline, initial low-cost concept, and the low-cost next-generation FC assembly processes.

fillers required would not need precision bimodal distributions of spherical particles. Precision filler particles are major cost drivers in conventional capillary flow underfills. Capital equipment and engineering support costs also tend to be lower for the next-generation process due to the reduction of process steps and the ability to use underfill application techniques (stencil printing and dispensing) requiring less precision than for conventional processing. This ultimately reduces the engineering support required for the underfill process.

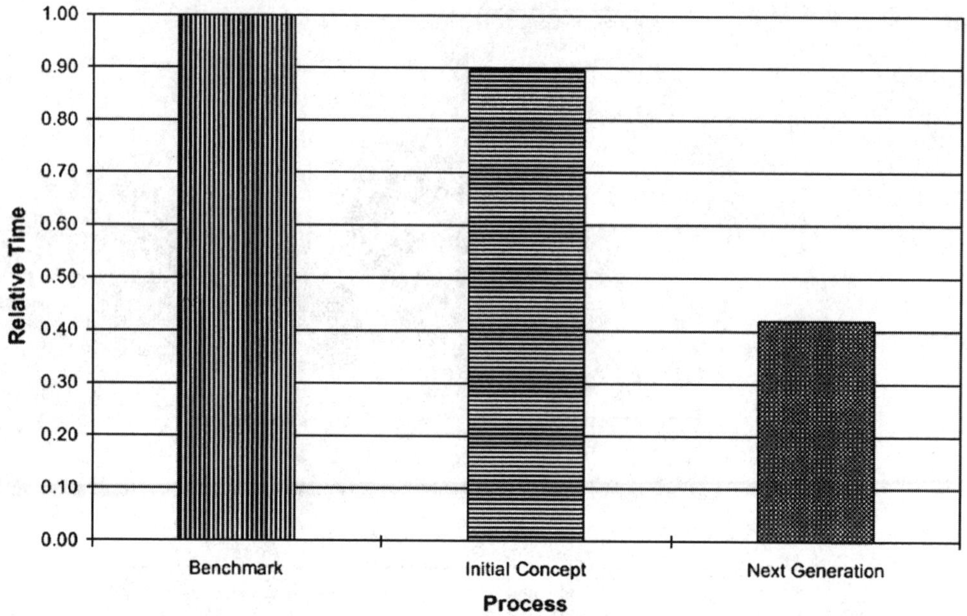

FIGURE 1.10 Comparison of the relative wafer to assembly process cycle times of baseline, initial low-cost concept, and the low-cost, next-generation FC assembly processes.

1.2.3 Cost Implications of Throughput, Chip Size, and Number of Chips per Panel

In order to further understand the cost benefits of the low-cost, high-throughput FC process, cost analysis was used to analyze and compare the conventional and high-throughput processes.[1,25–27,29,30] Parametric studies are presented in Figs. 1.11 and 1.12 illustrating the relative costs of assembly from the bare wafer-level to the final board assembly. The parameters varied included throughput, chip size, and number of chips per panel. The modeling parameters included a 5000-panel-per-day throughput with 30.5 cm (12 in) square substrates populated with ten 1.27 cm (0.5 in) square devices.

The effects of throughput on assembly cost are shown in Fig.1.11. As the throughput increases, the assembly cost decreases. At throughputs below 3000 panels per day, the assembly cost is dramatically decreased with increasing throughput. At throughput levels above 3000 panels per day, relatively little cost benefit is derived with increases in throughput. Moreover, notice that the high-throughput FC process achieves nearly a 60 percent cost reduction over the conventional process over the entire throughput range studied. The substantial reduction in assembly cost for the next-generation FC process is largely due to its characteristically low cycle times. These low cycle times are achieved by the elimination of process steps (relative to the conventional process) and transformation of the FC assembly process into an area array process. While not directly obvious from Fig. 1.11, it is interesting to note that throughput has by far the greatest impact on profitability. Even though the assembly cost decreases relatively little at higher throughputs, relatively small increases in throughput can significantly increase profit margins.

The size of the FC devices used in the large area panel assembly also has a significant effect on assembly cost. Increased device sizes will be characteristic of future device designs and performance demands, as predicted by industry roadmaps. Figure 1.12 shows the results for relative assembly costs with increased device size. The curves for the conventional and high-throughput assembly processes diverge in this plot. Divergence signifies that the next-generation process is very cost effective over the baseline process for increased device size. For smaller chip sizes, a 5 times cost reduction is achieved. As chip size increases, the cost reduction increases to 6 times. Elimination of the time-consuming capillary flow process and transformation of the assembly process from a point-to-point process to an area-based process are the major factors leading to this dramatic cost reduction. In general, printing of the underfill materials enables processing of the entire panel in a single process step.

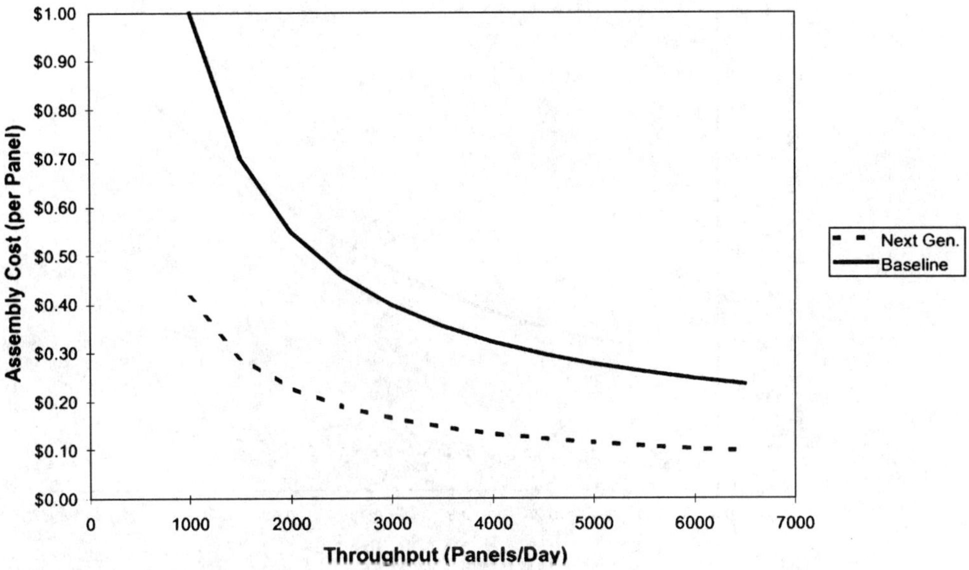

FIGURE 1.11 Effect of throughput on wafer to assembly cost.

In contrast, underfill processing based on dispensing processes the individual chip sites on a point-to-point basis, requiring excessive cycle times. In addition, the time-consuming underfill process is essentially eliminated in the next-generation FC process. This follows because compression flow of the underfill occurs simultaneously with chip placement, a required element of the FC assembly process regardless of the underfill process used. The compression flow underfill process also eliminates the chip size effect on underfill processing time. Typically for standard capillary flow underfill processing, underfill flow times increase with the square of the chip size. For compression flow underfill processing, this size effect is eliminated.

Figure 1.13 shows the effect of die per panel on assembly cost. Again, the curves for the two processes diverge, suggesting that the next-generation process will be more cost effective in meeting future miniaturization requirements. In this plot, cost reduction ranged from 2 to 3.3 times, and will continue to increase as more die are added to the panel. Area-oriented processing makes it so that the number of devices per panel has a minimal effect on cost in the next-generation process. The entire panel is processed in one step. This leads to reduced time and lower cost compared with the baseline's point-to-point processing. In general, increased cost due to additional die per panel can be attributed to the additional material required.

1.3 HIGH-THROUGHPUT FC PROCESSING USING NO-FLOW UNDERFILLS

Underfill processing is a time-consuming step in conventional FC assembly processes due to the flow mechanism used to fill the standoff gap between the chip and substrate. The current technique uses a dispensing method where underfill material is dispensed along one edge (or dual adjacent edges) of the chip and allowed to flow under the chip by capillary action. A schematic of the conventional FC assembly process is shown in Fig. 1.14. Based on first-order principles, the fill time can be estimated using Eq. (1.1) assuming the underfill materials are approximately Newtonian, where L is the length of the chip, h is the standoff height, μ is the viscosity, σ is the surface tension, and α is the wetting angle of the underfill, as shown in Fig. 1.15. A comprehensive analysis of underfill processing techniques is presented below.[1] Equation (1.1), illustrates that the underfill flow

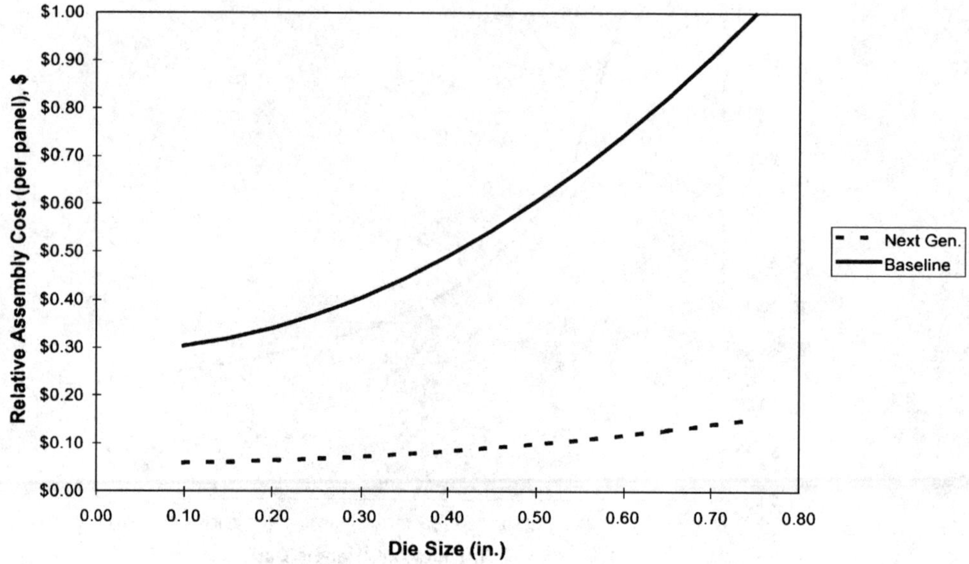

FIGURE 1.12 Effect of die size on relative wafer to assembly cost.

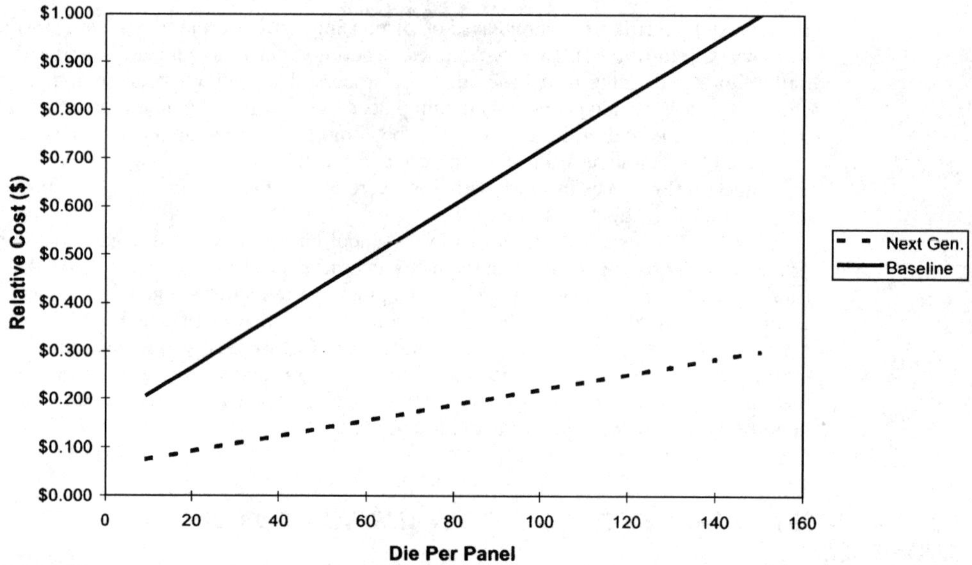

FIGURE 1.13 Effect of number of die per panel on relative wafer to assembly cost.

time for a given material increases with the square of the chip size and the inverse of the standoff height. As chip sizes increase and standoff heights decrease (due to the decreasing pitch requirements), indicated by the semiconductor packaging roadmaps, underfill flow times will only become compounded, inhibiting high-throughput production.[1,13,24–30,34]

$$t_{fill} = \frac{3\mu L^2}{\sigma h \cos(\alpha)} \tag{1.1}$$

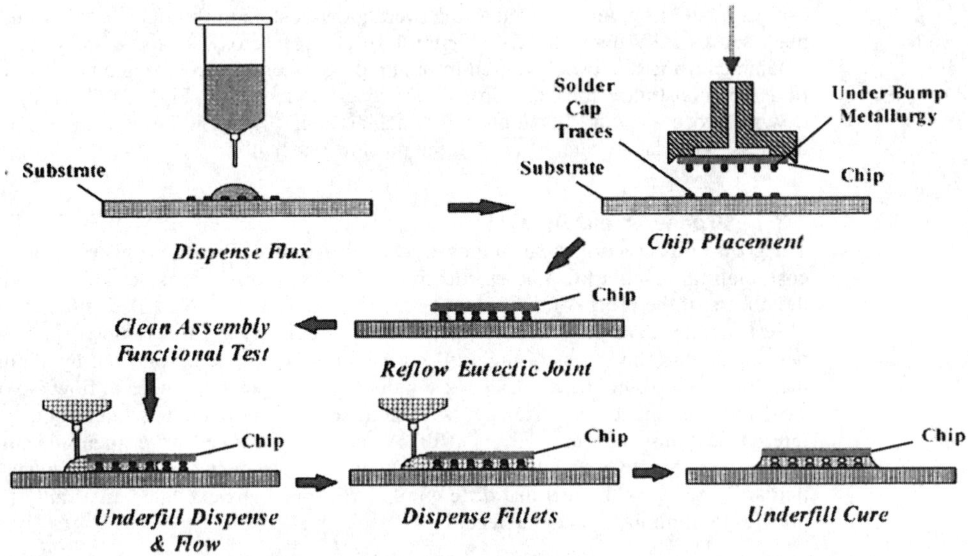

FIGURE 1.14 Conventional FCOB assembly process using capillary flow underfills.

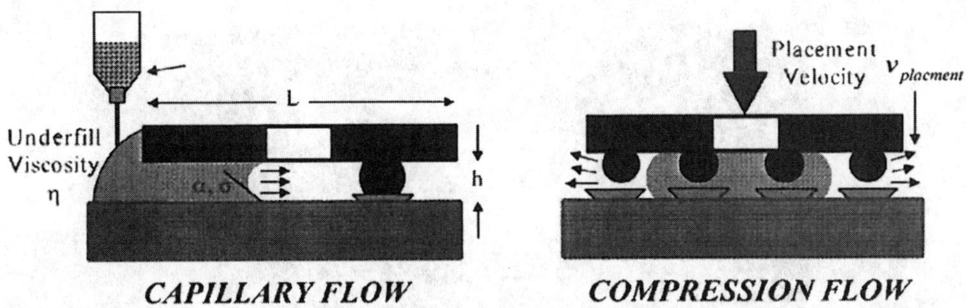

FIGURE 1.15 Underfill flow processing techniques including standard capillary flow and high-throughput compression flow.

With the new low-cost process, the flow time is reduced to the chip placement time, which is given by Eq. (1.2), where h_o is the initial distance of the chip to the substrate, h_f is the final distance of the chip to the substrate once assembled, and $v_{placement}$ is the chip placement velocity[1] (see Fig. 1.15). Equation (1.2) illustrates that in the new process the time required to underfill the chip is no longer dependent on the chip size, chip standoff, and underfill properties. In reality, the underfill flow time is eliminated in that the chip placement time would be incurred regardless of whether the underfill is processed or not.

$$t_{placement} = \frac{(h_o - h_f)}{v_{placement}} \tag{1.2}$$

The new low-cost FC assembly process has several critical requirements. Critical process steps include chip placement and simultaneous solder reflow and underfill cure. With respect to chip placement and reflow, the bumps on the chip must be able to flow through the underfill material, the underfill material must be able to act like a flux when reflowing the solder interconnections, the underfill must maintain a sufficiently low viscosity to enable chip collapse during reflow, and the solder must be able to flow and wet the pad within the same process cycle as the underfill cure. These critical process elements as well as the overall low-cost, high-throughput FC assembly process have been successfully demonstrated. Figure 1.16 shows a cross section of a typical FC test vehicle assembled using the low-cost, high-throughput FC assembly process and no-flow underfill. In addition to demonstrating the feasibility of the process and material, Fig. 1.16 illustrates the ability of the new FC process and no-flow underfill to achieve self-alignment during reflow, indicating the effectiveness of the latent catalyst used in the no-flow underfills.

1.3.1 Process Yield Analysis

The goals of this section are to demonstrate the feasibility of no-flow underfill materials and the low-cost, high-throughput FC process, identify the critical process variables affecting yield, and analyze the effects of the process on different underfill materials.[19,39,40] Reported is the assembly of a series of test vehicles to assess process yield and process defects. The test vehicles were assembled by dispensing a controlled mass of underfill material on the chip site, followed by alignment and placement of the chip onto the substrate pads, inducing a compression underfill flow. Next, the assemblies were reflowed in a commercial reflow oven in an air atmosphere to simultaneously form the solder interconnections and cure the underfill. A series of designed experiments identified the critical process variables to be underfill mass, reflow profile, placement velocity, and underfill type. Of particular interest was the fact that different underfill materials exhibited an affinity for unique reflow profiles to minimize process defects.

FIGURE 1.16 Cross section of a typical FC assembly using the low-cost high-throughput FC assembly process and no-flow underfill.

1.3.1.1 Test Vehicle Description. Six test vehicles were used in various combinations in the interconnection yield analysis experiments. The test die were supplied by FC Technologies and consisted of eutectic lead-tin (37Pb/63Sn) solder-bump FC devices with daisy-chain structures using a proprietary under-bump metallization (UBM) consisting of aluminum-nickel/vanadium-copper. All substrates were designed at Georgia Tech and consisted of high temperature tetrafunctional FR-4 having a glass transition temperature T_g of 170°C with thicknesses ranging from 0.79 to 1.57 mm. All substrate pads had metallizations consisting of copper, electroplated nickel, and immersion gold. Solder mask thicknesses for each substrate configuration ranged from 20 to 30 μm. Table 1.1 presents the various test vehicles and parameters used in the interconnection yield analysis experiments.

1.3.1.2 Experimental Procedures. This study focuses on evaluating process characteristics over a range of FC test vehicles, assessing the critical process characteristics and design parameters affecting yield, and identifying key factors contributing to process defects. A set of five designed experiments[21] is executed to examine the major process parameters affecting yield. These design experiments seek to determine the effects of key process variables and design parameters on assembly yield. Each experiment is designed to systematically investigate each of the following factors on assembly process yield: FC process variables, chip design variables, substrate design factors, and underfill surface geometry factors. The first experiment evaluates the effect of assembly process variables including chip placement force, chip placement velocity, reflow profile, underfill mass, and no-flow underfill material type on assembly yield. Due to the known interaction between chip placement force and chip placement velocity, the second experiment is intended to evaluate the extent of the interaction between these variables based on the hydrodynamic forces produced during assembly and assess their impact on assembly yield. The third experiment is designed to characterize the effects of FC design parameters including bump geometry layout (area and perimeter array), chip size, and solder bump density (number of bumps per unit area) on assembly yield. The fourth experiment investigates the impact of a key substrate design factor, specifically board thickness, on the assembly yield for FC processing based on no-flow underfills. The final experiment is designed to explore the interaction between no-flow underfill surface geometries and the formation of voids during chip placement. In this way, no-flow underfill surface geometries can be controlled and optimized to ensure minimum voiding during FC assembly.

In whole, these experiments provide a characterization of the low-cost, high-throughput FC assembly process using no-flow underfills, specifically providing knowledge of the feasible process windows and understanding of the key factors that control the process yield.

1.3.1.2.1 Process Variable Main Effects Analysis (Experiment 1). The first design experiment is used to determine the main effects of five process variables on the interconnection yield in test vehicles manufactured using the high-throughput FC assembly process. The experiment examines chip placement force, chip placement velocity, reflow profile, mass of underfill deposited, and type of no-flow underfill material. The intent is to examine the impact of these process variables on the number of solder interconnections that successfully reflow and wet the substrate pads. The PB8-4-0.79 test vehicles are used for this set of experiments (see Table 1.1). Two replicates of the experiment

TABLE 1.1 FC Test Vehicles and Associated Design Parameters

Test vehicle ID	Test chip	Chip size* (mm)	Bump pitch (μm)	I/O count	Bump layout	Substrate thickness (mm)	Nickel thickness (μm)	Copper thickness (μm)	Gold thickness (μm)
PB8-2-1.57	PB8	5.08	203	88	Perimeter	1.57	43	3.8	0.15
PB8-4-0.79	PB8	10.2	203	352	Perimeter	0.79	50	3.8	0.15
PB8-4-1.57	PB8	10.2	203	352	Perimeter	1.57	50	3.8	0.15
FA10-2-1.57	FA10	5.08	254	317	Area	1.57	37	4.4	0.17
FA10-2-0.79	FA10	5.08	254	317	Area	0.79	37	4.4	0.17
FA10-4-0.79	FA10	10.2	254	1268	Area	1.57	34	4.3	0.17

*Chip size is the side measurement of the square test chips.

are performed to allow for calculations of error in the assembly process and to validate the results. The chips are placed at four placement velocities, 0.1, 0.2, 0.3, and 0.4 mm/s and two placement forces, 400 and 900 g. Four underfill masses are investigated 15, 22, 28, and 35 mg. Two sources of no-flow underfill materials are investigated: a commercial no-flow underfill, underfill A, and a Georgia Tech developed no-flow underfill, underfill B,[17,18] along with an adhesion-promoted version of underfill B, underfill C.

Two reflow profiles are studied. One profile consists of a typical eutectic solder profile, termed the *step profile,* having a ramp, soak, peak spike, and cool-down. The second profile is a *rapid ramp profile* consisting of a ramp to the peak temperature (i.e., no soak stage) followed by a slow cool-down and finally a rapid cool down. The standard reflow profile, shown in Fig. 1.17, models a reflow profile used for eutectic solder pastes. It begins with a temperature ramp at 3.0°C/s to a soak temperature between 150 and 160°C, where it dwells for 60 s. The soak activates flux in typical solder pastes. Then a second temperature ramp at, 3.0°C/s, to the peak temperature of 220°C reflows the solder. The spike at the end of the profile keeps the board above 183°C for 80 s while the solder reflows and wets the pads. The rapid ramp profile, shown in Fig. 1.18, increases the temperature of the assembly at a constant rate of 1.5°C/sec up to a peak temperature of 215°C. This keeps the assembly above 183°C for 80 s while the solder reflows and wets the pads.

1.3.1.2.2 Interaction of Placement Force and Placement Velocity (Experiment 2). The hydrodynamic force that the underfill exerts on the chip during placement is proportional to the placement velocity. Therefore, as the placement velocity increases, the force the underfill exerts on the chip, and therefore the placement force, must increase in order to drive the chip through the underfill material to make contact with the substrate. This experiment isolates the placement force and velocity factors to determine if the increased hydrodynamic force of the underfill on the chip has an effect on interconnection yield. The rationale is that if the programmed placement force limit is reached prior to the chip bumps making contact with the substrate pads, the chip will be released some distance above the substrate. Hence contact with the substrate pads is not guaranteed, which could adversely affect interconnection yield. The placement velocities investigated are 0.4, 2.7, and 5.0 mm/s. The placement forces investigated are 400, 1000, and 1600 g. By design, the experiment helps to determine if increasing the programmed placement force allows the chip to penetrate the underfill while traveling at high velocities. The PB8-4-0.79 test vehicles are used for this set of experiments (see Table 1.1). Three replications of the experiments are run to allow for a calculation of error.

1.3.1.2.3 Chip Parameters Main Effects Analysis (Experiment 3). The goal of this experiment is to determine the effect of the chip parameters, such as size and bump pattern, in combination with placement force and velocity on interconnection yield. Four factors are investigated including chip size, bump geometry, placement force, and placement velocity. Two chip sizes, 5.08 by 5.08 mm and 10.16 by 10.16 mm, are used. The bump patterns investigated are perimeter and full area array. The placement velocities studied are 0.1 and 5 mm/s, and the placement forces were 400 and 1600 g. Four types of no-flow underfill materials are investigated: a Georgia Tech developed no-flow underfill, underfill B, an adhesion promoted version of underfill B, underfill C, a 20 percent silica filled version of underfill B, underfill D; and a 40 percent silica-filled version of underfill B, underfill E. The PB8-2-1.57, PB8-4-0.79, FA10-2-1.57, and FA10-4-1.57 test vehicles are used for this set of experiments (see Table 1.1).

1.3.1.2.4 Board Thickness Analysis (Experiment 4). To determine the impact of board thickness on interconnection yield, a simple test is performed. This experiment uses a single FC test vehicle design with two FR-4 substrate thicknesses of 0.79 to 1.57 mm. The primary metrics used for analysis include transmission x-ray microscopy and solder-joint interconnection yield verified by electrical continuity measurements. The FA10-2-0.79 and FA10-2-1.57 test vehicles are used for this set of experiments (see Table 1.1).

1.3.1.2.5 Placement Void Analysis (Experiment 5). Early experiments demonstrating the low-cost, high-throughput FC process[7] exhibited voiding in the underfill attributed to three factors: outgasing of the substrate and chip passivation during reflow, outgasing of the no-flow underfill during reflow, and capture voids formed during placement. It is theorized that as a chip is placed onto a substrate, it is possible for air to become entrapped between the chip and the underfill material on the board. A possible cause of the voids would be the capture of air pockets or bubbles at the chip-underfill interface due to depressions in the underfill surface and/or multiple points of contact between the chip surface and underfill. In support of this theory, acoustic microscopy images of FC test vehicles used to demonstrate process feasibility show large voids in the underfill at the center of die (these initial feasibility studies were performed by McGovern and Baldwin[19]).

To examine whether air is entrapped in the underfill during chip placement, a designed experiment is used to determine if the phenomenon occurs and what factors influence it. Four factors identified for possible interaction are studied to determine their impact on void formation: placement velocity, placement force, underfill viscosity, and bump pattern. Void analysis is performed using a Sonoscan D6000 Scanning Acoustic Microscope after chip placement. Calculation of the percentage voiding is performed using a commercial image analysis software package. The PB8-2-1.57 and FA10-2-1.57 test vehicles are used for this set of experiments (see Table 1.1).

1.3.1.3 Assembly Process. The test vehicle assemblies are produced using a consistent process from experiment to experiment adjusting the process parameters as stated in the experiments section. All substrates are baked for 2 h at 100°C to remove dissolved solvents and moisture. A fixed mass of underfill is dispensed by syringe onto the substrate bond site. Accuracy of the dispensed underfill mass is determined by weighing the samples in a precision balance. The substrate is then placed in a Kulicke and Soffa 6900 FC bonder. The bonder picks the chip, visions the corner bump array, aligns it to the substrate pads based on fiducials, and places it using programmed parameters, including placement velocity, placement force, and bond time. The assembly is then reflowed using a BTU Paragon 98N convection reflow furnace in an air atmosphere. The reflow profiles are optimized for each test vehicle using a KIC Thermal Profiling Prophet system. Acoustic microscopy is performed using a Sonoscan D6000 scanning acoustic microscope to analyze internal defects in the assemblies such as voids, cracks, etc. Interconnection continuity measurements are performed using a Kiethley 210 digital multimeter based on a two-point measurement technique.

A baseline set of process parameters is used for test vehicle assembly, except where the design experiments specify alternate parameters. Baseline placement process parameters include a 1000 g placement force, a 5 mm/s placement velocity, and a 300 ms bond time. Reflow is based on the rapid ramp profile shown in Fig. 1.17.

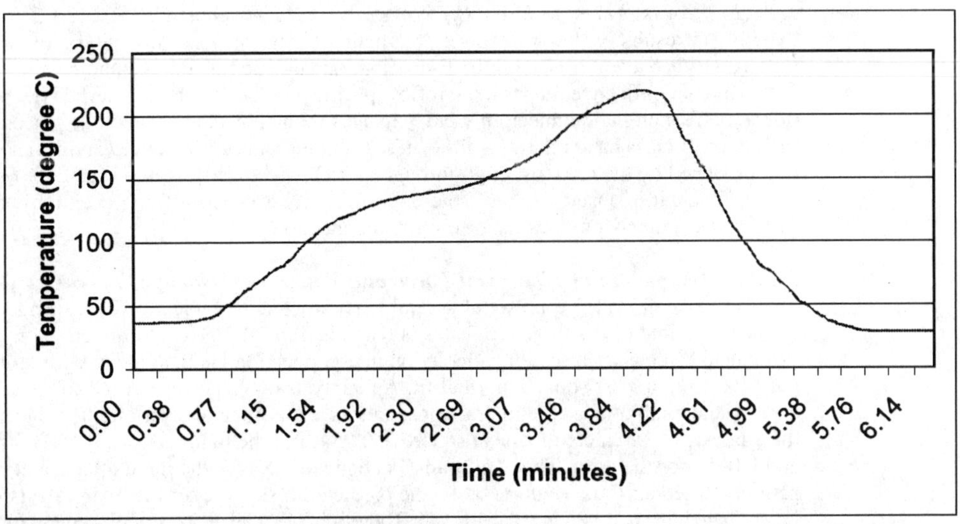

FIGURE 1.17 Conventional step eutectic solder reflow profile.

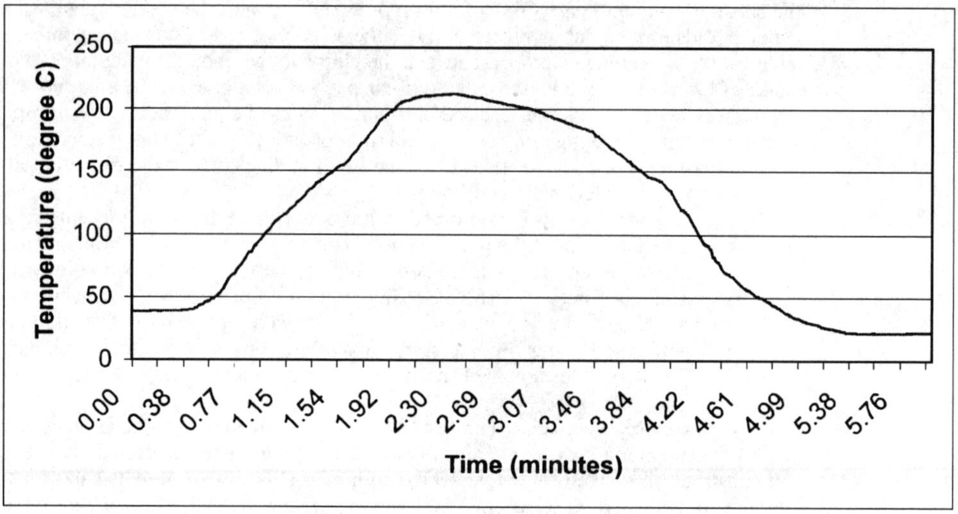

FIGURE 1.18 Rapid ramp reflow profile.

1.3.2 Yield Analysis Results and Discussion

Interconnection yield is determined by electrical resistance continuity measurements of the FC test vehicles via their daisy chain structures. The criterion for interconnection yield failure is a ±100 mΩ continuity loop resistance difference over baseline resistance values previously determined for the test vehicles. Interconnection yield is reported as a percentage of interconnection chains passing the resistance criterion relative to the total number of interconnection chains possible for the particular test vehicle. Depending on the test vehicles used, interconnection failures are based on groups of interconnections or interconnection chains that can be tested. For example, the PB8-2 test vehicle has test points allowing interconnection pairs to be continuity tested. Hence each PB8-2 has 44 interconnection chains from which the yield can be reported.

1.3.2.1 Process Variable Main Effects Analysis (Experiment 1). Table 1.2 presents the design experiment results of the process variable main effects analysis. An analysis of variance is done on the main effects data. Based on the F-distribution statistic for the five parameters investigated using a 95 percent confidence interval, the reflow profile and the underfill material type have the most significant effect on interconnection yield. Changes in the placement force, the placement velocity, and the underfill mass have relatively little effect on interconnection yield. A typical solder interconnection produced by the low-cost, high-throughput FC process is shown in Fig. 1.19. For the PB8-4-0.79 test vehicles, it is observed that if the underfill mass exceedes 35 mg (i.e., the maximum underfill mass tested), the assemblies have zero interconnection yield.

1.3.2.2 Interaction of Placement Force and Placement Velocity (Experiment 2). Three PB8-4 chips are assembled for each set of assembly parameters in Table 1.3. The average percentage interconnection yield of the three replicates is listed in Table 1.3. To determine the impact of changes in placement force and placement velocity on interconnection yield, a linear regression is performed on the data using a least squares normal fit. An analysis of variances is used to calculate the contributions of the main effects, the interactions, and the experimental error on interconnection yield. From the analysis of variance, the mean square is 0.0541 for the main effects, 0.1083 for the interactions, and 0.0650 for the error. This yields an F statistic of 0.83 for the main effects and 1.67 for the interactions. Based an F-distribution table, the predicted F statistic for the main effects $F_{0.01,2,10} = 7.56$ is larger than the experimental value of 0.83. Therefore, the main effects of placement force and placement

TABLE 1.2 Yield Results of Assemblies in Experiment 1

Experiment no.	Underfill mass (mg) (mm/s)	Placement velocity profile	Reflow type	Underfill force (g)	Placement replicates	Average percent yield for two
1	15	0.1	Step	B	400	98.9
2	22	0.1	Ramp	B	900	99.0
3	28	0.1	Ramp	A	400	100
4	35	0.1	Step	A	900	98.9
5	15	0.2	Ramp	A	900	100
6	22	0.2	Step	A	400	81.8
7	28	0.2	Step	B	900	96.9
8	35	0.2	Ramp	B	400	100
9	15	0.3	Step	A	900	83.8
10	22	0.3	Ramp	A	400	100
11	28	0.3	Ramp	B	900	99.7
12	35	0.3	Step	B	400	100
13	15	0.4	Ramp	B	400	100
14	22	0.4	Step	B	900	98.9
15	28	0.4	Step	A	400	92.8
16	35	0.4	Ramp	A	900	99.7

FIGURE 1.19 Typical eutectic solder interconnection produced using the low-cost, high-throughput FC process based on no-flow underfills.

TABLE 1.3 Yield Results of Assemblies in Experiment 2

Experiment no.	Velocity (mm/s)	Placement force (g)	Average percent yield for three replicates
1	0.4	400	99.8
2	2.7	400	99.6
3	5.0	400	99.8
4	0.4	1000	100
5	2.7	1000	100
6	5.0	1000	99.8
7	0.4	1600	100
8	2.7	1600	99.8
9	5.0	1600	99.6

velocity do not impact yield statistically. Similarly, for the interaction, $F_{0.01,1,10} = 10.04$, which is greater than the experimental value of 1.67. Therefore, the interaction between the placement force and placement velocity does not contribute to the change in yield.[8]

1.3.2.3 Chip Parameters Main Effects Analysis (Experiment 3). The average interconnection yield measured for the two replicate test vehicle assemblies and the corresponding process variables are presented in Table 1.4. This design experiment shows that only one factor is important relative to interconnection yield under conditions of varying underfill type, chip size, bump layout, placement force, and placement velocity. More specifically, the filler content of the no-flow underfill dominated the interconnection results. The unfilled no-flow underfills yield 100 percent interconnection regardless of process parameters, chip parameters, or underfill chemistry. In contrast, the filled no-flow underfills yield 0 percent interconnection under all conditions. A typical FC solder interconnection cross section using filled no-flow underfill is shown in Fig. 1.20.

1.3.2.4 Board Thickness Analysis (Experiment 4). All 20 chips assembled had 100 percent interconnection yield indicating that substrate thickness over the range of 0.79 to 1.57 mm had no appreciable effect on interconnection yield based on a 90 percent confidence interval.

1.3.2.5 Placement Void Analysis (Experiment 5). Test vehicle assemblies are made using the parameters in Table 1.1. Two assemblies are made for each experimental condition. After assembly, each die is inspected using scanning acoustic microscopy with a 180 MHz transducer to inspect underfill voiding due to the placement process. None of the images show evidence of void generation in the no-flow underfill for the assemblies produced. Typical images from the 5.08 by 5.08 mm PB8 and 5.08 by 5.08 mm FA10 test vehicles are shown in Fig. 1.21. In these images, voiding appears as white regions within the gray area of the chip.

FC assembly using the low-cost, high-throughput process based on no-flow underfills is capable of achieving high interconnection yields provided key process design and control measures are integrated into production. The results show that reflow is the most important stage of the process where the reflow profile has the strongest influence on first-pass interconnection yield. Underfill filler content also has a dramatic impact on interconnection yield. The silica filled no-flow underfills studied impede solder interconnection yield. Physically, this result stems from geometric stability. Since the chip is rigid and the surface tension of solder high, the chip will have a tendency not to yield (i.e., chip yield is predicated on all interconnections on the chip yielding) if any three solder bumps, sufficiently separated, capture a filler particle at the interface between a bump and pad. This follows because three points define a plane, and if any three bumps are not able to collapse and contact the substrate pad, the chip will not yield. This phenomenon is reduced if the filler particles are reduced in size. While the probability that a particle is entrapped between a bump and substrate pad is reduced with a reduction is particle size, the overall effect on first-pass chip yield is still likely to exceed acceptable first-pass assembly yields (typically 90 to 99.5 percent at the chip level).

TABLE 1.4 Yield Results of Assemblies in Experiment 3

Experiment no.	Underfill type	Chip size (mm)	Chip type	Placement force (g)	Placement velocity (mm/s)	Average percent yield for two replicates
1	C	5.08	FA10	400	0.1	100
2	B	5.08	FA10	1600	0.1	100
3	D (20 percent filled)	5.08	FA10	1600	5.0	0
4	E (40 percent filled)	5.08	FA10	400	5.0	0
5	C	10.16	FA10	400	5.0	100
6	B	10.16	FA10	1600	5.0	100
7	D (20 percent filled)	10.16	FA10	1600	0.1	0
8	E (40 percent filled)	10.16	FA10	400	0.1	0
9	C	5.08	PB8	400	0.1	100
10	B	5.08	PB8	1600	0.1	100
11	D (20 percent filled)	5.08	PB8	1600	5.0	0
12	E (40 percent filled)	5.08	PB8	400	5.0	0
13	C	10.16	PB8	400	5.0	100
14	B	10.16	PB8	1600	5.0	100
15	D (20 percent filled)	10.16	PB8	1600	0.1	0
16	E (40 percent filled)	10.16	PB8	400	0.1	0

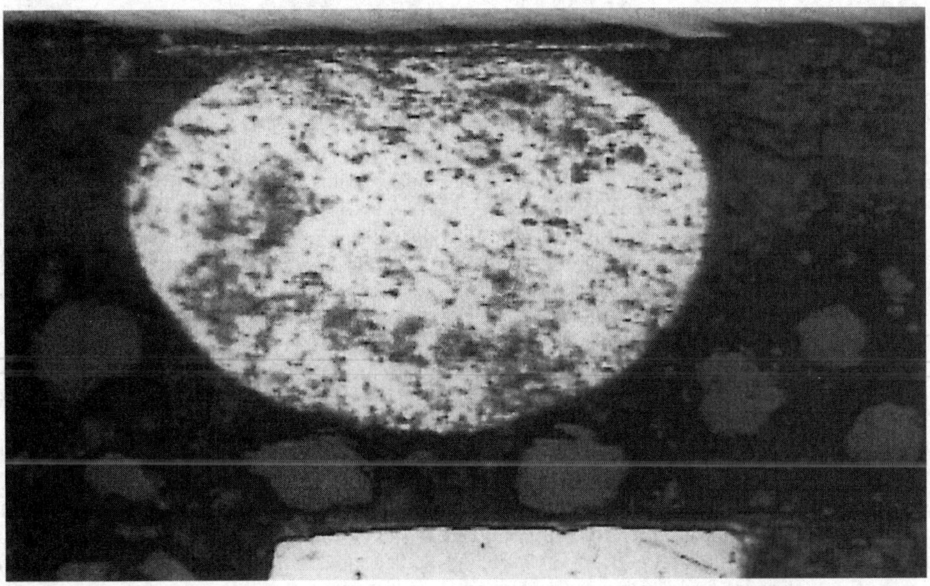

FIGURE 1.20 Solder bump impeded by filler particles from collapsing and forming an interconnection with the substrate pad.

Process parameters such as placement force and placement velocity have relatively little impact on yield over the ranges studied. Underfill mass or volume also has relatively little impact on yield, provided it was keep below a critical value that depends on the chip geometry and bump configuration. Under conditions where the underfill mass/volume is excessive (above a critical value), the assembled FCs are observed to "float" compromising alignment accuracy before and after reflow. The basic physics driving the chip floating phenomenon is discussed below, and a process design model is presented.

FIGURE 1.21 Scanning acoustic microscope image of under fill layer in a perimeter and area array die after chip placement.

1.3.3 Placement Voids

The results of Experiment 5 show no voids in the underfill after the chip placement process. This likely results from the initial underfill geometry being hemispherical in shape. If the underfill forms a hemispherical geometry, as in Fig. 1.22, no air pockets will tend to form during the placement process. As the chip moves into contact with the underfill, it wets at only one point toward the center of the chip. On further motion of the chip, the underfill continues to wet the chip surface from the chip center outward. Locally, the surface tension of the underfill can facilitate the wetting of the chip, driving out the air and eliminating capture voids.

Using the optical microscope, the geometry of the underfill was examined. Visual inspection of the underfill deposits for fluids having viscosities of 0.4, 3.4, 10, and 3400 Pa·s shows no depressions in the surface of the deposited underfills. Without depressions in the underfill deposits, the chips will tend not to have multiple contact points of the underfill with the chip surface. The chip starts with an initial point of contact at the top of the underfill deposit and locally wets the chip surface and bumps as it is moved toward the substrate. Lack of air pocket capture is one of the primary reasons voids are not formed.

The capability to assemble FCs using the low-cost, high-throughput assembly process incorporating no-flow underfill materials was demonstrated. The ability to implement the process on standard SMT assembly equipment, including placement machines and reflow ovens, was shown experimentally. The placement process prefers a low-viscosity, no-flow underfill material. For a given placement force, as underfill viscosity decreases, the allowable placement velocity increases, improving throughput. The low-cost, high-throughput FC assembly process also prefers an underfill with a low surface tension. The yield analysis experiments demonstrated a number of process recipes capable of producing high first-pass yields. It was found that the placement force, the placement velocity, and the underfill mass had relatively little effect on interconnection yield. Excessive underfill volume resulted in chip floating. Chip bump pattern and chip size were also found to have relatively little effect on assembly yield. Capture and assembly-induced void formation during assembly was also found to be minimal when the appropriate underfill geometry is deposited on the substrate.

1.3.4 Summary of Process Yield Analysis

FC technology represents a rapidly advancing area in commercial electronics. To enable low-cost, high-throughput processing compatible with high-speed SMT assembly and high-volume electronics packaging, a new assembly process was developed. This new process eliminates fluxing operations, secondary thermal curing steps, and the need for time-consuming capillary flow underfill processing using a compression flow technique where the underfill is applied prior to chip placement. The innovative process integrates the chip placement and polymer underfill processes using a compression or squeeze flow technique. It results in significantly lower assembly costs and reduced cycle time.

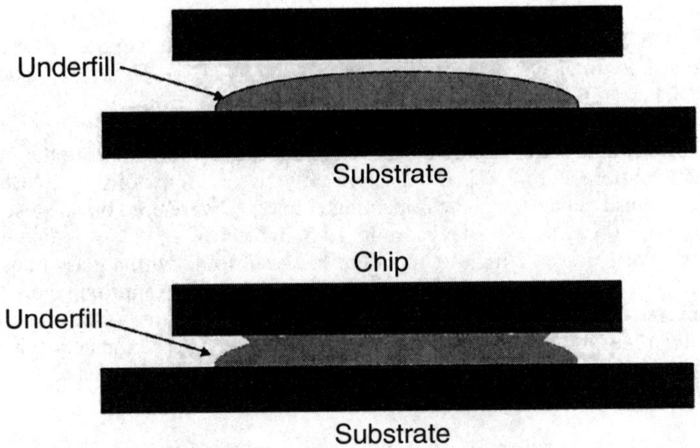

FIGURE 1.22 Chip contacting a spherically shaped underfill.

This work has demonstrated FC assembly using the low-cost, high-throughput process based on no-flow underfills and its ability to achieve high interconnection yields, provided key process design and process control measures were integrated into production. Critical process parameters affecting yield included reflow profile and no-flow underfill filler content. The results also suggest that adding fillers to no-flow underfill systems tended to inhibit robust interconnection formation, significantly lowering chip level yield. Process parameters such as placement force and placement velocity had relatively little impact on yield over the ranges studied.

Two process design models were developed that provide additional insights into designing the process parameters for the low-cost, high-throughput FC process. The compression flow placement model enabled prediction of key placement process parameters such as placement velocity, placement force, and bond time based on the FC design parameters. Moreover, the compression flow placement model was instrumental in analyzing the interaction between placement force and placement velocity. In particular, it was found that bond time was a key process design parameter in order to ensure high yield processing if high placement speeds are required or if placement forces are limited. A chip floating model was developed to determine the key process variables governing floating, including underfill mass, underfill surface tension, and surface wetting characteristics. Chip floating was analyzed to explain chip shifting and motion after placement that significantly affected yield when excess underfill volumes were used. Chip floating was found to depend on the steady-state forces acting on the chip, namely, chip weight, underfill surface tension, hydrostatic pressure on the chip, and pressure differential across the fillet meniscus. In terms of process variables, underfill mass, surface energy, and underfill wetted length had the strongest impact on chip floating. The process design model was developed to understand the basic physics of the floating phenomenon and predict process variables to eliminate process defects. A floating parameter was developed to predict the critical underfill volume resulting in chip floating. The model predictions agreed well with experimental measurements of chip floating with varying underfill masses.

1.3.5 Reliability Analysis

This section will evaluate the reliability of several commercial no-flow underfills and demonstrate the affects of board design and chip design parameters on reliability. Section 1.3.6 will provide a comprehensive failure-mode analysis of the resulting FC assemblies.[36,38–40]

1.3.5.1 Experimental Procedures

1.3.5.1.1 Test Die. Four different test die were used to evaluate the effect of bump pitch, bump layout, and chip size on yield and reliability. All test die were daisy chained devices supplied by FC

Technologies with tin-lead eutectic solder bumps and silicon nitride passivation. The under bump metallization (UBM) consisted of an aluminum, nickel-vanadium, copper system. The daisy-chain structure allowed continuity resistance measurements to be taken. The specifications of the test die are shown in Table 1.5.

1.3.5.1.2 Test Vehicles. Six different test vehicles were used to evaluate the effect of chip size, bond pad metallization, and bond pad definition on yield and reliability. The PB8-4 chip was used on three different test vehicles, and the remaining chip types were used on one test vehicle each. The important variables for each test vehicle are listed in Table 1.6.

All the test vehicles used a high-temperature FR-4 substrate with a glass transition temperature T_g of 180°C. The substrate thickness for the six test vehicles was approximately 0.78 mm. For the organic passivated substrates, the organic solderability preservative (OSP) was 0.05 to 0.13 μm thick. The Cu/Ni/Au boards had an electrolytic nickel layer approximately 3.8 μm thick and an immersion gold layer between 0.05 and 0.30 μm thick. Figure 1.23 shows a schematic of the circular defined bond pad and the rectangular defined bond pad.

1.3.5.1.3 No-Flow Underfill Materials. Seven no-flow underfills were evaluated in this study. All but one were supplied by commercial vendors. All these materials were dispensed during assembly with a controlled geometry and volume. Two of the materials studied required a postcure of 30 min at 160°C (Table 1.7).

1.3.5.2 Assembly Process.

Before assembly, the test vehicles were baked for 1.5 h at 125°C to drive off any entrapped moisture and solvents. Both substrates and die were stored in a desiccant chamber prior to use. A controlled amount of underfill was dispensed on each FC bond site using a CAM/ALOT 3800 automated dispense system with a 22 gauge needle. Two different dispense patterns were used during processing. The underfill amount necessary was determined by the chip size and bump layout, as well as the bump height. Previous work has modeled the relationship between underfill amount and die yield with respect to chip floating and placement parameters.[7,8]

Following underfill dispense, the FCs were placed on the boards using either a K&S 6900 FC bonder or a high-speed Siemens SIPLACE F5 with DCA placement system. Previous work has shown that placement force and dwell (bond) time can have an effect on yield when processing with a no-flow underfill.[6] Table 1.8 lists the placement forces used for the different FC test chips optimized for high yield.

TABLE 1.5 Test Die Used for Reliability Testing

Chip type	Chip size (mm)	Bump count	Bump pitch (μm)	Bump layout
PB8-4	10.2	352	203	Perimeter
PB8-2	5.1	88	203	Perimeter
FA10-4	10.2	1268	254	Area
FA10-2	5.1	317	254	Area

TABLE 1.6 FC Test Vehicles

Test vehicle	Chip type	Metallization	Bond pad definition
TV1	PB8-4	Cu/Ni/Au	Circular Mask
TV2	PB8-4	Cu/Ni/Au	Rectangular Mask
TV3	PB8-4	Cu/OSP	Circular Mask
TV4	PB8-2	Cu/Ni/Au	Rectangular Mask
TV5	FA10-4	Cu/Ni/Au	Rectangular Mask
TV6	FA10-2	Cu/Ni/Au	Rectangular Mask
TV7	PB8-2	Cu/OSP	Rectangular Mask

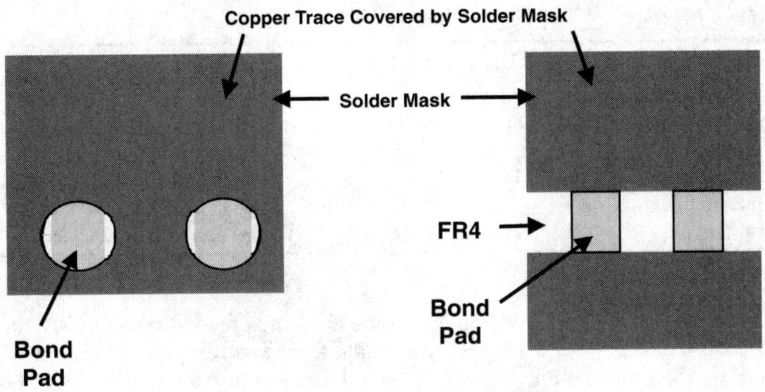

FIGURE 1.23 Circular mask opening (*left*) and rectangular mask opening (*right*).

TABLE 1.7 No-Flow Underfills Analyzed

Underfill	CTE	T_g	Post cure used
A	90ppm/°C	100°C	No
B	70ppm/°C	120°C	Yes
C	75ppm/°C	125°C	No
D	70ppm/°C	90°C	No
E	65ppm/°C	95°C	No
F	70ppm/°C	110°C	No
G	90ppm/°C	95°C	Yes

TABLE 1.8 Placement Forces Used for FC Assembly with No-flow Underfills

Test die	Placement force
PB8-4	800–1600 g
PB8-2	400–1000 g
FA10-4	800–1600 g
FA10-2	400–1000 g

After placement, the substrates were reflowed, simultaneously curing the underfill and forming the solder interconnections. The reflow profile used was determined by the chemistry of the underfill. For this study, three different profiles were used: rapid ramp profile, a conventional step eutectic style profile, and low soak temperature eutectic profile. All reflow profiles used an ambient air environment without nitrogen inerting. A postcure step for 30 min at 160°C was used for materials B and G.

1.3.5.3 Test Methods.
Several different standard accelerated tests were used to evaluate the no-flow underfills. Table 1.9 summarizes the series of tests used.

Liquid-to-liquid thermal shock testing (LLTS) was performed using an ESPEC TSB5 environmental test chamber. Test conditions were from −55 to +125°C with 5 min dwells in each bath. Air to air thermal cycling (AATC) was performed using a Thermotron ATS-320-DD test chamber. Test conditions were from −55 to +125°C with 10 min dwells at each temperature. Transfer times between temperature chambers were less than 10 s in each case. Temperature-humidity (TH) aging was performed in a Thermotron SM-8C test chamber. Test conditions were 85°C and 85 percent relative humidity for a total of 1000 h. Moisture preconditioning tests were done at levels 1 and 3

TABLE 1.9 Reliability Test Methods and Conditions

Reliability test	Industry standard	Test conditions	Cycle time
Air-to-air thermal cycling	JESD22 A104-A	+125°C to −55°C	20 min
Liquid-to-liquid thermal shock	JESD22 A106-A	+125°C to −55°C	10 min
Temperature humidity	JESD22 A101-B	85°C 85 percent RH	100 h
Level 1 preconditioning	J-STD 020-A	85°C 85 percent RH	168 h
Level 3 preconditioning	J-STD 020-A	30°C 60 percent RH	192 h
Autoclave	JESD22-A102	121°C, 100 percent RH, 2.09 atm	192 h

according to J-Standard-020 Revision A. For the moisture preconditioning tests, the assemblies were baked for 24 h at 125 +5/−0°C. They were then aged according to the test level. Level 1 required 168 h at 85°C/85 percent relative humidity and level 3 required 192 h at 30OC/60 percentRH. The test assemblies were then reflowed according to the J-STD and inspected. Autoclave testing was performed in an ESPEC TPC-422M test chamber. Test conditions were 121°C, 100 percent relative humidity, and 2.09 atm absolute pressure for a total of 120 h.

1.3.5.4 C-SAM Analysis, Continuity Testing, and Failure Criteria. During accelerated life testing, test vehicles were tested for electrical continuity and examined using acoustic microscopy. Every 100 h or 100 cycles, electrical continuity was checked. A change in resistance of greater than ±5 percent was classified as a failure. For the autoclaved test vehicles, electrical continuity was measured every 24 h.

Every 200 cycles, a sample of the test vehicles was examined with a Sonoscan D6000 acoustic microscope using a 180- or 230-MHz transducer. Scans were taken at the die-underfill interface to investigate delamination, void growth, die cracking, and underfill cracking. The underfill-to-substrate interface was also scanned for defects. After scanning, test vehicles being subjected to AATC and LLTS were baked out for a minimum of 2 h at 80°C before testing continued. This was done to drive off trapped moisture introduced by the C-SAM, which could induce premature failures.

1.3.6 No-Flow Underfill Reliability Results

All reliability data were analyzed using the Weibull distribution. The Weibull distribution can be used to model early life, intrinsic, and wear-out failures. Weibull distributions return two parameters, the theta parameter (Θ) and the alpha parameter (α) as shown in Eq. (1.3). The theta parameter corresponds to the time to 63.2 percent failure in the sample and is commonly referred to as the mean time to failure (MTTF). The alpha parameter is the shape factor and refers to the slope of the best-fit line. The alpha parameter is associated with the failure rate of the assemblies. An increase in the alpha parameter corresponds to an increase in failure rate. Reliability data are presented in terms of these two values, as well as the number of cycles to first failure.

$$F(t) = 1 - e^{-(t/\Theta)^{a}} \tag{1.3}$$

1.3.6.1 Liquid-to-Liquid Thermal Shock. Table 1.10 summarizes the reliability data for four no-flow underfills subjected to LLTS using the PB8-4 test die with Cu/Ni/Au metallization and circular solder-mask-defined pads (TV1). Under this set of conditions, underfill E performed the best with the largest MTTF (Fig. 1.24).

Table 1.11 summarizes data for four no-flow underfills on test vehicle 2 (TV2) with a rectangular solder mask opening consisting of a hybrid mask and metal-defined pad (Fig. 1.25). In this case, underfill C performed worse on boards with rectangular openings compared with boards with circular openings (i.e., comparing TV1 and TV2).

Table 1.12 summarizes data for four no-flow underfills tested on test vehicle 3 with a circular solder mask opening and an organic solderability preservative (Fig. 1.26). Comparing the effect of

TABLE 1.10 Weibull Parameters for TV1 (PB84-Au-Circular) Subject to LLTS

Underfill	Θ	α	Cycles to first failure
C	1990	4.4	1000
D	1930	3.9	1000
E	2040	3.5	1000
F	1770	5.5	1100

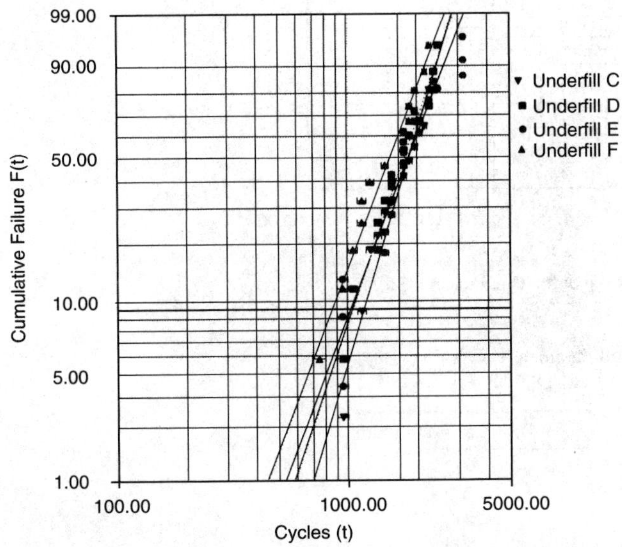

FIGURE 1.24 Weibull plot for TV1 subject to LLTS.

TABLE 1.11 Weibull Parameters for TV2 (PB84-Au-Rectangular) Subject to LLTS

Underfill	Θ	α	Cycles to first failure
A	1550	3.9	600
B	970	3.0	100
C	1180	5.4	600
G	2210	5.1	1200

substrate metallization based on Tables 1.10–1.12 shows that underfill C performed slightly worse on Cu/OSP relative to Cu/Ni/Au, whereas underfills E and F performed slightly better. Again, underfill E was the best performer, followed by underfill C.

Underfills E and F were subject to LLTS testing on TV7 with the results presented in Table 1.13 and Weibull plot in Fig. 1.27. Underfill E performed slightly worse on TV7 compared with TV3 with the first failure occurring 400 cycles earlier and an MTTF approximately 100 cycles lower. Underfill F performed slightly better with first failure occurring 200 cycles later on TV7 and the MTTF was approximately 300 cycles higher than on TV3.

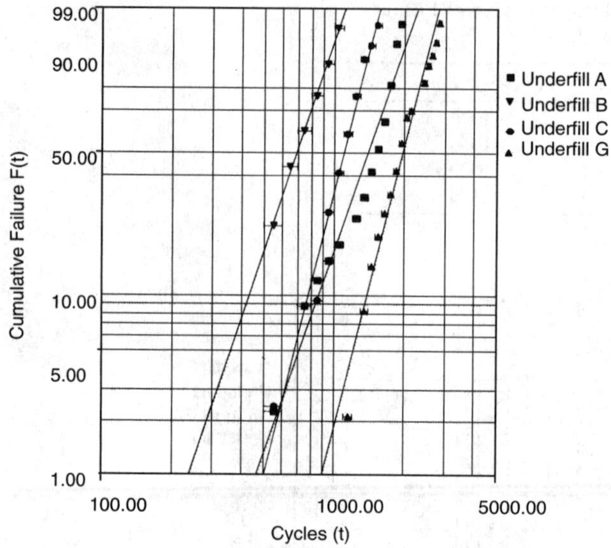

FIGURE 1.25 Weibull plot for TV2 subject to LLTS.

TABLE 1.12 Weibull Parameters for TV3 (PB84-OSP-Circular) Subject to LLTS.

Underfill	Θ	α	Cycles to first failure
A	1380	4.8	600
C	1830	5.5	1000
E	2170	5.2	1100
F	1840	6.0	800

Underfill C was tested on six of the seven test vehicles to examine the effects of bump geometry, chip size, metallization, and pad definition on reliability performance, with the results summarized in Table 1.14 and Fig. 1.28. The MTTF for the FA10-2 was approximately 400 cycles longer than the FA10-4. Similar results hold for the PB8 test die as well, with an approximate difference in MTTF of 300 cycles, with the smaller die having higher reliability. Underfill C was also more reliable on the PB8-4 format with circular-solder-mask-defined pads relative to a hybrid defined pad, with a difference in MTTF of approximately 800 cycles.

1.3.6.2 Air to Air Thermal Cycling. AATC results on TV2 (PB8-4-Au-rectangular) are shown in Table 1.15 and Fig. 1.29. The MTTF in AATC on TV2 was substantially lower for all the underfills tested except for underfill B. Underfills C, A, and G had first failure 300 cycles earlier in AATC, and their MTTF was reduced by 300 to 900 cycles. The more rapid failure rate can be attributed in part to the longer dwell time in AATC, which accelerates solder creep, and solder extrusion and increases solder fatigue. Hybrid pads impart a larger stress concentration, which could also promote more rapid joint fatigue.

Three additional commercial no-flow underfills were tested on TV1 (PB8-4-Au-circular) (Fig. 1.30), and two additional commercial no-flow underfills were tested on TV3 (PB8-4-OSP-circular) (Fig. 1.31). As seen with the other underfill chemistries tested, the time to first failure was lower than it was in LLTS for underfills D, E, and F, as shown in Table 1.16. In addition, the MTTFs for underfills E and F were approximately 400 cycles lower in AATC than LLTS testing. Underfill D had a, MTTF that was approximately 1000 cycles lower in AATC.

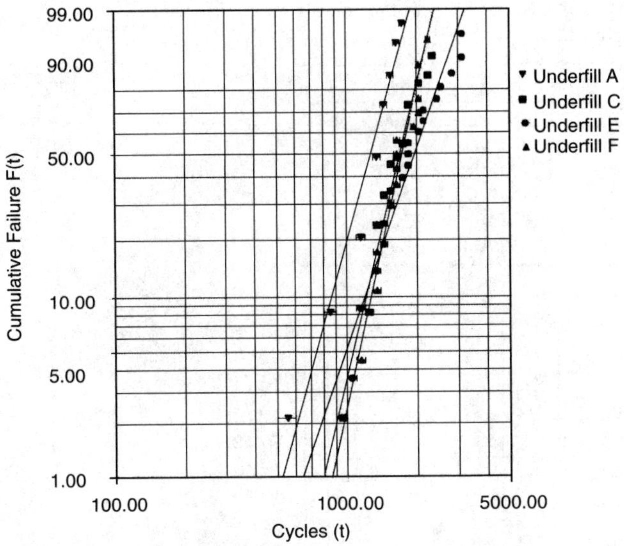

FIGURE 1.26 Weibull plot for TV3 subject to LLTS.

TABLE 1.13 Weibull Parameters for TV7 (PB82-OSP-Rectangular) Subject to LLTS.

Underfill	Θ	α	Cycles to first failure
E	2030	3.1	800
F	2170	4.9	1000

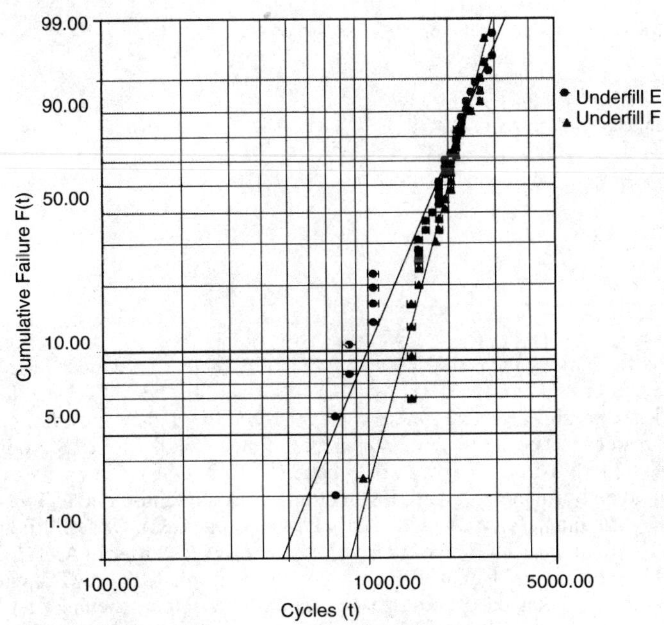

FIGURE 1.27 Weibull plot for TV7 subject to LLTS.

TABLE 1.14 Weibull Parameters for Underfill C on Six Test Vehicles Subject to LLTS.

Chip	Definition	Metallization	Θ	α
PB8-4	Circular	Cu/Ni/Au	1990	4.4
PB8-4	Rectangular	Cu/Ni/Au	1180	5.4
PB8-4	Circular	Cu/OSP	1830	5.5
PB8-2	Rectangular	Cu/Ni/Au	1510	3.3
FA10-4	Rectangular	Cu/Ni/Au	1300	7.8
FA10-2	Rectangular	Cu/Ni/Au	1690	6.0

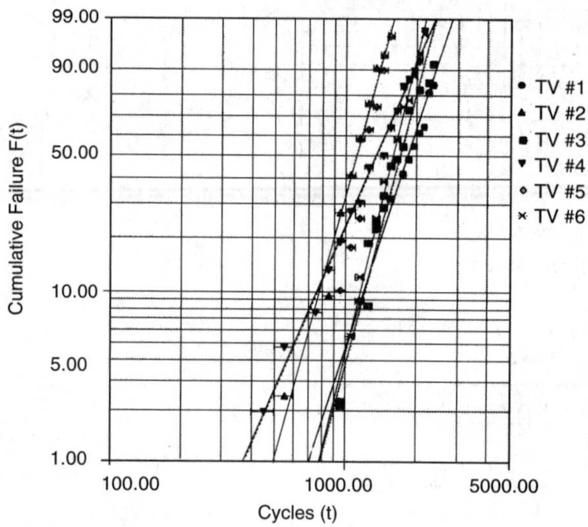

FIGURE 1.28 Weibull plot for underfill C on TV1–TV6 subject to LLTS.

TABLE 1.15 Weibull Parameters for TV2 (PB8-4-Au-Rectangular) Subject to AATC

Underfill	Θ	α	Cycles to first failure
A	800	3.8	300
B	1040	2.8	100
C	820	3.3	300
G	1340	7.8	900

AATC results for underfill E and F are summarized in Table 1.17. Underfill E passed the same number of cycles to first failure in AATC as in LLTS. However, the MTTF was lower by approximately 300 cycles. Underfill F required more cycles to reach first failure in AATC and had an MTTF that was approximately 100 cycles higher than the mean time to failure in LLTS. Additionally, the performance of both underfills E and F was slightly better on TV3 (OSP) than on TV1 (Cu/Ni/Au). This may be due to gold embrittlement in the UBM that increases the failure rate in TV1 during AATC.

On TV7, underfills E and F were tested (Table 1.18 and Fig. 1.32). Underfill F had a much lower reliability in AATC than in LLTS. The MTTF was over 900 cycles lower in AATC, and the first failure occurred 800 cycles earlier. Underfill E had superior performance in AATC, with an increase in MTTF of approximately 300 cycles, and it passed 1500 cycles before reaching first failure.

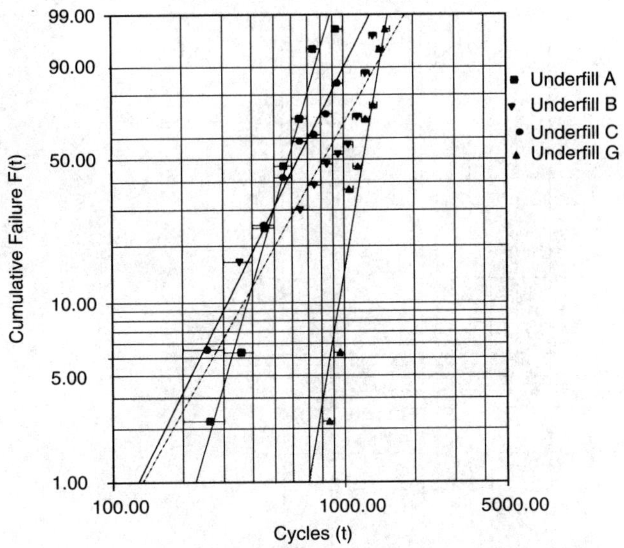

FIGURE 1.29 Weibull plot for TV2 subject to AATC.

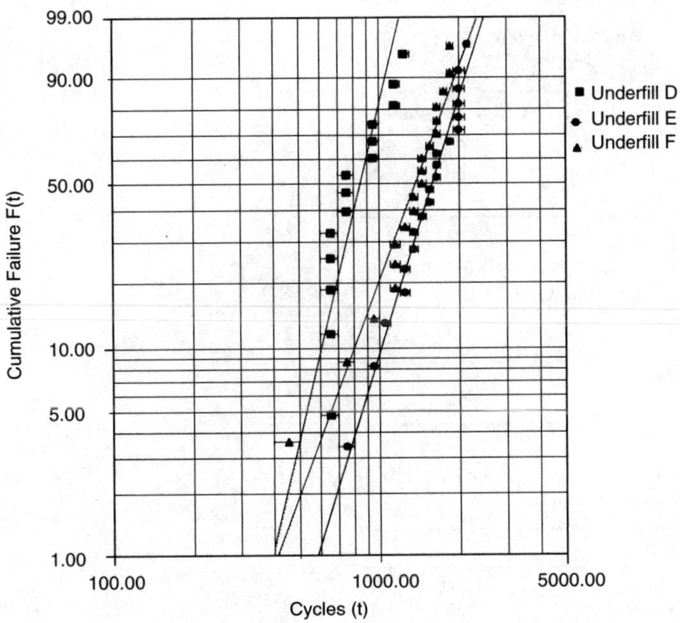

FIGURE 1.30 Weibull plot for TV1 subject to AATC.

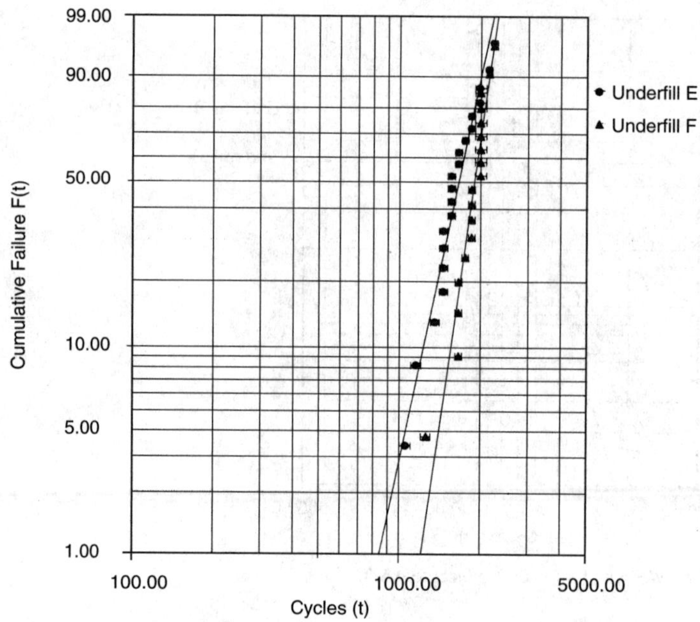

FIGURE 1.31 Weibull plot for TV3 subject to AATC.

TABLE 1.16 Weibull Parameters for TV1 (PB8-4-Au-Circular) Subject to AATC

Underfill	θ	α	Cycles to first failure
D	910	5.5	600
E	1720	4.2	800
F	1510	3.5	600

TABLE 1.17 Weibull Parameters for TV3 (PB8-4-OSP-Circular) Subject to AATC

Underfill	θ	α	Cycles to first failure
E	1750	6.3	1100
F	1970	9.4	1300

TABLE 1.18 Weibull Parameters for TV7 (PB82-OSP-Rectangular) Subject to AATC

Underfill	θ	α	Cycles to first failure
E	2330	6.2	1500
F	1250	1.8	200

1.3.6.3 Temperature Humidity Aging and Autoclave. In addition to LLTS and AATC testing, TH testing was also performed. The commercial no-flow underfills tested did not appear to be susceptible to the 85°C/85 percent relative humidity test condition as shown in Tables 1.19 and 1.20.

Underfills C and A were also subjected to moisture preconditioning based on J-STD-020A standard levels 1 and 3 preconditioning. As the data in Table 1.21 show, both underfills passed level 3

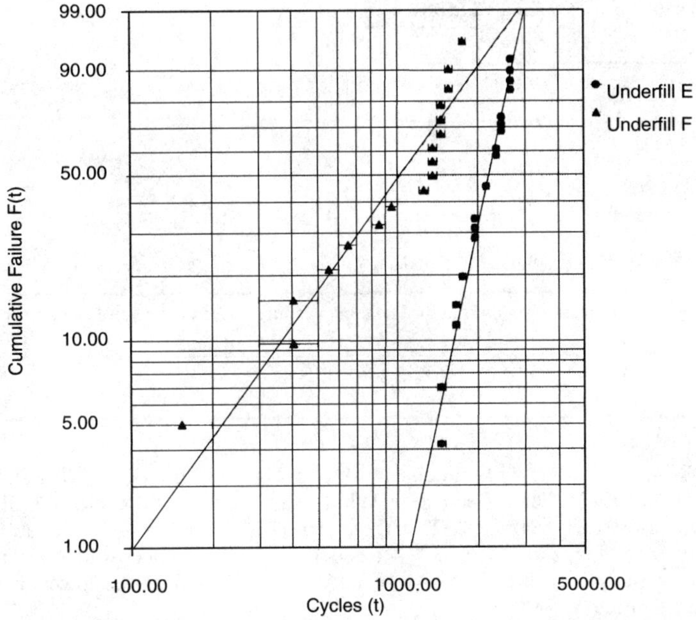

FIGURE 1.32 Weibull plot for TV7 subject to AATC.

TABLE 1.19 TH Reliability Data for TV3 (PB8-4-OSP-Circular)

Underfill	Failures at 1000 h/total sample size
A	0/32
C	0/32
D	0/20
E	0/20
F	0/20

TABLE 1.20 TH Reliability Data for TV1 (PB8-4-Au-Circular)

Underfill	Failures at 1000 h/total sample size
D	0/20
E	0/20
F	0/20

preconditioning. Underfill A came very close to passing level 1 preconditioning, with only one chip failing after reflow. Underfill C had over 50 percent of the assemblies pass level 1 preconditioning.

Underfills A and C were also subject to autoclave testing on TV3. The times to percentage failure are summarized in Table 1.22.

1.3.7 Reliability Data Discussion

In liquid-to-liquid thermal shock (LLTS) testing, using Cu/Ni/Au or Cu/OSP appears to have no significant impact on reliability when all other variables are held constant. Underfills C, E and F were

TABLE 1.21 J-STD 020 Preconditioning Data for TV3 (PB8-4-OSP-Circular)

Underfill	Preconditioning level	Passed chips/total tested
A	1	19/20
A	3	32/32
C	1	11/20
C	3	32/32

TABLE 1.22 Autoclave Test Data for TV3 (PB8-4-OSP-Circular)

Underfill	Percent failure 0 h	Percent failure 24 h	Percent failure 48 h	Percent failure 72 h	Percent failure 96 h	Percent failure 120 h
A	0	0	0	10	45	100
C	0	0	19	93	100	100

tested on PB8-4 test boards with circular mask openings with both Cu/Ni/Au and Cu/OSP metallization (TV1 and TV3). The difference in MTTF was less than 200 cycles for all three underfills. During air to air thermal cycling (AATC), the effect of the metallization on reliability appears to be material-dependent. For underfill E on TV1 and TV3, the difference in MTTF was 25 cycles. For underfill F, the MTTF for the Cu/Ni/Au metallization was 460 cycles lower than the MTTF for Cu/OSP metallization.

For all but two test board configurations, AATC testing resulted in lower MTTF than LLTS testing. This difference ranged from 250 cycles with underfill F on TV1, to 870 cycles on TV2. The two exceptions to this trend were underfill E on TV7 and underfill F on TV3. In these two cases, the MTTF was higher during AATC. It is interesting to note that both these test vehicles used Cu/OSP metallization. For the material systems tested, the Cu/Ni/Au substrates always had lower reliability in AATC testing than in LLTS.

Across all seven underfills tested, there appears to be no direct relationship between the number of cycles to first failure and the MTTF. For example, both underfill A and underfill C experienced their first failure at 600 cycles on TV2 during LLTS testing. However, the difference in MTTF between the two materials was approximately 370 cycles. Another example of this can be seen when comparing underfills D and F on TV1 during AATC testing. Both materials experienced first failure at 600 cycles, but the difference in MTTF was over 600 cycles.

The data obtained using underfill C indicate that chip size and mask design may affect LLTS reliability. When comparing TV2 and TV4, the only difference between the two substrates is the size of the die (10 versus 5 mm), both having the same bump density. The metallization (Cu/Ni/Au) and the mask opening (rectangular) are identical. The MTTF for the small test die was greater by 325 cycles. This difference is also seen when comparing TV5 and TV6. Again, the only difference in board/die configuration is the size of the die (10 versus 5 mm). In this case, the MTTF values differed by 390 cycles in LLTS, with the smaller die being more reliable.

When comparing the reliability data for underfill C on TV1 and TV2, the MTTF was 800 cycles higher for TV1. This suggests that underfill C is more reliable when used on a circular mask opening as opposed to rectangular. However, there are insufficient data to determine whether this large difference is material-dependent.

When comparing the reliability data for underfill C on TV4 (perimeter array) and TV6 (area array) during LLTS, the MTTF was 200 cycles higher for TV6. This difference is not statistically significant enough to indicate that bump density has an effect on reliability.

Of the six no-flow underfills tested, two required a postcure (underfills B and G). Overall, no-flow underfill B was shown to have poor reliability in both LLTS and AATC testing, with first failure occurring after 100 cycles and MTTF values equal to or less than 1000 cycles. No-flow underfill G, conversely, had the highest reliability of all the materials tested on the same test vehicles. These data do not indicate whether a postcure is beneficial to no-flow underfill reliability.

1.3.8 Summary of No-Flow Underfill Reliability

Different test vehicles were used to evaluate the effect of varying chip size, interconnection density, pad surface finish metallization, and solder mask opening design. Accelerated reliability tests performed included liquid/liquid thermal shock (LLTS), air/air thermal cycling (AATC), moisture sensitivity preconditioning, and temperature-humidity aging (TH). This discussion has shown that low-cost FC technology based on select commercial no-flow underfills is reliable, surviving 1000+ LLTS cycles and having MTTFs in excess of 2000 cycles. It has also been demonstrated that bond pad definition can have a large effect on underfill reliability. Small die also appear to be more reliable than large die when all other design factors are equal, while bump density appears to not have a significant affect. No correlation between cycles to first failure and mean time to failure was found. The effect board metallization has on reliability depends on both the test method used and the board/die configuration. There was also no direct correlation between the use of a postcure and overall reliability.

1.3.9 Failure-Mode Analysis

This section will describe several failure modes of commercial no-flow underfills and describe the progression of failure.[36,38]

1.3.9.1 Solder Fatigue. The dominant failure mode for FC assemblies using no-flow underfills is solder fatigue cracking. Ultimately, solder fatigue cracking was responsible for all electrical failures observed in test vehicles subjected to LLTS and AATC. All the electrical failures produced during AATC occurred prior to delamination of the chip-underfill or the board-underfill interfaces. Additionally, all electrical failures observed in assemblies subjected to LLTS occurred either prior to delamination, or, if after delamination showed no significant correlation with the locations of delamination. That is, fatigue cracks causing chip failure were not typically located in regions of the chip experiencing delamination. Based on the AATC and LLTS electrical failure observations, it can be assumed that the mode of interconnection failure was solder fatigue leading to fatigue cracking. Figure 1.33 shows an example of solder fatigue cracking occurring through the UBM side of the bump propagating through the solder and causing an electrical open.

In order to substantiate the experimental fatigue life results, a simple analytical approach was used, based on the Suhir trilayer model.[10] The analytical formulation is based on trilayer structures and can be used to estimate the interfacial shear stresses in a FCOB assembly subjected to a thermal load (such as temperature cycling). Under the assumption of interfacial adhesion, the shear stress can be used to approximate the shear strain in the solder interconnection. The Coffin-Manson relationship can be used to estimate the mean time to solder fatigue failure, based on the plastic strain on the solder interconnection.[11,23,24] Figure 1.34 shows an FCOB trilayer assembly used for the solder fatigue analysis.

The top and bottom materials are represented as materials 1 and 3, which refer to the silicon die and the FR-4 substrate. Suhir showed that the maximum shearing stress (σ_{xz}) occurs at the die-underfill interface and is given by Eq. (1.4).[37]

$$\sigma_{xz} = \frac{\Delta\alpha\Delta T}{\sqrt{\lambda\kappa}} \frac{\sinh kx}{\cosh kl}$$

$$\Delta\alpha = \alpha_3 - \alpha_1$$

$$\Delta T = \text{cycling temperature differential}$$

$$\lambda = \frac{h_1^2}{12D_1} + \frac{h_3^2}{12D_3} + \left(\frac{h_1 + h_3}{2\sqrt{D}}\right)^2$$

$$D_1 = \frac{E_i h_i^3}{12(1 - \nu_i^2)}; i = 1,3$$

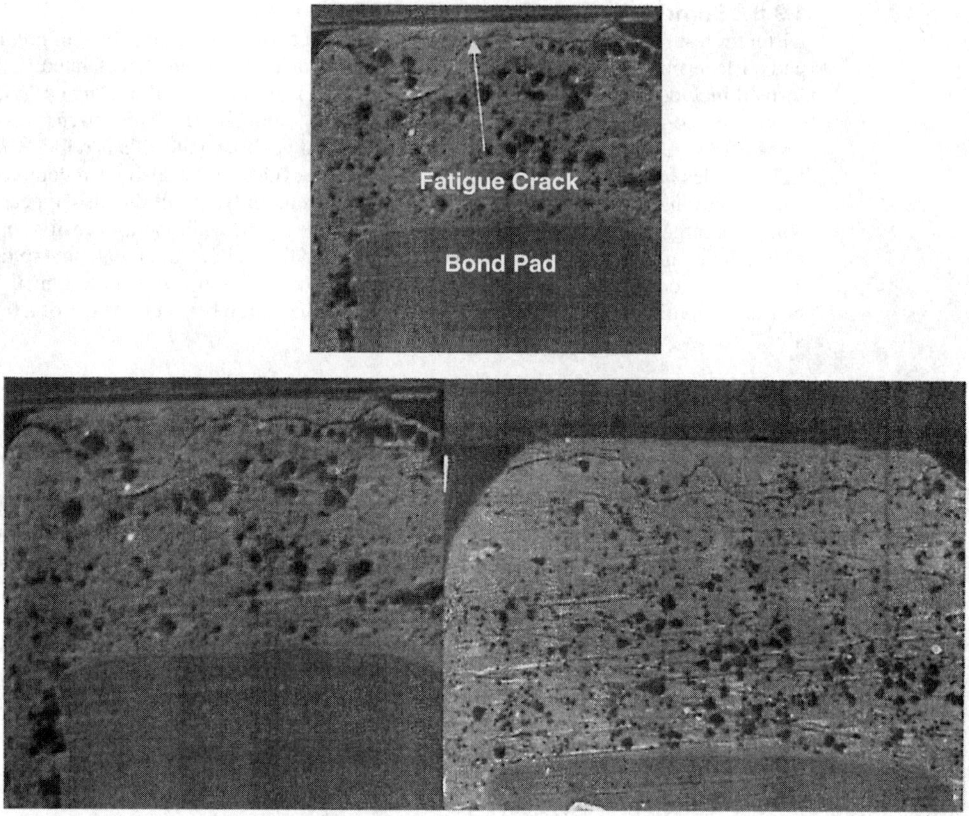

FIGURE 1.33 Scanning electron micrographs of a typical solder fatigue crack.

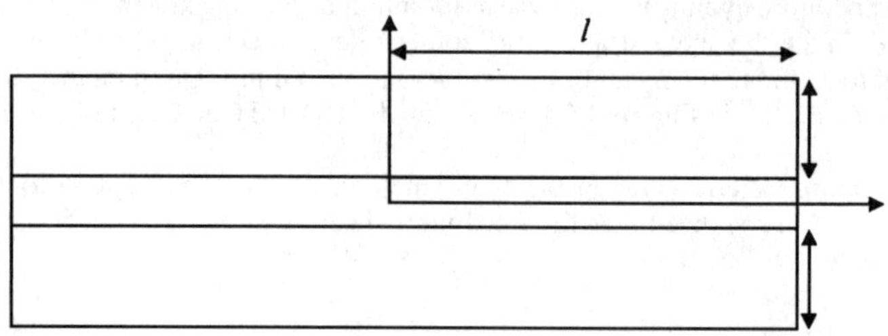

FIGURE 1.34 Structure used for solder fatigue analysis.

$$\mathrm{D} = \mathrm{D}_1 + \mathrm{D}_3$$

$$\kappa = \frac{h_1}{3G_1} + \frac{h_3}{3G_3} + \frac{2h_2}{3G_2}$$

$$\kappa = \sqrt{\frac{\lambda}{\kappa}} \qquad (1.4)$$

where h_i = thickness of the ith material

v_i = Poisson's ratio of the *i*th material
E_i = elastic modulus of the *i*th material
G_i = shear modulus of the *i*th material

A Coffin-Manson relationship (Eq. 1.5) is then used to estimate solder joint fatigue life corresponding to the MTTF.[33,35,37]

$$N_p = 1.2928\,(\epsilon_{solder})^{-1.96} \qquad (1.5)$$

Based on the strain values calculated from Eq. (1.4), the Coffin-Manson equation predicts a fatigue life of 2160 cycles. Table 1.23 shows a comparison of the predicted MTTF and the LLTS and AATC experimental results. The analytical prediction of the MTTF provides an order of magnitude approximation of the true fatigue life of an FCOB assembly using no-flow underfills. In the best case for TV1 subjected to LLTS, the predicted life is within 10 percent of the experimental life.

1.3.9.2 Delamination. Delamination refers to a loss of adhesion between surfaces. In the case of an underfilled FC, the adhesion loss is at the chip passivation-underfill interface, the underfill-solder interface, or the underfill-solder mask interface. Delamination in the test vehicles subjected to LLTS testing occurred at the chip-underfill interface. In most cases, the delamination initiated at the corners of the test die. This results from the large stresses generated by the CTE mismatch between the silicon chip, the FR-4 substrate and the no-flow underfill. Underfill delamination typically does not occur until after solder fatigue failure has begun to occur.

Underfill A exhibited chip-underfill delamination in all LLTS assemblies subject to testing and only one assembly failed J-STD-020A level 1 preconditioning due to chip-underfill delamination. Delamination in LLTS testing was very similar between underfill A and underfill C. Typically the delamination was minor and originated predominantly in the chip corners and chip perimeter. Figure 1.35 shows a typical delamination progression found in TV2 assembled with underfill A. In all cases, the first instances of delamination occurred after 1200 cycles. From the time of delamination initiation to chip failure, the regions of delamination remained small and experienced very little growth. Assemblies subjected to AATC did not show any evidence of delamination prior to chip failure.

Underfill B exhibited chip-underfill delamination in test vehicles subjected to LLTS and AATC. Underfill B delamination occurred the earliest and was the most severe of all underfill materials tested. Figure 1.36 shows typical delamination progression for AATC. The first signs of delamination appeared at 700 cycles, with delamination spreading to all four sides of the chip by 1200 cycles.

Underfill C exhibited chip-underfill delamination in all test vehicles subjected to LLTS. Over half of underfill C assemblies subjected to J-STD-020A level 1 preconditioning failed due to chip-underfill delamination. Delamination occurring in LLTS was minor and originated in the chip corner or along the chip perimeter. Figure 1.37 shows typical delamination progression for TV1 assemblies subjected to LLTS.

Underfill D, when subjected to LLTS for TV2 showed very small amounts of chip-underfill delamination. Delamination began around 1000 cycles and progressed through 1800 cycles. Again, the delamination can be found in the chip corners and perimeter (Fig. 1.38).

TABLE 1.23 Summary of Experimental and Predicted Mean Times to Failure

Reliability Test	Test Vehicle	Experimental MTTF	Predicted MTTF
AATC	2	650	2160
LLTS	2	1150	2160
LLTS	1	2000	2160
LLTS	3	1700	2160

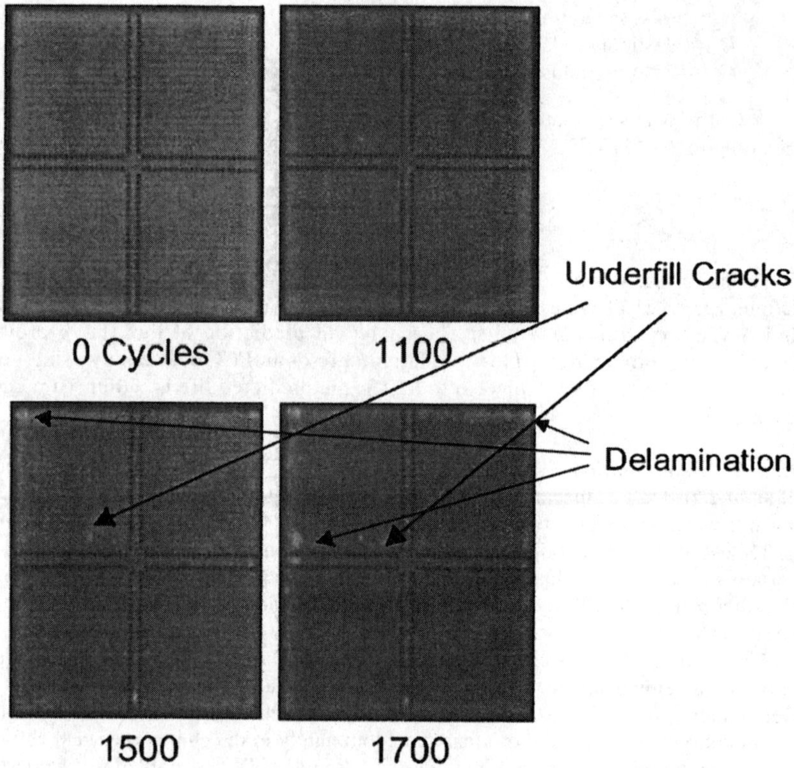

FIGURE 1.35 Delamination progression for underfill A-TV2, subjected to LLTS.

Underfill E showed little to no chip-underfill delamination in LLTS through 1800 cycles. Figure 1.39 shows a progression of acoustic micrographs through 1800 cycles showing no visible delamination. However, underfill cracking was visible at 1400 cycles.

Underfill F showed chip-underfill delamination in assemblies subjected to LLTS. Figure 1.40 shows typical delamination progression occurring along the perimeter of the chip for LLTS assemblies. Delamination initiated around 600 cycles and grew as cycling continued.

Underfill G exhibited chip-underfill delamination only in assemblies subjected to LLTS. Assemblies subjected to AATC did not show any delamination prior to electrical failure. Figure 1.41 shows typical delamination progression for the LLTS assemblies. Delamination first occurred around 1900 cycles and exhibited only minor growth through the next 900 cycles.

1.3.9.3 Underfill Fillet Cracking. Underfill fillet cracking was a common underfill material failure for no-flow underfills, although it can also occur in filled capillary underfills. Fillet cracking occurs mainly in assemblies subjected to LLTS and AATC cycling. Fillet cracks may be classified as one of three types: chip-side cracks, board-side cracks, and complete cracks. Figure 1.42 shows examples of the three crack types.

Chip-side cracks are small cracks that originate at the chip-underfill interface and usually form a V-shaped pattern that extends partially down the fillet. Board-side cracks are short cracks that extend up the fillet or run parallel to the fillet edge. Chip-side cracks can lead to localized areas of delamination between the underfill and the side of the chip, whereas board-side cracks can lead to localized areas of delamination between the underfill and substrate-solder mask. A complete fillet crack extends from the top of the fillet to the substrate and propagates completely through the fillet. In cases of extreme cracking, the underfill fillet separates from the chip, eliminating the stress-relief properties of the underfill.

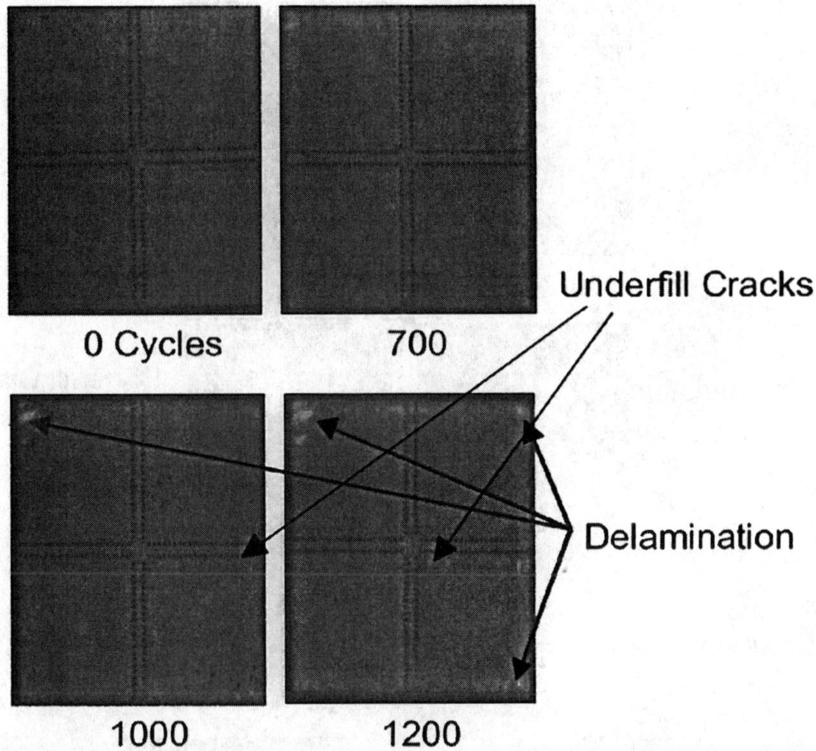

FIGURE 1.36 Delamination progression for underfill B-TV2, subjected to AATC.

Fillet cracking is driven by the large hoop stress produced by the underfill generated during the cooling cycle of LLTS. The net difference between the CTE of the underfill (~70ppm/°C) and the silicon (~3ppm/°C) produces very large hoop tensile and interface shear stress in the fillets.

All underfills tested fillet cracked during cycling. The number of cycles before cracking occurred and the severity of the cracking varied from material to material. For example, underfill B began to fillet crack after 200 cycles LLTS, and underfill E began to fillet crack after 1300 cycles LLTS.

1.3.9.4 Bulk Underfill Cracking. In addition to fillet cracking, cracks may be seen in the bulk underfill underneath the test die occurring during AATC and LLTS testing. Typical bulk underfill cracks in no-flow underfills are shown in Fig. 1.44. It is important to notice that the cracks typically occur adjacent to the solder joints, propagating between the joints. In addition to cracks propagating between solder joints, cracks were also seen to initiate and propagate along the metallization pattern on the test die, as seen in the right-hand second image in Fig. 1.44. Moreover, it has been observed that solder from the joints tends to extrude into underfill cracks during AATC and LLTS cycling, which has the potential to cause electrical shorting, as shown in Fig. 1.45.

Figure 1.46 shows a series of C-SAM images at the chip-underfill interface where fine lines near the solder joints appear and expand as cycling continues. Cross sections show that these fine lines correspond to cracks in the bulk underfill, as illustrated in the figure, indicating that C-SAM analysis can be used to detect bulk underfill cracks.

Bulk underfill cracking has been seen in no-flow underfills as well as in capillary underfills. In addition, for no-flow underfills, bulk cracking occurs before chip failure due to solder fatigue cracking. The chip shown in the acoustic images in Fig. 1.46 failed at 2400 cycles of LLTS, yet underfill cracking is visible in acoustic images at 1400 cycles.

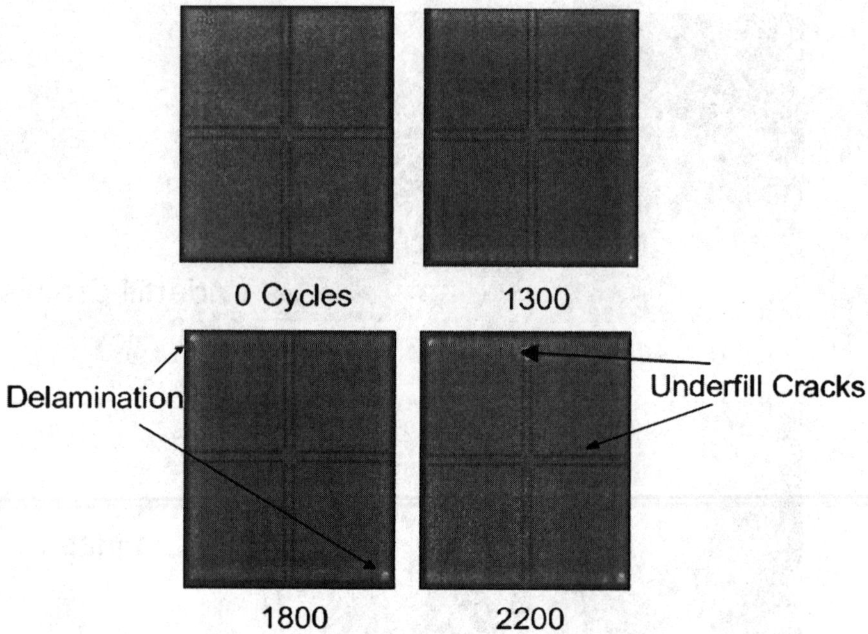

FIGURE 1.37 Delamination progression for underfill C-TV1, subjected to LLTS.

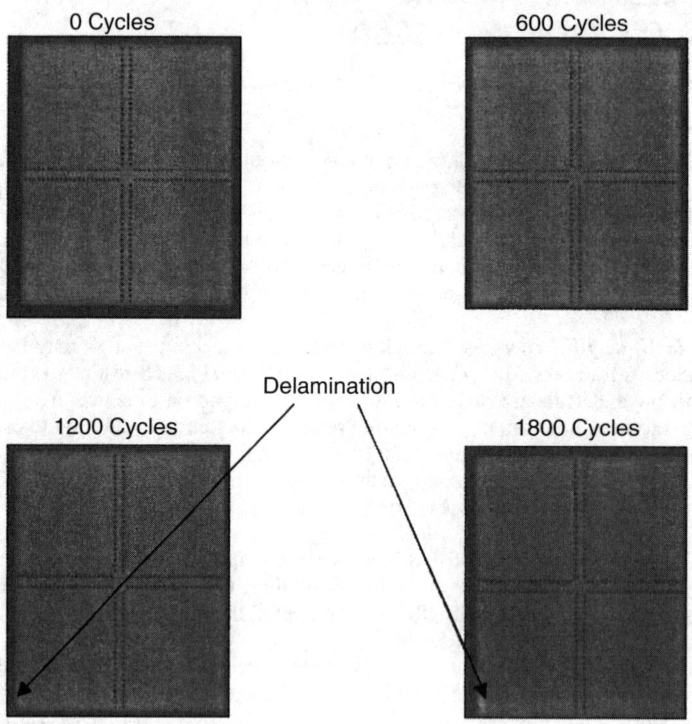

FIGURE 1.38 Delamination progression in underfill D-TV2, subjected to LLTS.

FIGURE 1.39 Cycling progression for underfill E-TV2, subjected to LLTS.

Figure 1.47 shows C-SAM images of a cycled chip showing underfill cracking at the chip-underfill interface and the underfill-board interface.[12]

Bulk underfill cracking was seen in all seven underfills tested to varying degrees. Cracking occurred the earliest and was the most severe in underfill B. Underfills A, C, E, and F all exhibited similar amounts of bulk cracking. Minimal underfill cracking was seen in underfills D and G.

Figure 1.48 shows a schematic of copper circuit trace on a substrate covered by solder mask and underfill. A simple analysis of this microstructure gives one explanation for the underfill and substrate cracks observed. During the reflow process, the underfill material hardens at a temperature of approximately 200°C and locks the assembly in a zero stress state at this temperature. As the assembly cools to room temperature, all four materials attempt to contract different distances based on their respective CTEs. The larger the CTE, the more the material wants to contract. Based on the approximate CTE values presented in Fig. 1.48, both the underfill material and the substrate will contract to a greater extent than the copper trace. The difference in CTE between the substrate and the underfill generates large in-plane stresses in the underfill layer. When the assembly is cycled, it experiences temperatures as low as $-55°C$, which represents a ΔT of 250°C from the zero stress state. The large temperature change produces large cyclic in-plane stresses that may promote underfill cracking. Sharp corners found on the copper circuitry generate local regions of stress concentration that increase the tendency for crack initiation near the corners of the copper trace. Cracks that form in the solder mask have a tendency to propagate into the bulk underfill layer. Figure 1.48 shows the areas of stress concentration that promote underfill and substrate crack initiation.

Figure 1.49 indicates that underfill cracks can also form between solder bumps. An analysis similar to the one just presented provides one explanation for underfill cracks forming between solder interconnections. Figure 1.49 shows a top-view schematic of solder interconnections surrounded by underfill material. The CTE of the underfill is nearly three times greater than the CTE of the solder. During cycling, the large difference in CTE generates a stress concentration along the circumference of the solder-underfill interface. Any micrcracks in the underfill that exist along the interface can be

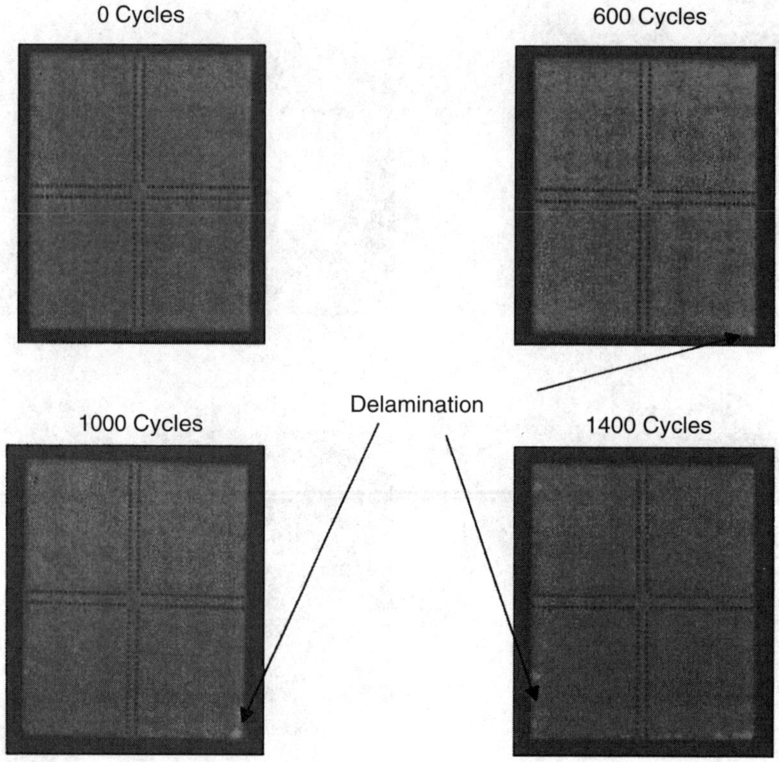

FIGURE 1.40 Delamination progression for underfill F-TV2, subjected to LLTS.

initiation points for underfill cracks. The most common cracks that initiated along the circumference of the solder occurred in the narrowest region between interconnections. This region has the smallest cross-sectional area and therefore is the most likely location for fatigue crack initiation.

1.3.10 Summary of Failure-Mode Analysis

The preceding discussion has shown that low-cost FC technology based on select commercial no-flow underfills is reliable, surviving 1000+ LLTS cycles and having MTTF in excess of 2000 cycles. It has also been demonstrated that bond pad definition can have a large effect on underfill reliability. Small die also appear to be more reliable than large die when all other design factors are equal. Solder interconnection density also has a positive effect on reliability, where higher solder interconnection densities tend to be more reliable.

In addition, this discussion has presented a comprehensive failure-mode analysis of FC systems using no-flow underfills. The dominant failure mode was solder fatigue cracking. Due to the high adhesion strength of the no-flow systems, underfill delamination was found to be a secondary failure mode generally occurring after fatigue failure. For the first time, bulk underfill cracking was reported as a dominant failure mode in no-flow materials, driven by local CTE mismatches. Fillet cracks are also reported and discussed.

1.4 WAFER-LEVEL FC PROCESSING

FC technology has been shown to be a reliable method for the direct electrical and mechanical attachment of silicon chips to polymeric substrates. However, the assembly of encapsulated FC circuit

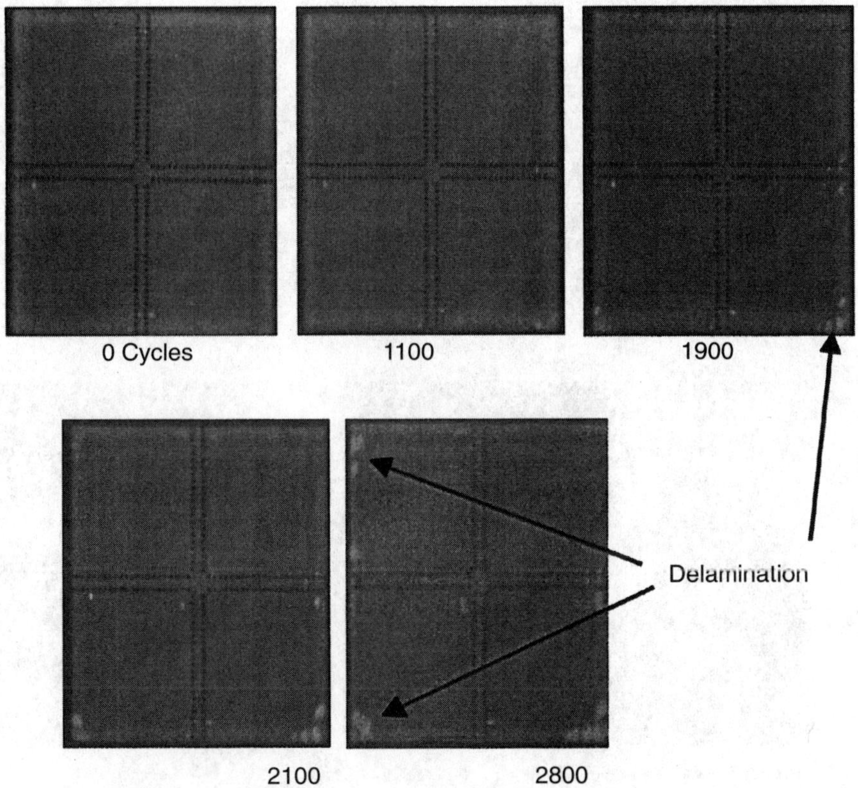

FIGURE 1.41 Delamination progression for TV2 underfill G, subjected to LLTS.

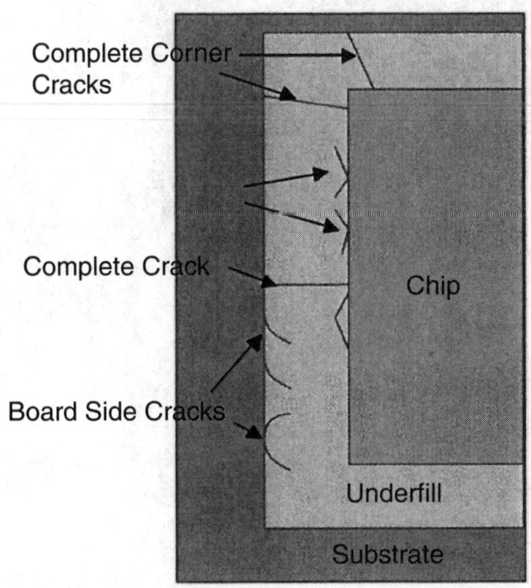

FIGURE 1.42 Underfill fillet crack types.

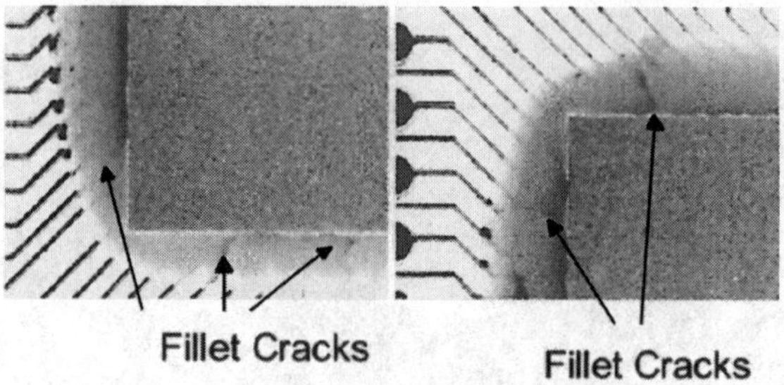

FIGURE 1.43 Typical fillet cracks found in underfill B assemblies subjected to LLTS.

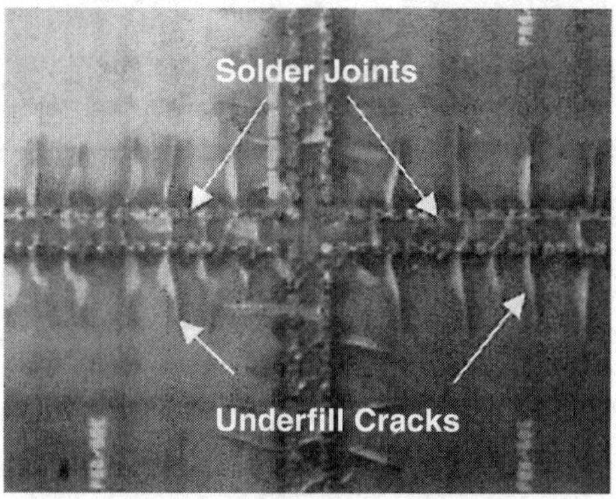

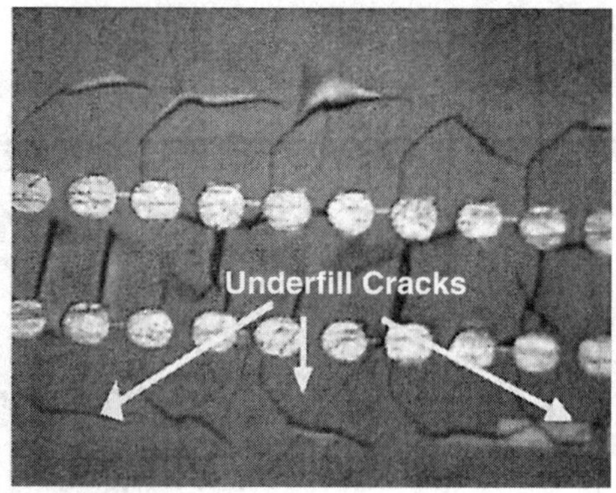

FIGURE 1.44 Examples of bulk underfill cracking.

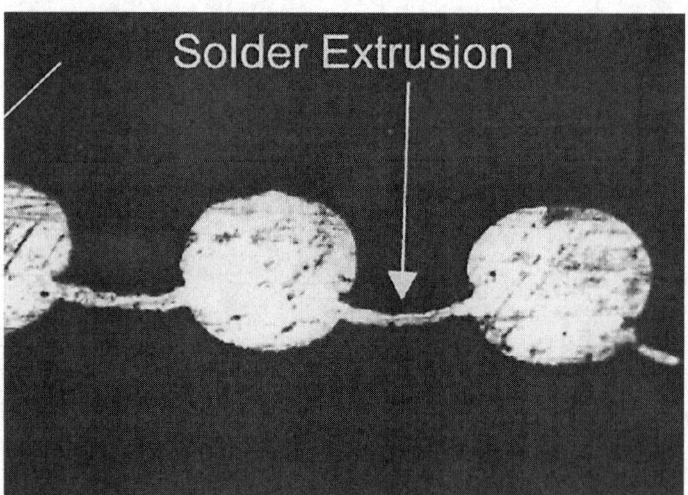

FIGURE 1.45 Example of solder extrusion into underfill cracks between adjacent bumps.

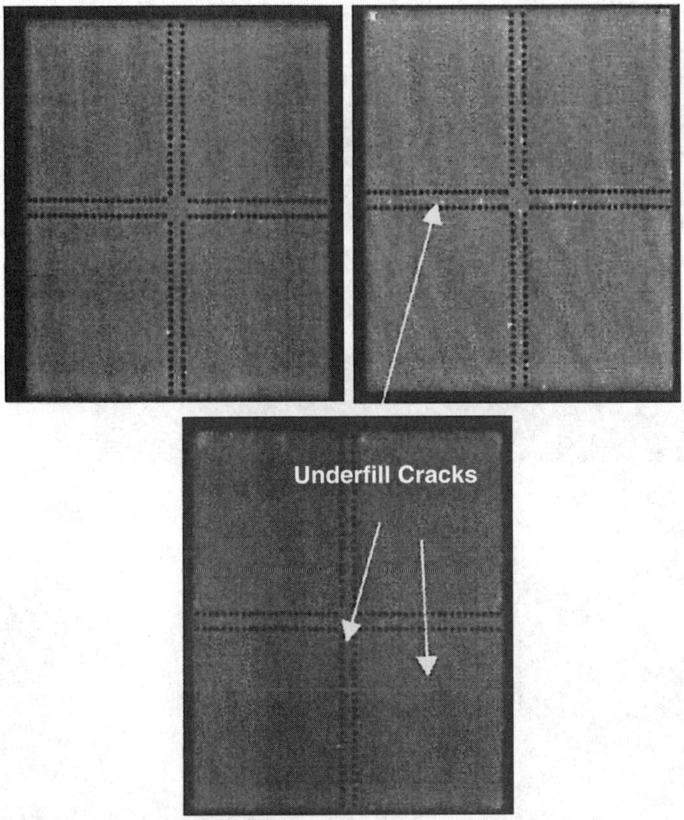

FIGURE 1.46 Underfill E C-SAM images at 0 cycles (*top left*), 1400 cycles (*top right*), and 2200 cycles (*bottom*).

FIGURE 1.47 C-SAM images of TV3 subject to LLTS at the chip-underfill interface (*left*) and the underfill-board interface (*right*).

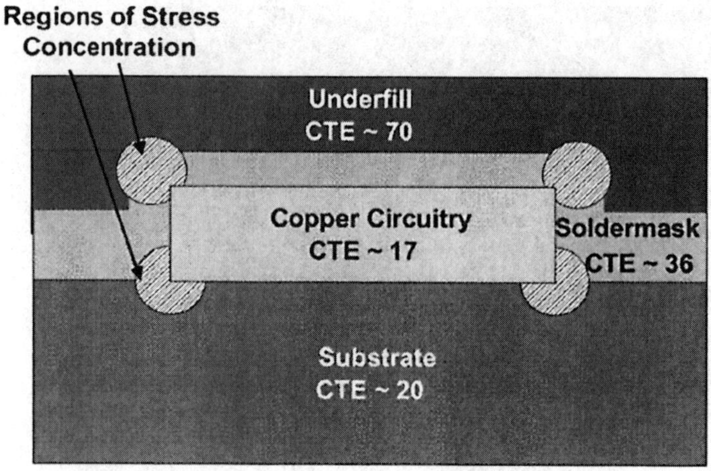

FIGURE 1.48 Stress concentration regions produced by substrate circuitry.

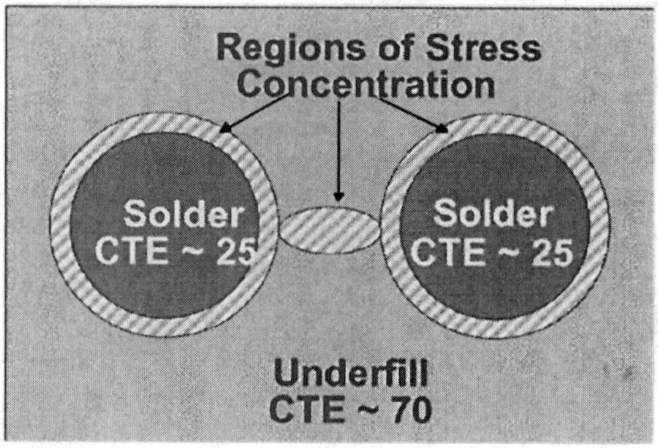

FIGURE 1.49 Stress concentrations caused by solder interconnections.

boards remains a time consuming and expensive process. This discussion focuses on the development of innovative material systems and process technology that can potentially make FC assembly transparent to SMT and high speed packaging lines by eliminating flux application, underfill application, and underfill cure from the assembly process.[15] The process under development is unique in that it uses a wafer-level underfill approach, a placement process comparable with CSPs (i.e., low placement forces), and underfill systems that enable simultaneous reflow of solder interconnections and cure of the underfill. The basic concept is to apply modified no-flow (reflowable) underfill encapsulants to FC bumped wafers as a form of wafer scale packaging. The wafer-applied underfills dry to a solid film, enabling packing, shipping, handling, and board assembly similar to CSPs. On reflow, the wafer-applied underfills liquefy, wet the substrate, and provide fluxing action on the solder-joints bond-pads, resulting in a high reliability underfilled FC. The wafer-applied fluxing underfill solid presents several non-trivial challenges, but none of the required properties are mutually exclusive from a materials standpoint.[11,15] The process is considerably different from other solid underfill film approaches such as substrate-applied dry-film laminated underfills[11] or wafer-applied solid-polymer film interposers and flux films.[11]

The new assembly process under development is shown in Fig. 1.4. It provides numerous advantages over conventional FC processes because it saves floor space and capital costs by removing underfill dispensers and cure ovens. Moreover, the wafer application of underfill is an enabling technology for fine-pitch FC processing (i.e., ≤150 μm). The new process could also increase worldwide FC implementation by reducing the process learning curve and by eliminating special training requirements necessary for current FC dedicated equipment. The program currently underway focuses on the development of a wafer-level application process, determination of the underfill material characteristics for application on bumped wafers, characterization of the FC assembly process, and reliability qualification of FC test vehicle assemblies. This discussion presents an overview of low-cost FC processing based on wafer-applied underfill systems, an analysis of several wafer-level deposition techniques on solder bumped wafers, an examination of critical factors and challenges involved therein, a feasibility analysis of high-speed placement of underfill coated bumped chips, and a characterization of the placement process based on bump and fiducial imaging using the vision system of a commercial placement system. Process challenges arise with the underfill application after bumping. This discussion investigates the challenges associated with polymer film thickness and uniformity, including edge effects, shadowing, and nonconformal coatings over the complex bump topologies of FC wafers. The effects of viscosity, wetting angle, and multiple underfill layers on coating uniformity are also discussed. The process technology, challenges, and preliminary yield analysis associated with the assembly of preunderfilled FCs directly to low-cost substrates is also presented. Of particular interest is how a placement machine handles, images, and places coated chips. Also addressed, but not analyzed, is the reflow of underfill coated wafers. This process step will be analyzed in future publications.

1.4.1 Desired Coating Characteristics

The reflowable underfill may be applied either before or after the prebumped wafer is diced. For the purposes of this discussion, only the case in which the wafer is coated before dicing is examined. In this case, the underfill is applied to a prebumped wafer over the complex surface topology of the bumps and active circuitry.

In order for the coated die to reflow and make a reliable connection, the coating should have certain desired geometric and surface properties. Some application methods produce shadowing effects behind the bumps as seen in Fig. 1.50*a*. This can lead to voiding in the underfill during FC assembly. Other methods promote a conformal coating as shown in Fig. 1.50*b*. This profile creates a thick film coating over the bumps, which can have an impact on the reflow process. A nonconformal geometry, shown in Fig. 1.50*c*, is the preferred coating geometry. During reflow, the underfill liquefies, the flux reduces the surface oxides, the solder wets the substrate pads, and the die collapses, forming the interconnections. As the die collapses, underfill flows under compression forces out past the die edges and wets up the sides of the die, forming fillets.

The coating geometries are determined in part by the zero-shear-rate viscosity of the underfill and by the coating method. If the zero-shear-rate viscosity is low enough, the material will tend to self-level and create the profile shown in Fig. 1.50*c*. The gravity-driven flow enabling self-leveling is governed by wafer-applied underfill rheology under low shear rates. The coating technique also has an impact on the nonconformal coating characteristics of the applied underfill film. Inherently, certain coating techniques promote nonconformal coats based on the physics of wetting. For example, surface tension driven coating techniques tend to promote nonconformal coatings, whereas pressure-driven coating techniques tend to promote conformal coating. Figure 1.51 shows the viscosity of a

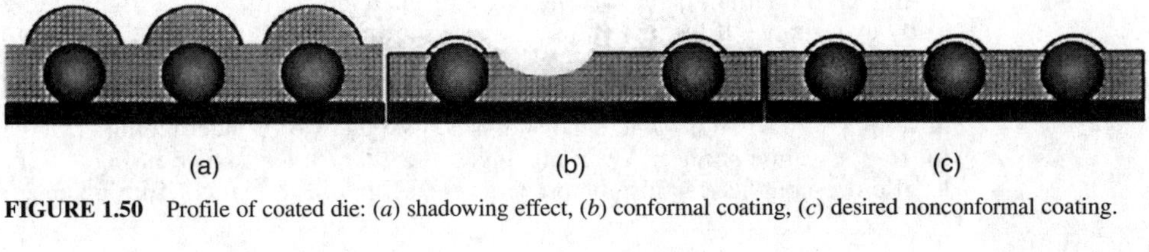

FIGURE 1.50 Profile of coated die: (*a*) shadowing effect, (*b*) conformal coating, (*c*) desired nonconformal coating.

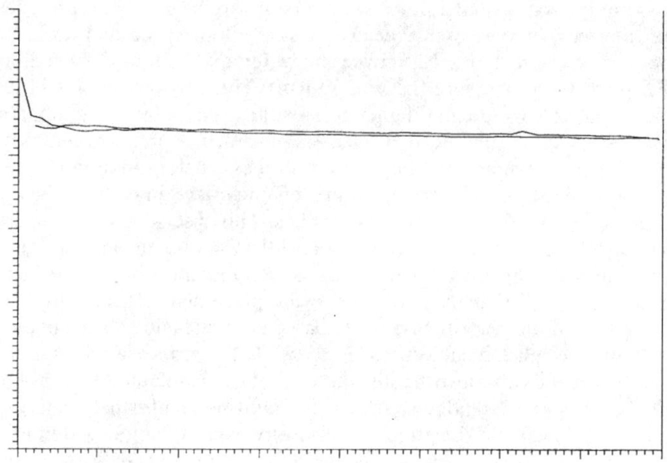

(a)

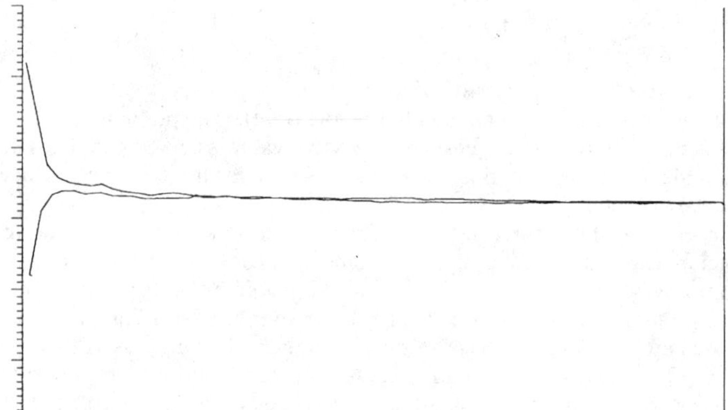

(b)

FIGURE 1.51 Apparent viscosity of material A under various shear rates: (*a*) material A and (*b*) material A diluted.

prototype wafer-level underfill as a function of varying shear rate. Both solvent-diluted and undiluted versions of a prototype underfill (material A in Table 1.24) were analyzed as representative materials. The prototype underfill rheology was measured using a TA Instruments rheometer at 25°C. Notice the order-of-magnitude decrease in the high-shear-rate viscosity when the underfill is diluted with 50 percent solvent.

Equally important is the wetting angle of the underfill material. Large wetting angles are especially undesirable. If the material behaves as shown in Fig. 1.52*a* with a wetting angle of larger than 90 degrees, then the underfill has a tendency not to wet the wafer surface or bump surfaces. Moreover, the underfill can be prone to dewetting the surface. Such large wetting angles can result in voiding in the regions where the bumps contact the wafer, undermining the integrity of the underfill to passivation surface. A wetting angle near 90 degrees, shown in Fig. 1.52*b*, improves wafer-level processing but can also be prone to underfill void formation. Wetting angles near 90 degrees also can cause the material to bead at the center of the die, resulting in placement problems during assembly. If the material is thickest at the center of the die, the die could roll or twist out of alignment after placement due to the instability from having a single contact point. Ideally, the wetting angle should be 50 degrees or less, as shown in Fig. 1.52c, for adequate surface quality. Processing coatings with this profile allows for more accurate volume control due to the planarity of the resulting coating geometry. Moreover, this profile exhibits more uniform coatings under conditions of multiple coating passes. Multiple coatings passes tend to exaggerate any previous coating nonuniformities. Numerous material rheologic studies have been performed to assess the required material properties for coating bumped wafers.

Wetting angles measured for prototype wafer-applied underfill materials on eutectic solder, nitride passivation, and solder mask are shown in Table 1.24 based on goniometer measurements taken at room temperature. The dilute versions of materials A and B are 50/50 mixtures by volume of material and solvent. As indicated in the table, the wetting angles remained relatively constant over the different surface materials that were coated. This is interesting in that one would tend to expect different wetting angles between the solder surface and the nitride surface due to their divergent surface energies. However, this does not seem to be the case. Diluting the prototype underfills with solvent resulted in a 50 percent decrease in wetting angle. Material B exhibited the largest wetting angle, approximately 45°. Based on the results to date, it would appear that wetting angles in the range of 35 to 50 degrees should provide adequate performance for coating bumped wafers.

1.4.2 Coating Methods and Materials

Determining the required material properties is the first step in developing the back-end wafer-level packaging process. The next step involves finding a consistent and accurate coating method by which the material may be deposited onto the bumped wafer. The need for a quantitative comparison of different coating methods exists. As a relative comparison of deposited underfill coating, a film quality factor was developed consisting of a *characteristic* film thickness divided by the standard deviation of the surface height variation given by Eq. (1.6). Note that this is a die-level comparison. The characteristic film thickness is taken as a representative film thickness at or near the die center. This was chosen because the maximum film thickness tends to occur at the center of the chip, and

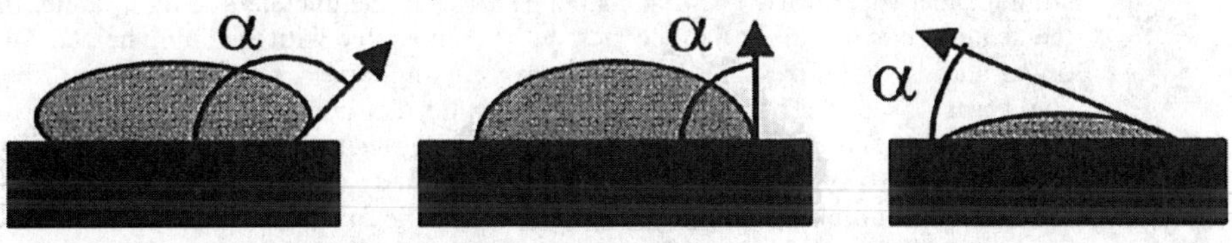

FIGURE 1.52 Profile of three different wetting angles. *a.* $\alpha > 90°$; *b.* $\alpha = 90°$; *c.* $\alpha < 90°$.

TABLE 1.24 Wetting Characteristics of Prototype Wafer-Level Underfills

	Wetting Angle (degrees)								
	Surface and Sample Diameter (cm)								
	Solder mask			Solder-covered substrate			Nitride passivation		
Sample	1	0.5	0.25	1	0.5	0.25	1	0.5	0.25
Material A	39	39	40	36	36	38	39	36	37
Material A Diluted	19	17	18	16	17	16	19	20	15
Material B	47	44	38	42	42	48	45	45	41
Material B Diluted	10	12	12	17	16	16	15	17	18
Material C	37	32	33	32	29	33	30	27	19

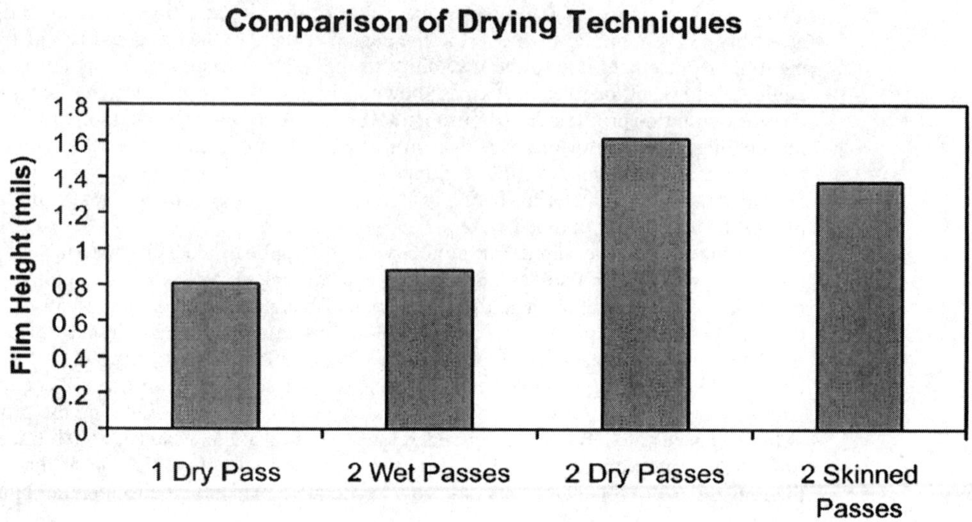

FIGURE 1.53 Film thickness for different drying techniques.

at this location film thickness is easily measured using a stylus contact surface profiling system (for perimeter I/O devices).

$$Q_{film} = \frac{h_{char}}{\sigma_{film}} \tag{1.6}$$

where h_{char} is the characteristic film thickness, and σ_{film} is the standard deviation of the surface height.

Each of the coating methods evaluated has advantages and disadvantages, with some demonstrating clear superiority. The first challenge arises in the thickness of the coating. In order to evaluate the coating, it is useful for the thickness to be comparable with the bump height. Most of the methods under analysis produce coats up to 25 μm in a single pass, whereas the wafer bump height in this study was 150 μm. Therefore, multiple coats are required to achieve the 150 μm value. Since the materials wet onto themselves differently than they wet the passivated wafer, a set of experiments was designed to investigate the differences. The first set compared the additive effects on surface thickness of multiple coatings by varying how dry the first coating was when the second was applied. The results in Fig. 1.53 show that a single coating thickness of 20 μm was achieved for material A after drying. The figure also shows two consecutive coatings with no drying between the first and second coats. In this case, a minimal increase in thickness over a single dried coating was observed. However, if the first coat was completely dry before the second was applied, the resuling coating thickness was essentially twice that of a single coat. If the second coating was applied after the first had dried partially, an intermediate thickness was achieved. This "skinning" process is of little production value because the amount of drying is difficult to control and dramatically affects the coating thickness. Note that all thickness measurements in Fig. 1.53 were taken after the material had been dried according to manufacturer's recommendations.

In the next phase of work, the effect of the different underfill materials and coating methods on surface quality and coating thickness was investigated. Two different materials and three different coating processes were studied. Fig. 1.54 shows the resulting coating thickness and standard deviations at the die level. Notice that material A Diluted yields a coating thickness much less than that of material A for the same coating method. This is largely due to the fact that material A is more viscous than its diluted counterpart. Material A diluted has a viscosity of 0.17 Pa·s, whereas material A has a viscosity of 2.2 Pa·s at room temperature. As expected, the coating thickness is highly sensitive to variations in viscosity. This observation holds true for most coating methods investigated. Also, note that method 2 yields a much thicker coating than method 1 for the same material. This result is not necessarily conclusive because there are unique ways to vary the thickness within each coating method, but it is representative for most experimental conditions studied. It is important to note that the surface quality obtained with method 2 was inferior to that obtained with method 1. The superior coating technique studied in this work was method 3 based on coating quality and thickness. It produces slightly thicker coats than method 1 and generates a surface with significantly less variation.

How the surface roughness and/or film uniformity changes as the number of coats increases is another characteristic of interest. Figure 1.55 shows the maximum film thickness h_{char} and standard deviation of the surface height σ_{film} for eight sites across a solder bumped wafer after a single coating using Method 3. Notice that the film thickness varies by as much as 10 percent and the height standard deviations are around 20 percent of the film thickness. Figure 1.56 shows the results for the film uniformity after five successive coatings completely dried between passes. As evidenced in this figure, the absolute variations in film thickness and standard deviations are the same as those seen with a single coating. This means that relative to the film thickness, the film surface variations have decreased. In other words, the quality factor has increased by a factor of 5. Closer observation of the data actually indicates that the quality factor has increased by a factor of 6, suggesting that this particular underfill material has self-leveling tendencies over the course of multiple coats. A disadvantage of the multiple coats is that they can result in a buildup of material on top of the solder bumps, which has the potential to compromise reflow. Fortunately, each coating pass deposits material on the tip of the solder bump having a thickness one or more orders of magnitude less than that deposited over the unbumped regions of the die.

Figure 1.57 shows a laser profilometer raster scan of a wafer that has been coated with a liquid prototype wafer-level underfill material. The image shows a single die in a wafer and adjacent die edges. The darkest areas are the bumps around the edge of the die adjacent to the saw streets, indicating the highest points in the measurement. Notice that the material is thickest in the central areas of the die, particularly in the upper right corner. The "beading" of this material was common due to

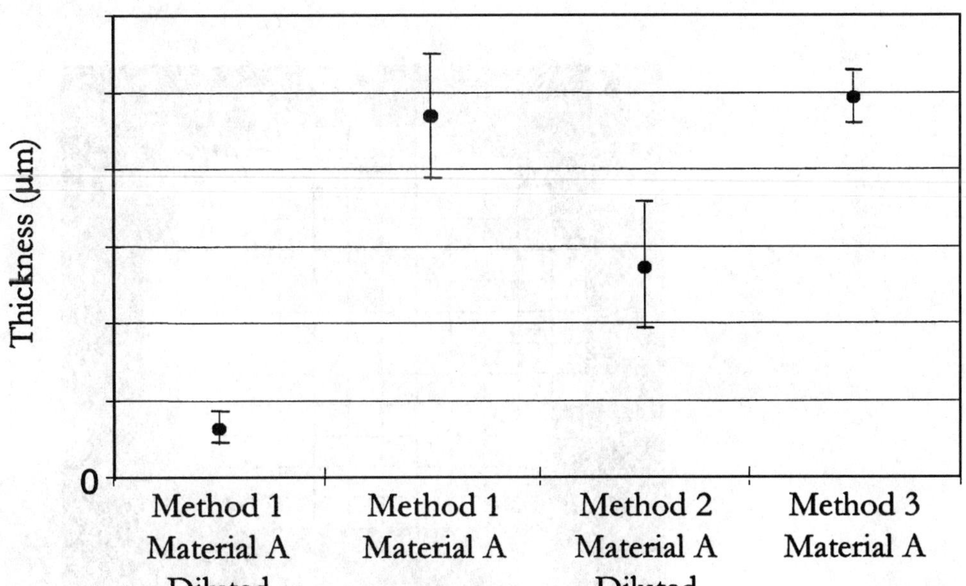

FIGURE 1.54 Comparison of different coating methods and materials.

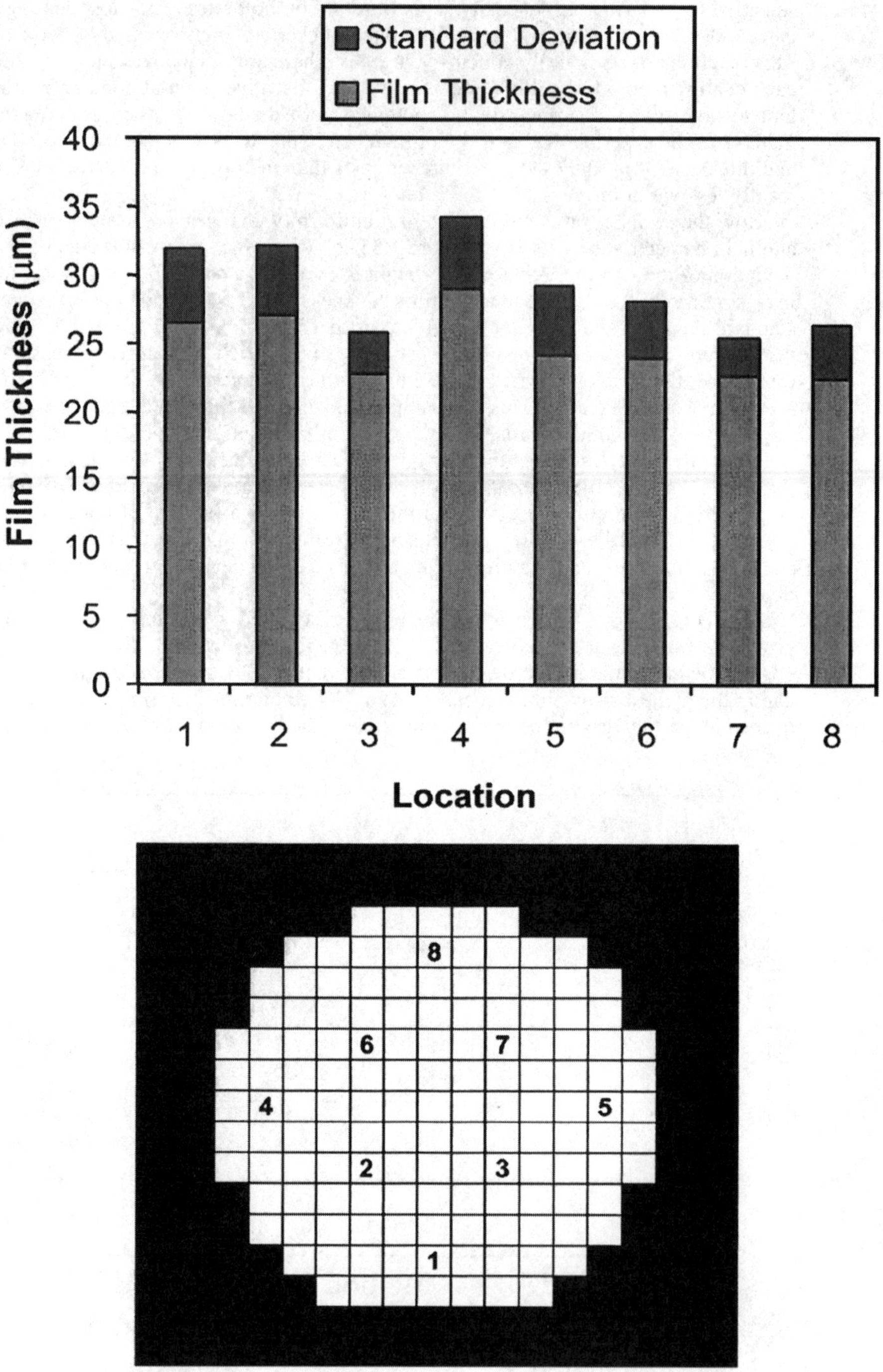

FIGURE 1.55 Film thickness results after one coat.

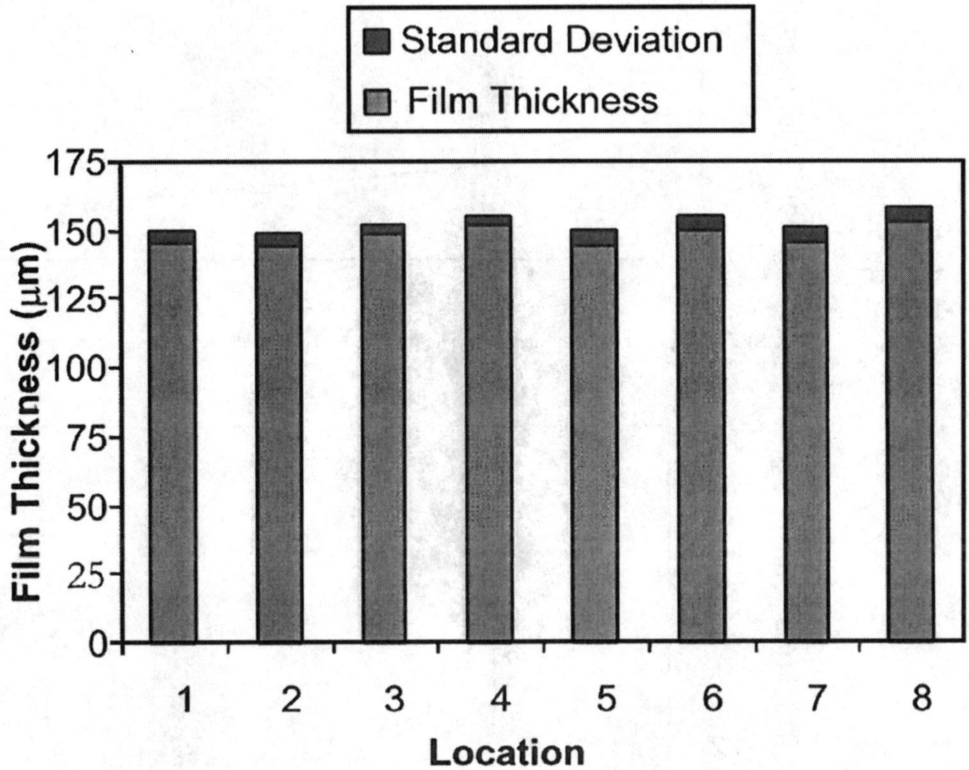

FIGURE 1.56 Film thickness results after five coats.

its rheology and wetting characteristics. This is undesirable because the curved surface caused by the beading reduces the amount of material that may be deposited on the die since the maximum film thickness should be maintained less than the bump height. Fortunately, most other candidate materials studied did not exhibit this beading behavior.

The last issue investigated with respect to coating methods was how thickness may be varied for a particular coating method. For most techniques, the deposition rate or speed was a critical factor for film thickness. Figure 1.58 shows the effect of increasing the application speed for coating method 1. For the range of this experiment, the correlation of increasing speed was nearly linear. As expected, the more viscous material (material A) yields a thicker coat than the less viscous one (material A dilute). The deposition method in Fig. 1.58 corresponds with method 3 in Fig. 1.54. This relationship between speed and thickness is unique in that the slope is positive. Most other methods of application have a negative slope, indicating an inverse relationship between speed and thickness. A number of process cycle time advantages can be derived from this positive coating speed versus thickness relationship.

1.4.3 Chip Imaging and Placement

In order for wafer-level underfill materials and process technology to be successful, placement machines must be capable of accurately aligning and placing the coated chips onto the substrate. Normally, key features on the chips and substrate are located using automated vision systems. These vision points are used as references in the alignment of the chip and the substrate. Key features on the chip are either fiducials or solder bumps. Key features on the substrate are either fiducials or bond

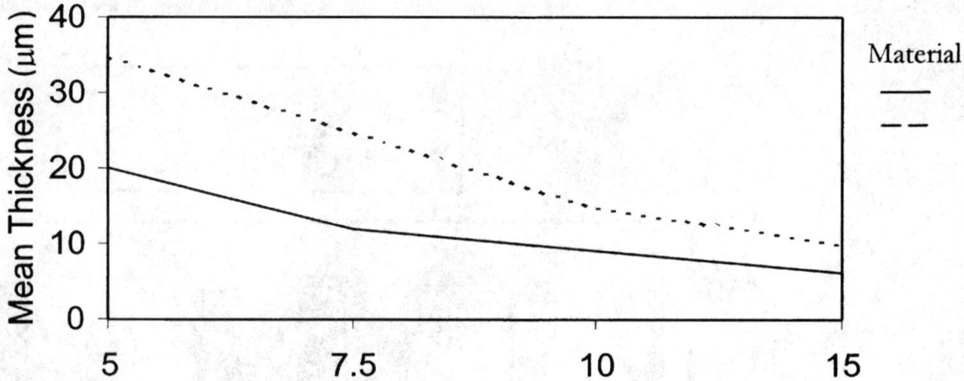

FIGURE 1.57 Raster scan of undiced solder bumped wafer at the die level. Image shows the coating thickness level versus gray scale.

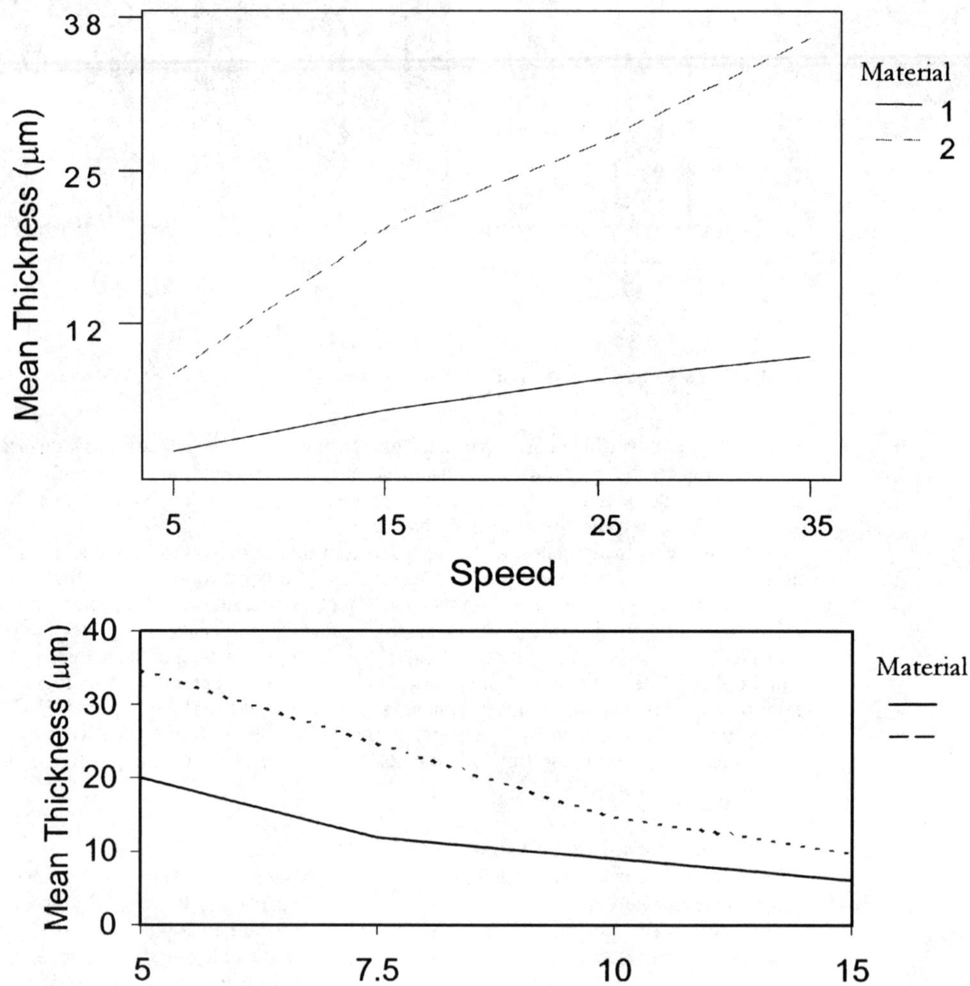

FIGURE 1.58 Film thickness results for different application speeds and materials: (*top*) method 1; and (*bottom*) method 2.

pads. Even if the coating is relatively uniform, the vision system may not be able to locate the alignment points on the chip due to the coating acting as an obstruction. With every coating method under consideration, some material covers the top of the solder balls. Fiducials are covered by the full thickness of the coated underfill material. A key issue to be investigated is whether a typical high-speed FC placement machine can vision and locate these alignment points in the presence of the underfill material coating. Transparent coatings would allow for the best imaging of bumps or fiducials. However, the wafer scale underfills being developed are filled materials. The high solid filler content makes current prototype materials translucent or opaque. Fortunately, many placement machines offer flexibility within their vision system to optimize imaging. Parameters that may be adjusted include illumination angle, light intensity, and image acceptance transform (level transforms).

Figure 1.59 presents an example of how the illumination angles affect feature brightness for coated and uncoated chips. The figure shows four illumination angles x-plane, plane, middle and steep based on the high speed FC placement system used to investigated FC placement of underfill-coated FCs. For each angle, the inside corners of four die are shown. Die 1 is a 200 μm pitch eutectic solder bumped FC device with silicon nitride passivation. Die 3 is a 356 μm pitch high-lead solder bumped FC device with polyimide passivation. For each illumination angle, the coated version of die 1 is shown in the upper left corner, the uncoated version of die 1 is shown in the upper right corner, the coated version of die 3 in the lower left corner, and the uncoated version of die 3 in the lower right corner. The placement machine vision system looks for contrast, which is usually provided by the high reflectivity of the bumps or fiducials compared with the low reflectivity of the die passivation and IC surface. Notice that the bumps on coated die 3 are most visible with steep illumination, whereas the uncoated die 3 bumps are more visible with the other three illumination angles. The coating is most visible in plane illumination for each die. For coated die 1, illumination angles x-plane, middle, and steep provided good contrast, with x-plane providing the highest contrast between the bumps and underfill coating.

1.4.4 Automated Vision Analysis of Underfill-Coated Chip

Design experiments were performed to determine the main effects and imaging system configuration that results in the placement machine successfully imaging the chips on a variety of chips and coatings.

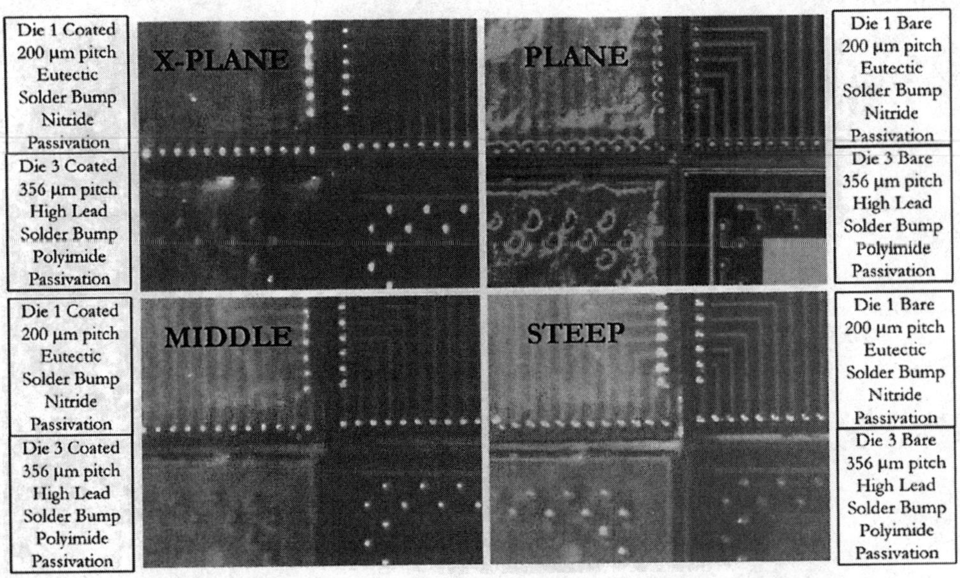

FIGURE 1.59 Comparison of illumination angles of x-plane, plane, middle, and steep for coated and uncoated version of die 1 and 3.

The high speed FC placement system used in this study enabled control of three imaging parameters: illumination angle, light intensity, and image acceptance transform (level transforms). Illumination angle refers to the angle of incidence of light striking the die. The placement system used has preset illumination angles referred to as x-plane, plane, middle, and steep. Each lighting angle used an array of red (680 nm) LEDs to illuminate the die. x-Plane lighting comes from a ring of LEDs located in the same horizontal plane as the die such that the light they emit illuminates the side of the die and the bumps. Plane is the next level of LEDs above x-plane. They are located in a circular array about 80 degrees above the die, but through a system of mirrors, the light strikes the die at an angle of approximately 30 degrees above the horizontal. The middle lighting level is a square array of LEDs. The light they emit strikes the die approximately 50 degrees above the horizontal. The final series of LEDs is steep. This is a vertical light source located much farther from the die than the other light sources and provides light perpendicular to the surface of the die via a mirror system. Light can be delivered from any combination of these angles simultaneously. Figure 1.60 shows a schematic of the lighting system, including the lighting angles, bounce mirrors, and CCD camera.

Another useful tool employed by vision systems is the acceptance transform function, or simple image digital filter. This helps the placement machine vision system recognize the features of the die; it is analogous to a light filter. By allowing certain colors to be emphasized or neglected, image acceptance transforms can create clear, well-defined images in which the important features of the die are recognizable to the vision software. Graphical representations of first-and second-order transforms are shown in Fig. 1.61. In selecting the proper combination of illumination angle, intensity, and transform, a greater percentage of image acceptance is obtained.

The test vehicles studied include five die of each of three types: die 1 with a single row of eutectic solder 200 mm pitch perimeter bumps and silicon nitride passivation, die 2 with two staggered rows of eutectic solder 356 mm pitch perimeter bumps and silicon nitride passivation, and die 3 with two staggered rows of high-lead solder 356 mm pitch perimeter bumps and polyimide passivation. Each of these die was coated with a prototype wafer-applied underfill system. Coating was achieved by stencil printing.

In order to design a relevant experiment, viable values for illumination levels and transform values had to be chosen. This is important because there are thousands of possible combinations, most of

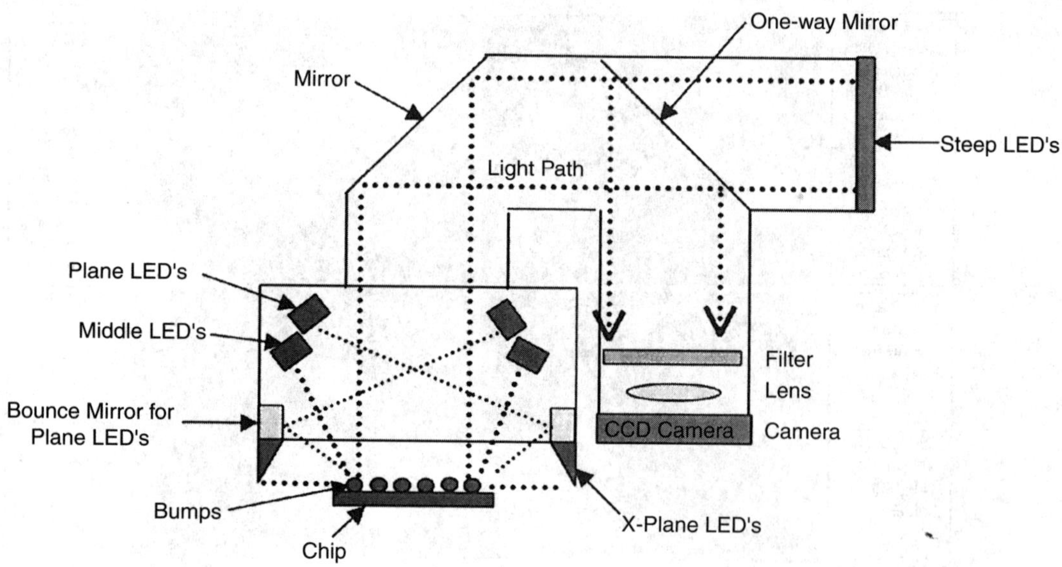

FIGURE 1.60 Commercial high-speed FC placement system vision system schematic.

which yield a null result. Preliminary tests were conducted to determine the combinations of illumination and transform values that resulted in feature recognition for the largest percentage of die. For each chip feature recognized and accepted, the vision system displays boxes around regions in the image that it considers to be the desired features. If the system targets the correct features for a particular type of die, such as corner bumps in the case of die 2, then that trial is deemed a success. The system either finds all features or none, so there is no ambiguity in declaring a particular test a success or a failure. Since each of the test die has different passivation, bump material, bump location, and bump size, different factor levels were chosen for each.

A five-factor design of experiments was performed to determine which illumination and transform values resulted in the placement machine recognizing bumps on the greatest number of underfill-coated chips. Each of the four illumination angles was a factor, with two levels of illumination each, one level being zero and the other selected such that when all other illumination angles are set to zero and a first-order transform is used, the best possible image of the die is provided. The fifth factor was the acceptance transform, which had four levels. The selected transforms represent both first-order, which is a linear adjustment of color, and three second-order cases, which allow piecewise manipulation of the spectrum (see Fig. 1.61). Each individual combination was tested ten times with each of the 15 test die, with the successful number of recognized images recorded for each combination.

Figures 1.62 through 1.64 show the main effects plots for illumination angles x-plane, plane, middle, and steep for die 1, 2, and 3, respectively. For die 1 shown in Fig. 1.62, plane and x-plane factors are instrumental to successful feature recognition. Middle and steep have little affect. Die 1 had 22 percent of the die recognized, the highest of the chip types tested, indicating that nearly one in four configurations tested resulted in successful bump recognition. The low acceptance percentage results from the fact that most configurations tested resulted in no recognition of the chip bumps or fiducials. A few of the illumination conditions studied yielded complete recognition and acceptance of the chips. For die 2, in Fig. 1.63, Middle was the main lighting angle providing effective illumination. Five percent of these die were recognized, indicating that only 1 in 20 configurations resulting in successful bump recognition. For die 3 shown in Fig. 1.64, plane and middle were the important effects. Eight percent of the high-lead solder, polyimide passivated die were recognized. It is important to note that for Die 3, the chip fiducials (covered with a prototype wafer-applied underfill coating) were imaged instead of the bumps that were used for die 1 and 2. The vision system was unable to image bumps on the bump-die-passivation-underfill combination of die 3.

From these results it can be concluded that certain illuminations made a positive impact on the imaging process, with the specified lighting angles being major factors enabling robust imaging for particular chip, bump, and passivation combinations. In addition, some transforms have proven to be unsuitable for the imaging of underfill-coated die. The overall percentage of successful image recognition configurations was low for this series of experiments. The design of experiments were set up to be primarily a main effects analysis, designed to selectively determine suitable combinations of image parameters and critical illumination parameters. Indeed, so many of the combinations were completely ineffective for imaging the underfill-coated die that their presence in the results lowers the overall success rate substantially. Fortunately, sufficient numbers of the individual combinations resulted in robust image recognition of the underfill-coated chips. Moreover, for the FC systems studied, it was found that bump layout, bump passivation, bump metallurgy, and chip passivation all played influential roles in FC imaging. For die 1 and 2, the eutectic solder bumps were imaged. For die 3 with the high lead-solder bumps and Polyimide passivation, the fiducials on the die were imaged.

Figure 1.65 shows the acceptance transform main effect plots, which was a four-level factor. Die 1 worked best with first-order transforms and second-order transforms with high cutoff values. Visioning of the FCs worked best using second-order transforms at specific illumination angles for die 2 and 3. In general, the vision reproducibility was very repeatable for the die configurations and underfill coatings studied. Consistency and uniformity of the coating were important factors that currently are under further investigation. Based on the current results, perimeter bumps near the edge of the die, such as die 1, were easier to image than perimeter bumps further from the die edge, as with die 2 and 3. High-lead bumps were not as easy to image due to their high lead content and consequent

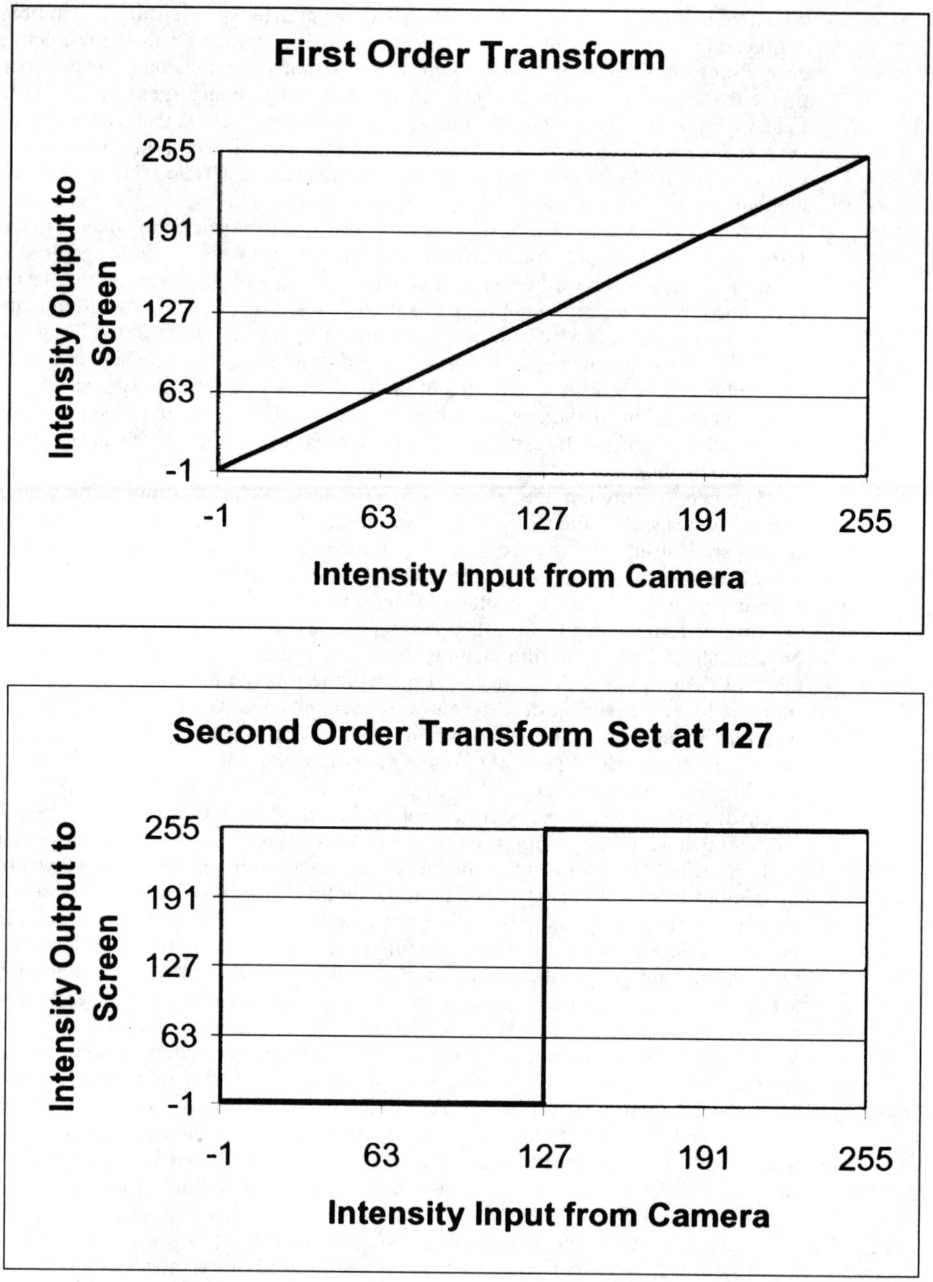

FIGURE 1.61 First- and second-order acceptance transforms for placement system vision.

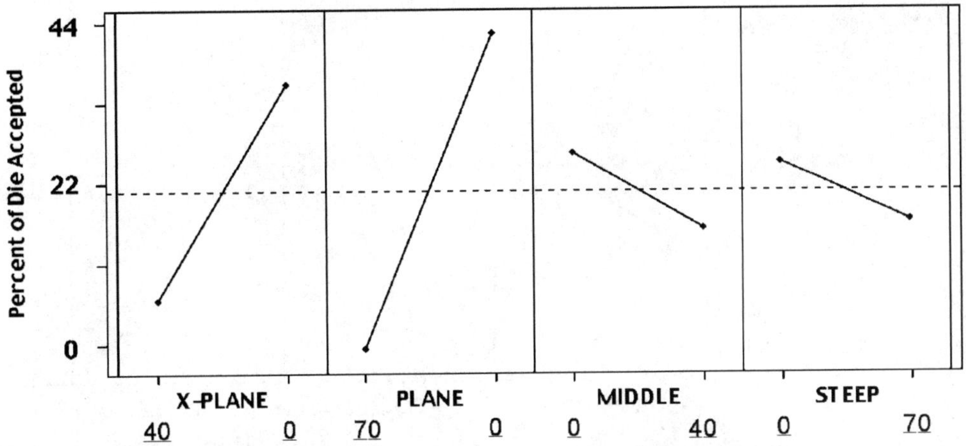

FIGURE 1.62 Main effects plot for die 1, eutectic solder, 200 μm pitch, silicon nitride passivated, underfill-coated die.

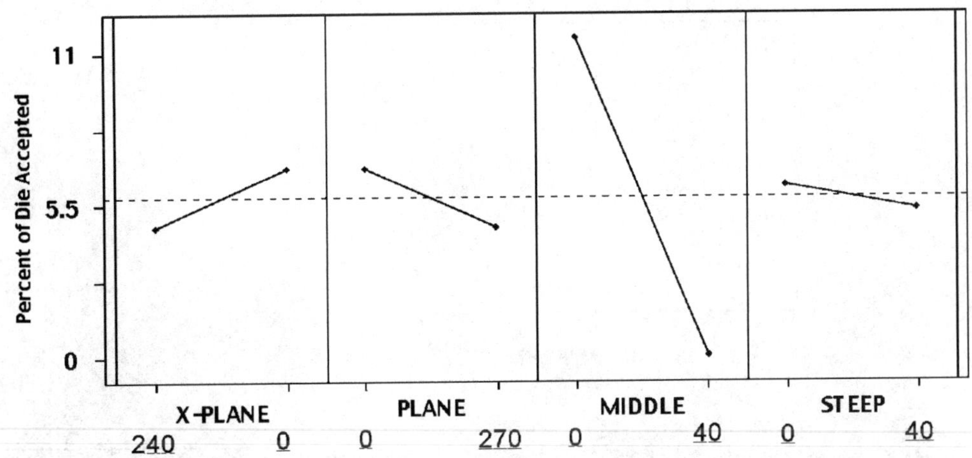

FIGURE 1.63 Main effects plot for die 2, eutectic solder, 356 μm pitch, silicon nitride passivated, underfill-coated die.

lack of reflectivity, evident by the die 3 results. Eutectic bumps provided better reflectivity, giving better contrast and between the passivation and the bumps. It appears that silicon nitride passivation provided high image contrast therefore better imaging of the underfill-coated die compared with polyimide passivation. In general, the silicon nitride was more reflective under the underfill coating, providing higher contrast with the solder bumps. This study indicated that given die with bump topologies and surfaces similar to those used with uniform underfill coatings, imaging of underfill-coated die for high-speed automated FC placement is feasible.

1.4.5 Analysis of Underfill-Coated Chip Placement Accuracy

This set of experiments was structured to examine how accurately underfill-coated die can be placed on a substrate. Potentially, the underfill coating can cause misalignment due to refraction through the coating material or shadows formed by the coating topology. Therefore, it is important to determine

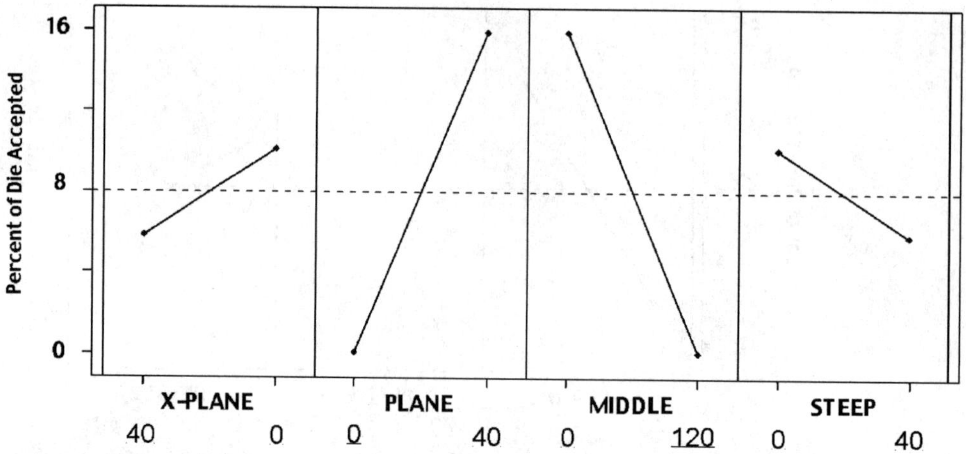

FIGURE 1.64 Main effects plot for die 3, high lead solder, 356 μm pitch, and polyimide passivated, underfill-coated die.

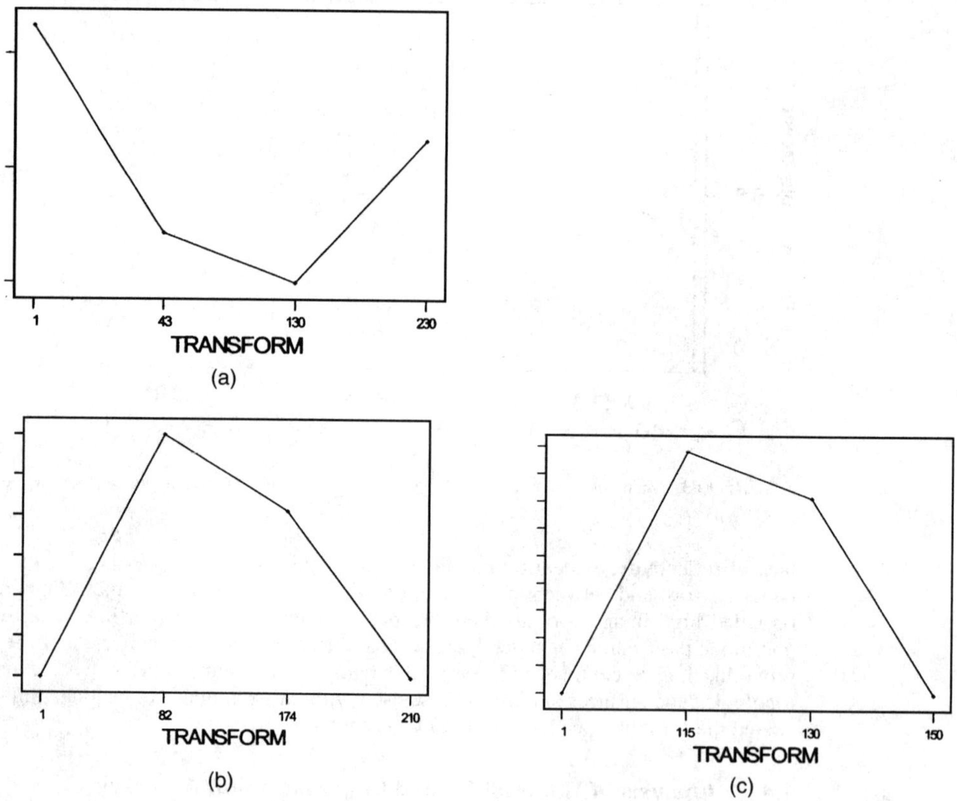

FIGURE 1.65 Main effects plots for (*a*) die 1, (*b*) die 2, and (*c*) die 3 between the vision system acceptance and the acceptance transform factors.

if the machine's image is a correct representation of actual die features because proper placement depends on the ability of the machine's vision system to correctly image bumps or fiducials on the die. Typically, global or local fiducials are used on the substrate. The goal of chip placement is to position the die such that each solder bump is aligned directly above its corresponding pad on the substrate site. This is a necessary but not sufficient condition for all the electrical connections to be formed during reflow. Although perfect alignment during placement is preferred, a die has been placed successfully if its bumps are sufficiently in contact with their respective pads to allow for proper wetting during reflow. This acceptable margin of error has been shown to be less than half the diameter of the minimum substrate pad.[15]

To determine the placement accuracy for die with preapplied underfill, a series of die-to-board placement experiments were performed using a Siemens SIPLACE 80 F5-DCA high-speed FC placement machine. Die 2 and die 3 were placed on matching substrates having six sites, with the die 3 substrates having a eutectic solder cap on the bond sites. Die 1 was placed on matching substrates having eight sites. A tacky adhesion promoter was applied on the substrate at each chip site to secure the chip after placement. X-ray inspection was performed using a Fein Focus Roentgen System providing a transmission x-ray image of the die bumps and substrate pads. The x-ray micrographs were used to determine the relative die-to-board offset. Using the bump's calibrated diameter as a reference distance, the distance between the center of the bump and the center of the pad was measured from the micrograph, providing an x, y, and θ offset measurement based on bump-to-pad distances measured at opposite corners of the die.

The results of the placement experiment indicated that chip placement accuracy and repeatability were equivalent to these of the same chip and substrate combination without preapplied underfill. All the die tested were consistently place within a 50 μm accuracy within the tolerance specifications for the placement machine. Table 1.25 shows the bump-to-pad offsets measured for 10 replicates of five coated FC devices of each die type. Based on these initial FC placement results, there was sufficient alignment of the underfill-coated FCs to allow for reflow. Of the die placed, die 3 showed the greatest offset. It is believed that this consistent placement offset is due to the lower illumination contrast features associated with the high-lead-solder bumps and polyimide passivation.

1.4.6 Wafer-Level FC Processing Summary

The wafer-level underfill FC packaging process, including underfill application, dicing, FC placement, and reflow, is complex, and a number of challenges remain. The investigation presented here has been designed to characterize the material properties necessary to build functional assemblies and demonstrate that the individual steps involved in processing are feasible. Future experiments will be aimed at increasing the surface quality of the underfill coating, evaluating the impact of coating quality on process yield, and increasing the yield of FC assemblies.

TABLE 1.25 Bump-to-Pad Placement Offsets for Six Sites (Measured in Microns)

		Die 1		Die 2		Die 3	
	Die	X Offset	Y Offset	X Offset	Y Offset	X Offset	Y Offset
Lower Right	1	25	25	−25	50	25	50
	2	25	50	−25	50	25	75
	3	0	50	−25	50	25	50
	4	0	50	−25	25	25	50
	5	25	50	−25	25	50	50
Upper Left	1	25	−25	25	25	50	0
	2	25	−25	25	0	50	0
	3	0	−25	0	25	25	25
	4	25	−50	25	25	25	0
	5	25	−25	0	25	50	0

In this study, several wafer-applied underfill coating techniques have been evaluated, and a number have been shown to be capable of depositing uniform nonconformal coatings over bumped wafers. Two preferred coating techniques were identified. A quality factor was defined as a metric to evaluate the various coating and drying techniques. Coating uniformity over 200 mm bumped wafers was also analyzed for select coating techniques exhibiting good uniformity over the wafers as well as locally over bumped die. A process characterization was presented for one of the coating methods studied. Various key material properties including underfill rheology, shear rate effects, and wetting angle were analyzed to determine the necessary material characteristics for uniform nonconformal coatings on bumped wafers.

Vision and illumination of underfill-coated bumped FC test die were also verified. Illumination angle and image acceptance transforms were the most critical factors having an impact on robust imaging of the coated die. Moreover, die configuration and topology were also important, particularly the bump material, the bump reflectivity, the die passivation, and the bump position. In addition to demonstrating the feasibility of imaging and accepting underfill coated FC die, placement accuracy of these innovative FC systems was also verified, exhibiting placement accuracy of underfill-coated die comparable with equivalent uncoated die.

1.5 ANALYSIS OF NEXT-GENERATION FC PROCESSES

Critical to the success of low-cost, high-throughput FC assembly is the ability to design process variables and predict the process window for a given incoming material stream. Several critical process steps have been analyzed and design models developed to enable process compatibility analysis and promote rapid implementation of the technology. The first analysis focuses on underfill encapsulant processing, looking at conventional capillary flow and compression flow underfill processing.

Two critical prereflow process steps are also analyzed. The first is the dynamic chip placement process. Based on computational fluid modeling of underfill compression flow during placement, there exists a theoretical limit to the placement velocity of a chip. The limiting placement speed exists due to the presence of underfill between the chip and the substrate during placement.

The second critical process step occurs after chip placement. Due to steady state forces acting on the chip, it is possible that the chip can "float" above the board if the underfill mass is not properly controlled. A process design model has been developed to understand the basic physics of the floating phenomenon and to predict process variables to eliminate process defects. Using these process design models, a reliable prereflow assembly process can be designed to ensure high first-pass yield.[1,2,4,12,19,22,24,28,31,32,38–40]

1.5.1 Potential Void Formation

In the low-cost FC process, voids in the underfill can form via three mechanisms. The first is due to outgasing during reflow. Common conditions promoting *outgasing voids* are dissolved moisture or solvents in the polymer films covering the chip and substrate surfaces. Outgasing is common from the solder mask, FR-4, polyimide chip passivation, etc. Heat treating (i.e., drying) the boards and chips prior to assembly eliminates this type of voiding. Outgasing can also occur within the underfill itself due to excessive ramp rates or temperature exposure during reflow. Higher ramp rates and elevated temperature exposure can cause the underfill constituents to volatilize during reflow, producing voids. Such voids can be extensive if the reflow profile is not designed correctly. Figure 1.66 shows scanning acoustic micrographs of FC assemblies produced using the new low-cost, high-throughput process at two different reflow profile ramp rates and peak temperatures.[24,25,28]

The second type of potential void formation occurs during compression flow chip placement and is due to *compression voids* forming at the chip surface. Simulation results predict the potential formation of compression voids at the intersection of the active face of the chip and the bumps, as shown in Fig. 1.67. Potential compression voids are at the base of the solder bumps adjacent to the chip. Note that the term potential compression voids is used because the simulations could not be run to the point where the solder bumps and the substrate make contact due to instabilities in the

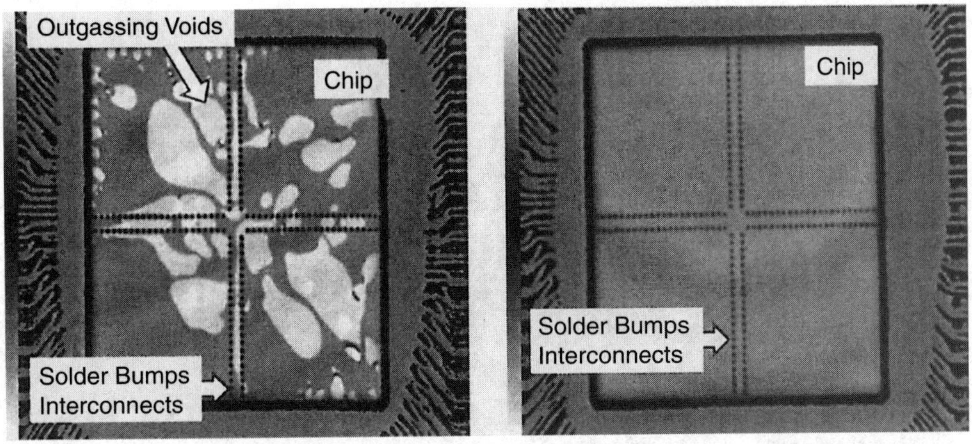

FIGURE 1.66 Scanning acoustic micrographs of low-cost next-generation fc assemblies demonstrating outgasing voids due to excessive reflow ramp rates. (*Left*) high ramp rate profile, (*right*) low ramp rate profile.

simulation software. Simulation results predict that compression voiding is minimal toward the center of the chip and that the number and size of voids increases toward the outer edge of the chip. Potential compression voids tend to be less prevalent at higher underfill compression flow pressures. By increasing the degree of cross-flow or squeeze flow, potential void formation is reduced. In general, void formation decreases with higher temperatures. Underfills with lower viscosity (i.e., lower temperatures in this case) are able to flow into the gaps and cavities more easily, reducing void formation. Exceptions to this trend occur when flow isolation and gravity flow take place. These cases result in "fill starvation" and large capture voids. Larger bumps and larger bump pitch reduce potential compression voiding. Simulation results also predict a strong interaction between the potential formation of compression voids and underfill print/deposition geometry. Lower ratios of the underfill print height to underfill print area result in predominantly compression flow and a higher probability of void formation at the bump-chip interface. Having a larger underfill print height to underfill print area ratio results in predominately squeeze flow and a lower probability of void formation.

Compression voids have been experimentally verified as occurring at the base of the solder bumps adjacent to the chip, as predicted by the flow simulations. Experimental results also show that the base underfill geometry deposited on the board strongly affects void formation. In agreement with the flow simulations, initial experiments indicate that compression and capture voids tend to occur under conditions where the fluid experiences compression flow in the z direction rather than a squeeze flow parallel to the chip and substrate surface. In general, lower print height to print area aspect ratios produce more voids. In contrast, higher print height to print area aspect ratios result in fewer compression and capture voids. The three-dimensional geometry of the deposited underfill is also found to be an important contributor to compression and capture void formation. This effect will be discussed further in future publications.

The third type of potential void formation is due to capture voids formed during placement of the chip where gas pockets (i.e., air in most cases) are captured between the chip and underfill surface. Capture voids, shown in Fig. 1.68 for an actual FC assembly after reflow, are typically larger than compression voids and are not localized to the bumps, like compression voids. The simulation results indicate the conditions under which such voids have the potential of forming.[23] For the lower underfill print height to underfill print area aspect ratio, the mean pressure and velocity profiles exhibit a saddle structure. The pressure profile is low at the center of the chip, increases forming a local maximum, decreases midway out the chip, increases to a second maximum, and decreases at the chip edge. Such a pressure profile implies that the underfill experiences a complex multidirectional flow where the underfill within the first pressure peak flows toward the chip center and the underfill outside the first pressure peak and within the second peak pressure flows toward the chip midpoint. Under these conditions, any capture and/or compression voids formed during chip placement in the

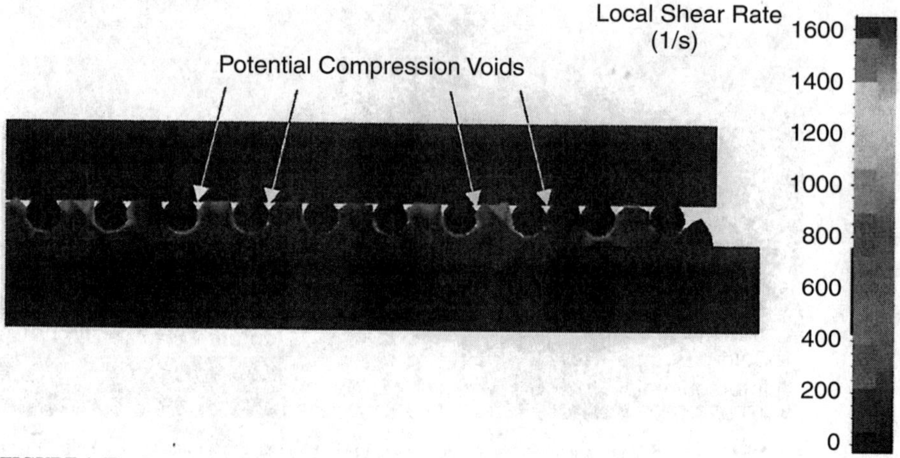

FIGURE 1.67 Compression flow simulation results showing potential void formation.

center or midpoint of the chip are isolated due to the pressure wells. The primary mechanism for eliminating such voids is entraining the interfacial voids in the cross flow of the underfill to expel the voids toward the chip edge. Controlling the underfill print/dispense geometry is also critical to eliminating compression and capture voids. The presence of pressure wells during assembly prevents entrainment of the voids in the cross flow resulting in the presence of capture and compression voids in the final assembly. In contrast, the flow simulation results show that a larger underfill print height to underfill print area aspect ratio results in a monotonically decreasing pressure profile from the chip center to the underfill-free surface at the chip edge. For such a profile, compression and capture void formation is minimized due to the ability of voids to be entrained in the cross-flow of the underfill drawing the void toward the underfill free surface at the chip edge. Preliminary experimental results support the simulation findings and suggest that capture voids are more prevalent in assemblies with a low underfill print height to underfill print area aspect ratio and that void formation decreases with increasing aspect ratio.

1.5.2 Design Guidelines

To reduce placement force requirements, larger bump size and bump pitch are preferred since smaller bump size and bump pitch resulted in higher placement forces. Selection of larger bump size and bump pitch also reduces the potential formation of voids during compression flow. Another design guideline concerns the underfill print height and viscosity as it relates to gravity flow. Results indicated that there is a clear tradeoff between the printed underfill height and the underfill viscosity. Lower viscosity materials in conjunction with higher print heights result in accelerated gravity flow and an underfill starved condition. Higher viscosity materials and lower print heights reduce gravity flow. However, lower underfill print heights reduce the extent of squeeze flow and increase compression flow ultimately leading the potential formation of more voids. Finally, due to its significant impact on pressure and placement force requirements, process temperature should be controlled to prevent gravity flow while providing for acceptable placement forces.[24,28]

The design guidelines derived from the simulation work are summarized below.

- Decreasing bump size and bump pitch results in larger placement forces.
- Potential voiding is reduced as bump size and bump pitch are increased.
- Very low viscosity underfills exhibit gravity flow, which leads to underfill starvation in the gap between the chip and substrate.
- Increased bump size and bump pitch result in smaller pressure gradients, thereby reducing the placement force requirements.

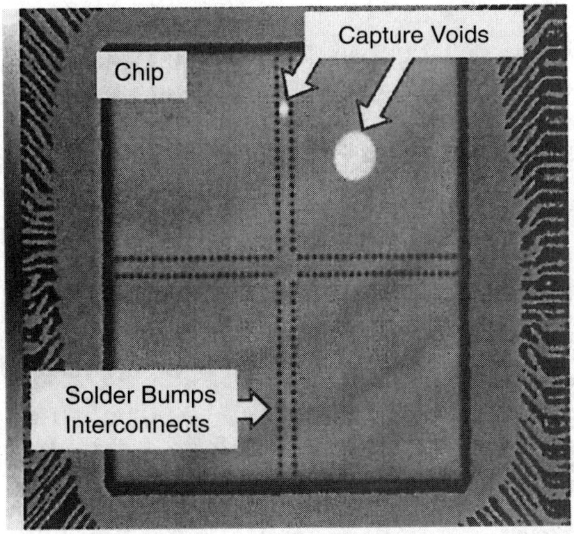

FIGURE 1.68 Scanning acoustic micrograph illustrating capture voids produced during compression flow chip placement.

- Increased temperature reduces underfill pressure and therefore placement forces.
- Placement forces of existing assembly equipment are sufficient for compression flow assembly.
- Higher-temperature processing during compression flow is not feasible done to the lower viscosity of the underfill, resulting in gravity flow and extensive void formation.
- Higher underfill print heights result in more squeezing flow, which provides the maximum pressure at the center of the chip and leads to better gap filling.

1.5.3 Placement Force Characterization of Compression Flow Underfill Processing

It is also important to analyze the flow pressure and placement force requirements for the compression flow underfill process to determine if standard placement systems can achieve the required placement forces. To accomplish this, a squeeze flow analysis between parallel plates can be used. Adopting Leider and Bird's[3,17,18] squeeze flow analysis with appropriate modifications for the flip chip compression flow case, the radial component of the general momentum equation (in cylindrical coordinates) (see Fig. 1.64) yields:

$$\rho\left(\frac{\partial v_r}{\partial t} + v_r\frac{\partial v_r}{\partial r} + \frac{v\theta}{r}\frac{\partial v_r}{\partial \theta} - \frac{v_\theta^2}{r} + v_z\frac{\partial v_r}{\partial z}\right) =$$

$$-\frac{\partial P}{\partial r} - \left(\frac{1}{r}\frac{\partial}{\partial r}(r\tau_{rr}) + \frac{1}{r}\frac{\partial \tau_{r\theta}}{\partial \theta} - \frac{\tau_{\theta\theta}}{r} + \frac{\partial \tau_{rz}}{\partial z}\right) + \rho g \tag{1.7}$$

For quasi-steady state, symmetric, inertia free flow and neglecting gravity effects, Eq. (1.7) simplifies to Eq. (1.8) where the shear stress gradient is proportional to the radial pressure gradient of the underfill. A schematic of the compression flow system under analysis is shown in Fig. 1.69.

$$\frac{\partial \tau_{rz}}{\partial z} = -\frac{\partial P}{\partial r} \tag{1.8}$$

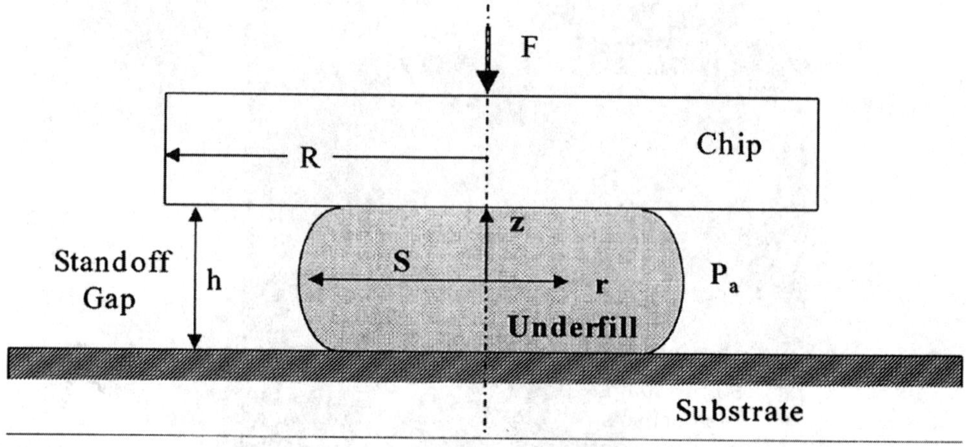

FIGURE 1.69 Compression flow analysis schematic for non-Newtonian squeeze flow between a chip and substrate.

Approximating the underfill flow as a shear thinning (or thickening) non-Newtonian fluid, yields an expression relating the shear stress and velocity gradient over the standoff gap.

$$\eta = m\,\dot{\gamma}_{rz}^{n-1} = m\left(-\frac{dv_r}{dz}\right)^{n-1} \tag{1.9}$$

$$\tau_{rz} = m\left(-\frac{dv_r}{dz}\right)^{n} \tag{1.10}$$

Integrating Eq. (1.8) with respect to z and combining with Eq. (1.10) results in an expression relating the radial pressure gradient and underfill rheology to the velocity gradient over the standoff gap.

$$-\frac{dv_r}{dz} = \left(-\frac{1}{m}\frac{dP}{dr}\right)^{1/n} z^{1/n} \tag{1.11}$$

Integrating Eq. (1.11) with respect to z yields an expression for the radial velocity profile over the standoff gap.

$$v_r = \left(\frac{h}{2}\right)^{1/n+1} \frac{1}{1/n+1}\left(-\frac{1}{m}\frac{dP}{dr}\right)^{1/n}\left[1-\left(\frac{2z}{h}\right)^{1+1/n}\right] \tag{1.12}$$

To relate the radial flow velocity to the chip placement speed, conservation of mass is applied to a cylindrical control volume of radius r encompassing the underfill material. Continuity then yields Eq. (1.13):

$$-\dot{h}\pi r^2 = 2\pi r\int_0^{h/2} v_r dz \tag{1.13}$$

Substituting into Eq. (1.13) the radial velocity profile (1.12) and integrating over the standoff gap height yields an expression for the pressure gradient over the chip surface:

$$-\dot{h}\pi r^2 = 2\pi r\left(\frac{h}{2}\right)^{1/n+1}\frac{1}{1/n+1}\left(\frac{1}{m}\frac{dP}{dr}\right)^{1/n}\int_0^{h/2}\left[1-\left(\frac{2}{h}\right)^{1+1/n} z^{1+1/n}\right]dz \tag{1.14}$$

$$\frac{dP}{dr} = -\frac{2^{n+1}m}{h^{2n+1}}\left[\frac{2n+1}{n}\right]^n\left(-\dot{h}\right)^n r^n \tag{1.15}$$

Integrating (1.15) over the wetted radius S gives an expression for the pressure distribution over the chip surface:

$$P = P_a + m\left(\frac{2n+1}{2n}\right)^n \frac{S^{1+n}}{(n+1)} \frac{(-\dot{h})^n}{\left(\frac{h}{2}\right)^{2n+1}}\left[1-\left(\frac{r}{S}\right)^{1+n}\right] \tag{1.16}$$

Next, it is important to estimate the placement forces necessary to achieve standoff heights typical of solder flip chip assembly prior to interconnect reflow (typically 75 to 100 (μm). This follows because for the compression flow underfill processing technique, chip placement and underfill flow occur simultaneously. The placement force can be estimated by integrating the pressure profile (Eq. 1.16) over the chip area. Assuming a constant volume of dispensed underfill V the placement force is given by Eq. (1.17) for non-Newtonian fluids. For Newtonian fluids ($n = 1$ and $m = \mu$), the placement force is given by Eq. (1.18).

$$F = m\pi\left(\frac{2n+1}{2n}\right)^n \frac{(-\dot{h})^n}{(h/2)^{2n+1}} \frac{\left(\frac{V}{\pi h}\right)^{\frac{3+n}{2}}}{3+n} \tag{1.17}$$

$$F = \frac{3\pi\mu\,(-\dot{h})\left(\frac{V}{\pi h}\right)^2}{h^3} \tag{1.18}$$

1.5.4 Simulation Analysis of Compression Flow Underfill Processing

This section focuses on flow simulation studies of the compression flow chip placement process using commercially available fluid flow analysis software.[24,28] It represents a fundamental advancement in compression flow simulation of polymers in its successful application to the complex geometries and surface topologies demanded by miniaturized FC assembly. Here a simulation methodology is developed and simulation studies are conducted to characterize the compression flow of the underfill, estimate required chip placement forces, evaluate the effect of underfill geometry, and assess the potential formation of voids. A major output of this work is process design guidelines for the next-generation FC process. Moreover, these studies give insight into a preliminary process window and outline the range of acceptable process parameters including underfill deposition geometry and underfill viscosity.

1.5.4.1 Finite Element Simulations. Analytical modeling of the compression flow process presented earlier serves as a starting point for the numerical simulations. Placement force estimates based on the analytical model demonstrate the feasibility of the process and give insight into acceptable process parameters (underfill viscosities, placement forces, etc.). Numerical modeling is chosen to address some of the potential yield and reliability issues (i.e., void formation, penetration of underfill by bumped chips, etc.) and gives insight into the flow characteristics of compression flow chip placement. The primary intent of the numerical simulation work is to characterize the compression flow process. The process characterization includes predicting the local shear rates, local velocity profiles, underfill pressure profiles, streamline bending effects, printed underfill shape (aspect ratio) effects, and potential void formation.

To implement numerical flow simulations, first, the parameters associated with compression flow chip placement are identified. The simulation parameters are divided into two categories: assembly and process parameters. *Assembly parameters* are those directly associated with the geometry of the compression flow chip placement process. *Process parameters* are those linked to the underfill material properties and process operating variables, such as material viscosity and process temperature.

The assembly parameters defined for the compression flow chip placement process include the chip geometry, bump geometry, and printed underfill geometry (i.e., the printed aspect ratio). The *aspect ratio* of the printed underfill is defined as the ratio of the print length to printed height (i.e., in two dimensions). These parameters are used to specify the finite-element mesh models for the numerical simulations. In the current work, only two-dimensional compression flow studies are performed.

Process parameters identified for compression flow chip placement include the isothermal process temperature (ultimately influencing the apparent viscosity of the underfill), chip placement speed, and the underfill material rheology. Figure 1.70 illustrates the various process parameters, where T is the temperature of the chip, underfill, and substrate; σ is the underfill surface tension; η is the underfill apparent viscosity; and $\dot{z}$ is the placement speed. Isothermal process temperatures are selected based on typical underfill dispense temperatures ranging from 25 to 80°C. In this preliminary investigation, the compression flow process is modeled as isothermal; however, this may not be the case for the actual assembly process. Future studies will explore the effects of nonisothermal process conditions in the compression flow process. The *chip placement speed* is defined as the z-axis speed of the placement machine's head during the final chip placement, sometimes called the *search speed.* In general, placement systems use high-speed motion to bring the chip close to the final placement position, followed by a slower search speed along the z-axis until a programmed placement force is reached. The placement speeds selected are based on reported specifications for commercial equipment ranging from 0.3 to 1.4 cm/s. Underfill materials are approximated as Newtonian or non-Newtonian fluids depending on their rheologic characteristics. Figure 1.71 presents the experimentally measured rheology of a typical commercial no-flow underfill material. Notice that this no-flow underfill exhibits Newtonian shear thinning and shear thickening characteristics over the shear rates and temperatures studied.

Two additional process parameters, shown in Fig. 1.70, exist for the compression flow process, placement force F_N and underfill surface tension γ. Unfortunately, the software chosen for analysis is unable to allow control of these quantities for a compression flow process. In general, compression flow modeling for polymer flows is extremely difficult due to the moving boundary and the free surface motion. For the software chosen, normal force control for free surfaces is not possible. The same scenario is true for control of the surface tension variable. Surface tension control for free surfaces is not possible for compression flow simulations, but nevertheless surface tension is held constant throughout the simulations. Specifying surface tension is important when comparing the performance of different underfill materials in compression flow processing.

Numerical modeling of the compression flow chip placement process is conducted using commercially available fluid flow software. POLYFLOW version 3.5.3 is chosen because of its advanced compression flow analysis capabilities for polymeric materials. Modeling of this process is essentially the simulation of compression molding with a very complex boundary. Although is another compression molding package is available, POLYFLOW offers the flexibility needed in terms of simulating the complex chip geometry in the compression flow chip placement problem. Compression molding is a relatively new feature in POLYFLOW, and its use in this study of compression flow chip placement is a unique application.

1.5.4.2 Parametric Studies. The response quantities selected for the simulation study include void formation, shear rates, fluid pressure distribution, local shear rates, fluid velocities, and streamlines. POLYFLOW allows for the visualization of both the mold and fluid throughout the simulations. Flow characterization includes color contour maps and cartesian plots of the pressure, velocity, and streamlines of the compression flow along defined sections. In addition, the maximum value of each quantity is recorded and plotted in order to characterize its behavior throughout an entire flow process. Visualization of the flow is achieved by plotting the deformed mesh at various time steps. The location and identification of potential voids are accomplished visually from the simulation output. The finite-element simulations are designed as two-dimensional, axisymmetric, time-dependent flow problems. Modeling is done axisymmetrically to minimize the CPU time required for a simulation.

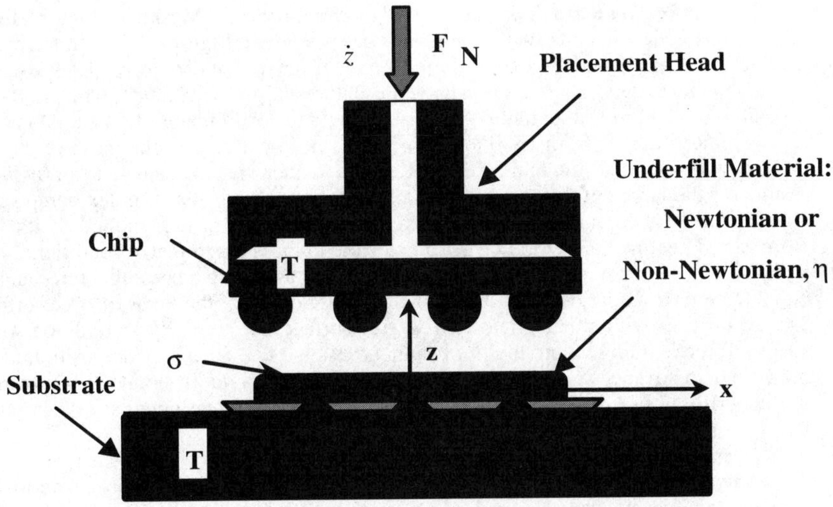

FIGURE 1.70 Process parameters for the simulation of compression flow chip placement.

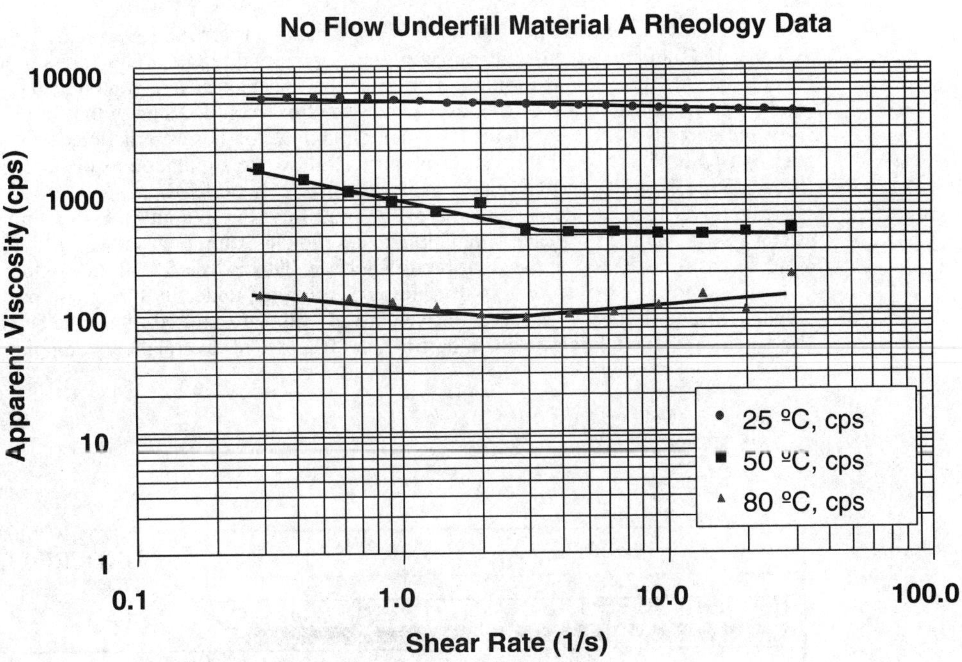

FIGURE 1.71 Rheology of a typical no-flow underfill at various temperatures.

1.5.4.3 Finite Element Mesh and Boundary Conditions. Mesh topology is divided into faces, edges, and vertices for the two-dimensional cases studied. Figure 1.72 is an example of one of the finite-element mesh models generated for the two FC devices studied. A mesh consists of 3 subdomains and 10 boundary sets. Two subdomains represent the chip and substrate, whereas the remaining subdomain (i.e., the fluid subdomain) models the liquid underfill material. The total number of elements for each mesh ranges from 1500 to 2300. Notice in Fig. 1.72 that the mesh of the underfill subdomain is very fine near the bump-fluid interface. A finer mesh is selected to provide more accurate results for the large deformations, especially in the regions of the complex bump geometry.

The fluid subdomain has three different boundary conditions applied to its four boundaries. Boundary 1, defined along the center axis of the chip, is a symmetry condition that allows for the simulation of the axisymmetric geometry. It is defined as having a tangential force and normal velocity equal to zero. Boundaries 2 and 3, the top and outer edges of the underfill-free surfaces, are defined as free surfaces with contact detection. A free surface is essentially a condition where a vanishing surface force is imposed and the fluid cannot cross the free surface. The position of the boundary is not known in advance and is therefore part of the solution. Finally, boundary 4, the interface between the underfill and substrate, is defined as a standard mechanical boundary where either the surface traction or velocity components were specified.

Two mesh models are created for the simulations performed in this work, model I and model II. These models represent two chip geometries with different underfill aspect ratios. Each simulation problem includes the density, gravity, and inertial terms of the underfill material during compression flow. By defining a value for the density, an incompressible fluid is specified. Inclusion of gravity allows for more accurate results and the prediction of gravity flows. Keeping inertial terms in the momentum equations makes the analysis more realistic and comprehensive with respect to fluid motion during rapid compression cycles.

1.5.4.4 Simulation Results. Numerical flow simulations are run for each mesh model. Two underfill materials and six viscosity parameters (i.e., isothermal temperatures) are simulated. Simulation results include the output of the mesh deformation, local shear rate contour, fluid pressure contour, and the fluid velocity profile. The results show the final recorded step of the simulation for the shear rate, pressure, and velocity outputs. The major highlights of these simulations are shown in Figs. 1.73 through 1.81. Figures 1.73 through 1.76 and 1.79 show the simulation results for mesh model I, and Figs. 1.77, 1.78, 1.80, and 1.81 show the results for mesh model II.

Figure 1.73 shows the progression of the fluid flow throughout an entire simulation. The local shear rate is recorded for each simulation, and a typical example is shown in Fig. 1.74 for model I. The shear rate is largest at the fluid-bump interface. This is true for all the compression flow simulations run and is expected due to the bumps piercing the underfill during compression. Looking at Figs. 1.75 and 1.76, the pressure and velocity profiles are plotted along the sections labeled AB. Notice for this model I case with underfill A at 50°C, we have a relatively complex pressure profile.

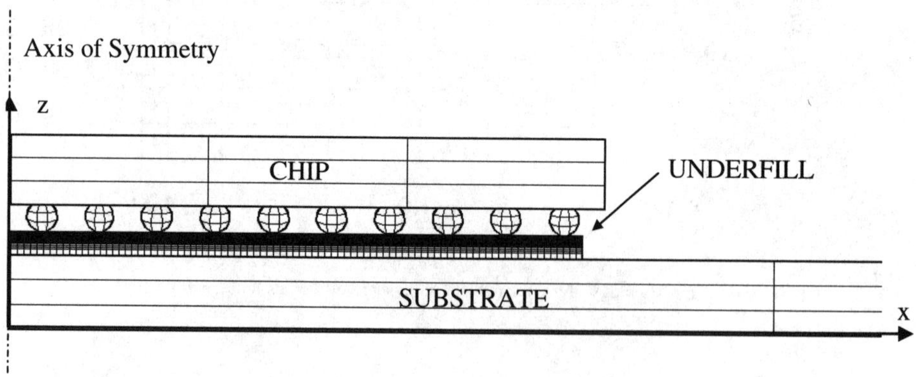

FIGURE 1.72 An example of a compression flow finite element mesh.

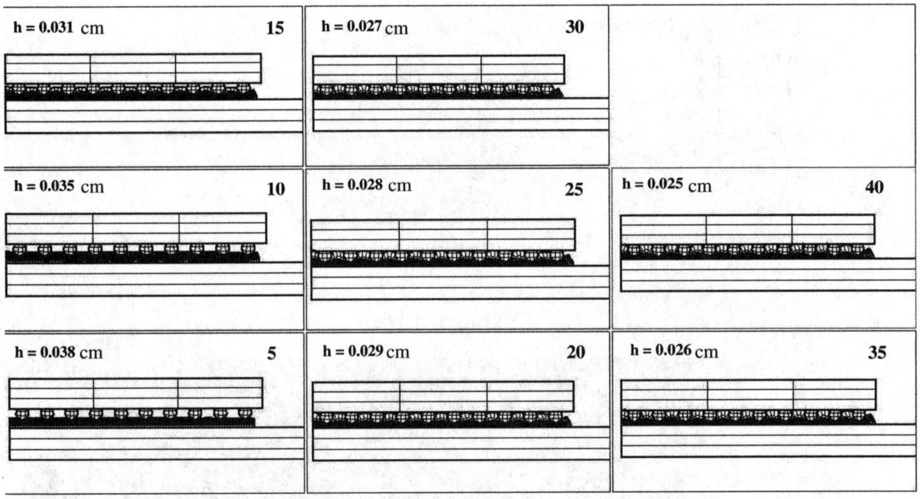

FIGURE 1.73 Progression of the compression flow process—vendor A, model I, T = 50°C.

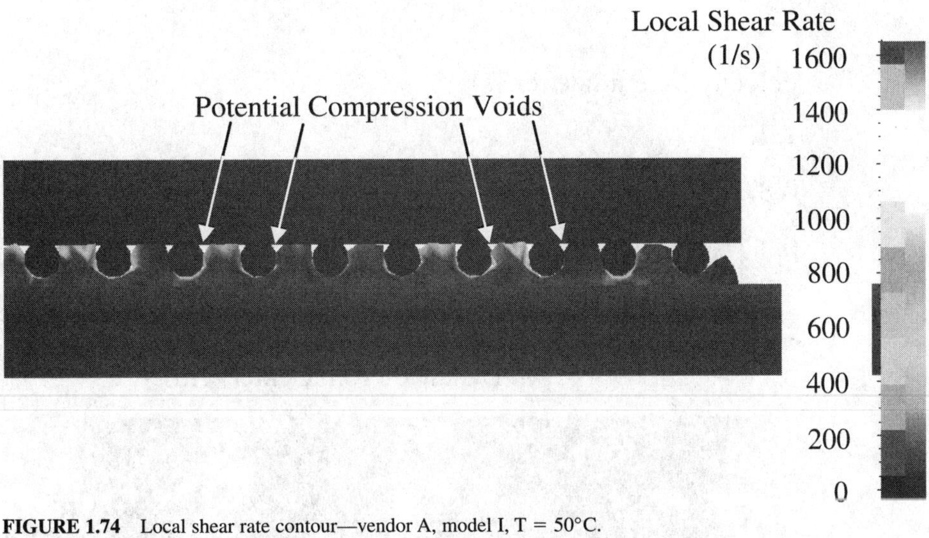

FIGURE 1.74 Local shear rate contour—vendor A, model I, T = 50°C.

Superimposed on the local pressure and velocity spikes under each bump, a saddle profile is seen where the mean pressure is low at the device center and again midway out the half chip length. Pressure peaks are seen just beyond the center and toward the outer edge of the chip. We see a similar trend in the underfill velocity in Fig. 1.76 with localized velocity spikes at the bumps, lower velocity at the center of the chip, lower velocity midway out the half chip length, and velocity peaks between the valleys.

Figure 1.77 shows model II with underfill B processed via compression flow at 80°C. Notice here that the underfill flow proceeds axially along the substrate surface prior to the chip and bumps making contact with the underfill. Under these conditions, the underfill is experiencing gravity-driven flow, where the material flows under its own weight. The velocity profile in Fig. 1.78 shows

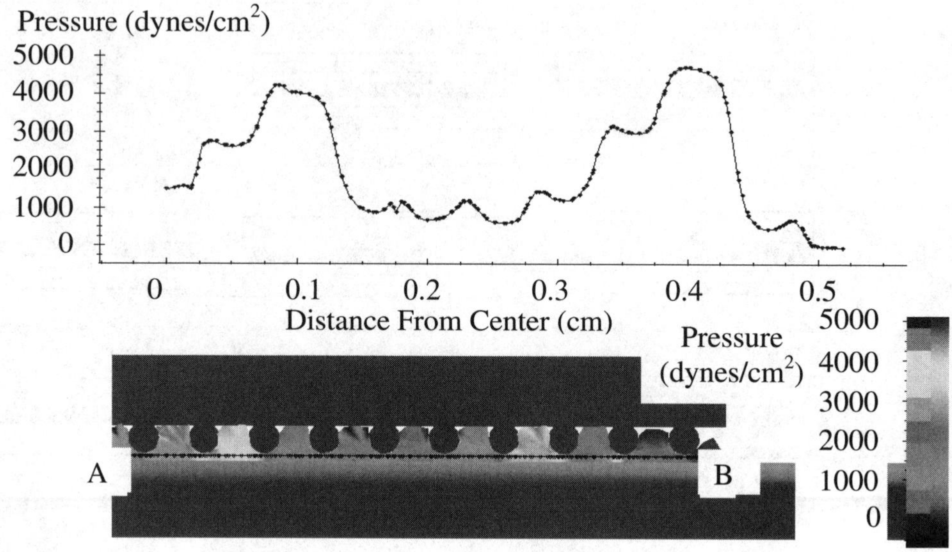

FIGURE 1.75 Fluid pressure contour and section profile (A-B)-vendor A, model I, T = 50°C.

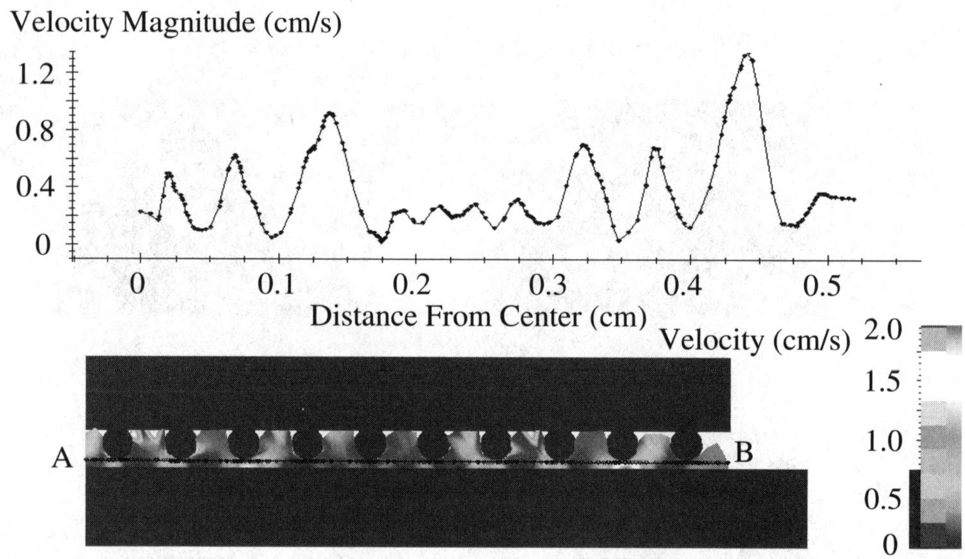

FIGURE 1.76 Velocity contour and section profile (A-B)—vendor A, model I, T = 50°C.

the underfill velocity significantly increasing toward the chip edge. Figure 1.79 also shows gravity flow in this case for model I with underfill A at 80°C. Gravity flow is characteristic of both underfill materials studied at higher temperature.

Figures 1.77, 1.78, 1.80, and 1.81 show the second chip model, model II, simulated. This model has a larger bump size and pitch compared with model I. Based on the simulation results of

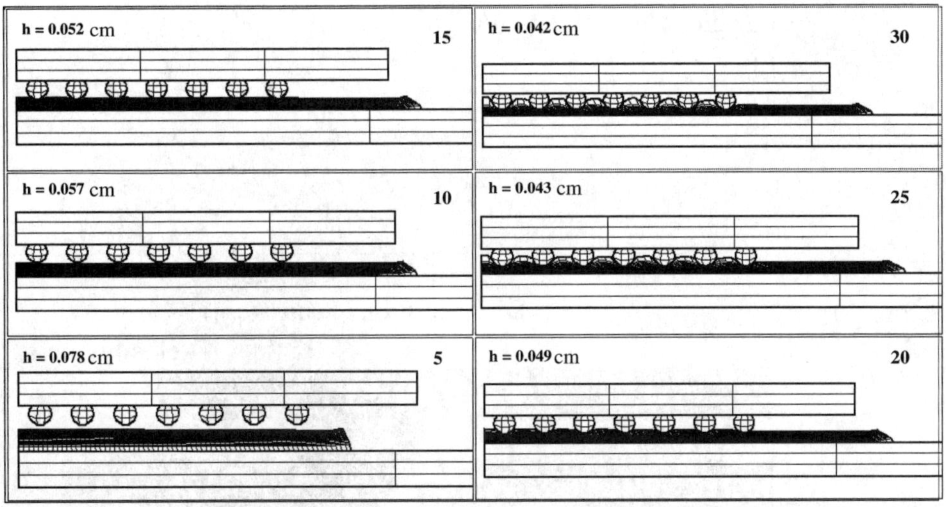

FIGURE 1.77 Gravity flow: progression of the compression flow process—vendor A, model II, T = 80°C.

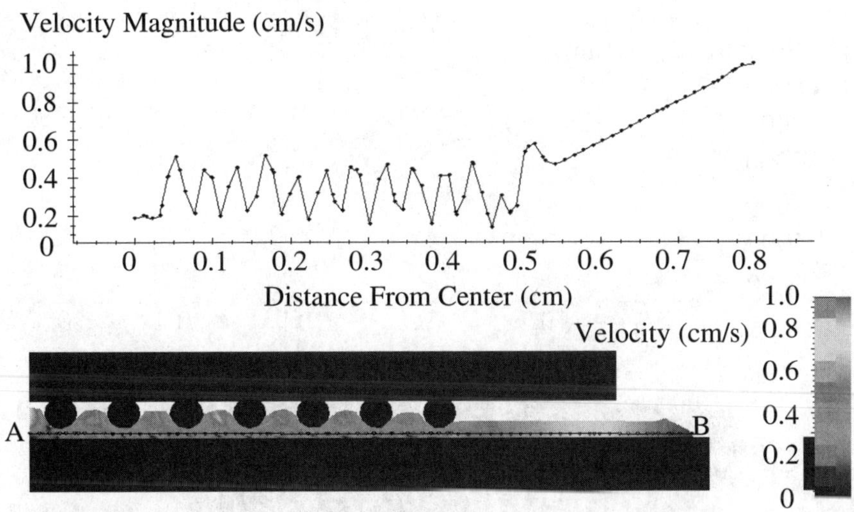

FIGURE 1.78 Velocity contour and section profile (A-B)—vendor A, model II, T = 80°C.

Model II, it is clear that squeeze flow dominated this case. The pressure contour of Fig. 1.80 indicates that the pressure decreases from the chip center to the edge. Similarly, the velocity profile of Fig. 1.81 shows a velocity contour tending to increase to the edge of the chip. It is interesting to compare this with the trend shown in Fig. 1.75 for the same material at the same temperature with a different underfill print height to area aspect ratio.

1.5.4.5 Discussion. Overall, the results of the compression flow simulations provide valuable information on the local shear rates, pressure distributions, and velocity profiles of the various chip configurations and underfill materials studied. The results also provide critical insights into the

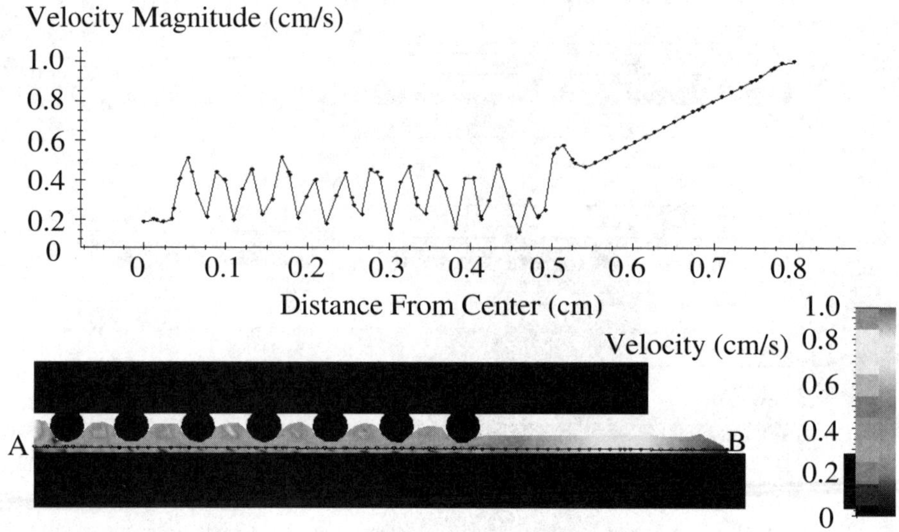

FIGURE 1.79 Velocity contour and section profile (A-B)—vendor A, model I, T = 80°C.

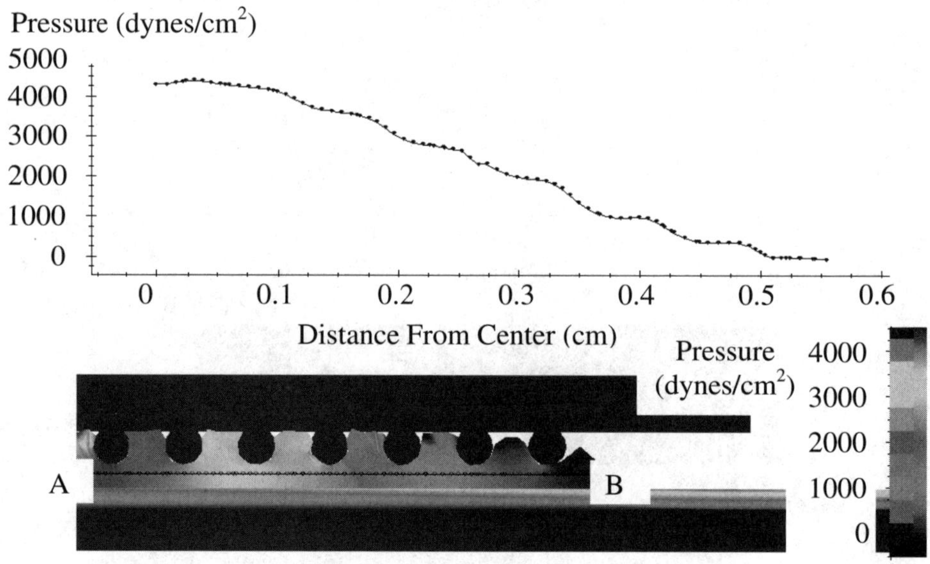

FIGURE 1.80 Fluid pressure contour and section profile (A-B)—vendor A, model II, T = 50°C.

potential formation of voids during next-generation FC processing using compression flow chip placement.

One of the critical elements of the new low-cost next-generation FC process is chip placement. Of particular importance is the ability to predict and understand the forces that must be applied to the chip to compress the underfill to relatively thin gaps (on the order of 25 to 100 μm). In this work, both analytical and numerical simulation results are used to estimate the forces required for the

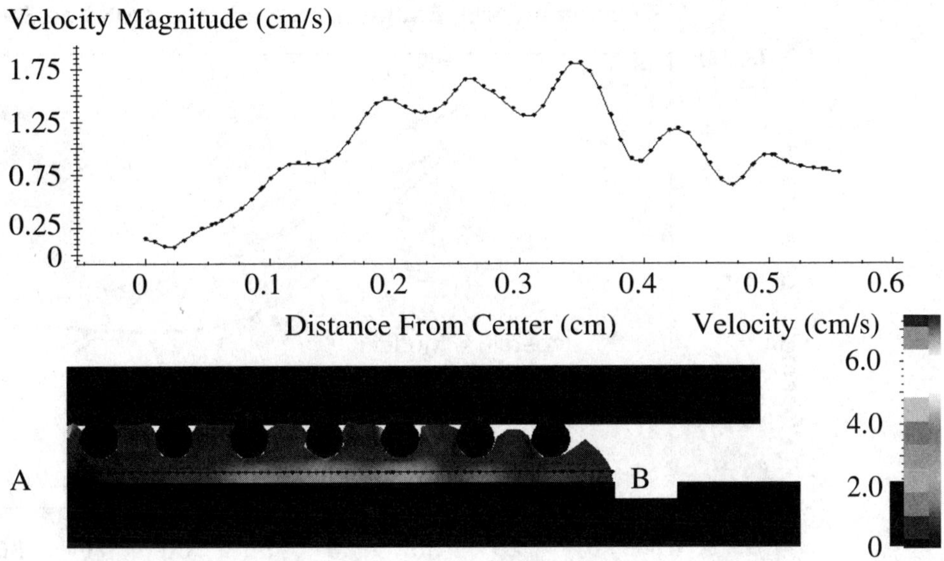

FIGURE 1.81 Velocity contour and section profile (A-B)—vendor A, model II, T = 50°C.

placement machines. While not directly comparable, both the analytical (Fig. 1.69) and numerical simulation (Figs. 1.75, 1.80, and 1.82) results suggest that the placement force capabilities of commercial placement machines are adequate for the next-generation FC process. Thus the analytical and numerical force predictions demonstrate that the process is feasible in terms of the assembly equipment capabilities. All the placement force results are within the capable ranges of existing placement equipment. Figure 1.82 shows the estimated placement forces predicted using the pressure profile simulation results of this study. Estimated placement forces are given for the underfill materials studied (A and B) for both models I and II. Notice in all cases for compression flow processing at 25°C that the required placement forces are less than 100 g well within the placement force capabilities of standard machines.

Another interesting phenomenon predicted by the simulation results is gravity driven flow of the underfill. In some cases, the viscosity of the material is sufficiently low that underfill starvation takes place within the chip-to-substrate gap. This effect is due to gravity flow of the underfill, as illustrated in Fig. 1.77. Gravity flow tends to be governed by the apparent viscosity of the underfill and ultimately temperature due the strong dependence of underfill viscosity on process temperature. Gravity flow is not observed for underfills A and B at 25°C. At 50°C, underfills A and B experience minimal gravity flow, as shown by Figs. 1.80 and 1.81. At 80°C, both underfills A and B experience significant gravity flow, as shown by Figs. 1.78 and 1.79. In particular, notice the relatively sporadic underfill velocities under the chips, as shown in Figs. 1.78 and 1.79. The gravity flow is particularly evident in the underfill segment outside the edge of the chip. In this region, Figs. 1.78 and 1.79 indicate linearly increasing velocity profiles, which are characteristic of uniform radial flow of the underfill due to its own weight. In general, the gravity flow results from the simulations indicate a limit in feasible process temperatures for the next-generation FC process.

Numerical simulations of the compression flow chip placement process are necessary to characterize the very complex flow characteristics of compression flow chip placement in low-cost next-generation FC processing. Moreover, the simulation results aid in identification of potential void formation during processing in order to eliminate this reliability concern. The numerical flow studies predict regions where voiding can occur and gave insight into the flow characteristics that will minimize void formation. Based on the simulation results, a series of design guidelines was formulated.

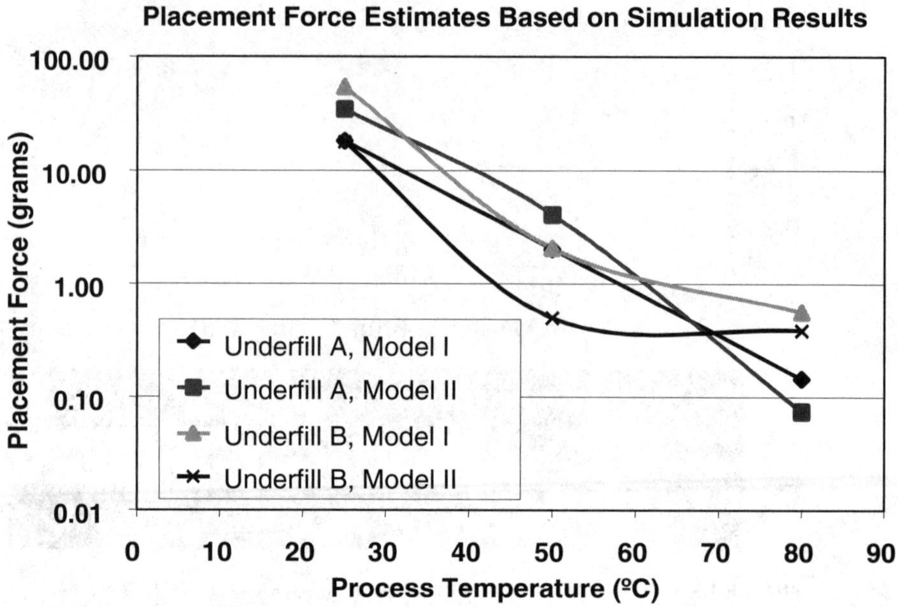

FIGURE 1.82 Placement forces predicted using simulation results for model I, model II, underfill A, and underfill B.

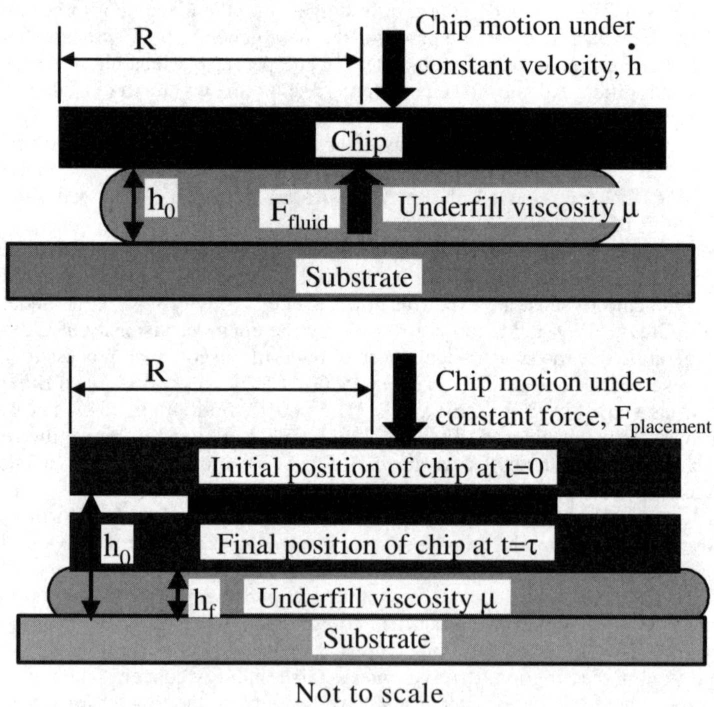

FIGURE 1.83 Schematic of compression flow processing and modeling parameters.

1.5.4.6 Summary of Compression Flow Simulation Analysis. FC technology represents a rapidly advancing area in commercial electronics. In order to ensure adequate reliability, FC assemblies undergo an underfill encapsulation process in which a polymer material is placed between the chip and the substrate. Conventional underfill processing is achieved through chip site to chip site dispensing and underfill flow via capillary action, making it a costly and time-consuming process, particularly as device sizes increase and standoff gaps decrease. To address the limitations associated with conventional FC processing, a low-cost next-generation FC process is proposed. This process eliminates the need for time consuming capillary flow processing using a compression flow technique where the underfill is applied prior to chip placement. The innovative process integrates the chip placement and polymer underfill processes using a compression or squeeze flow technique. It results in significantly lower assembly costs and reduced cycle time.

In general, the compression flow of the underfill material governed assembly yield and reliability for the next-generation process. Both analytical modeling and flow simulation studies were performed based on the compression flow chip placement process. In particular, the flow simulation studies represented a fundamental advancement in compression flow simulation of polymers in its successful application to the complex geometries and surface topologies demanded by miniaturized FC assembly. Simulation studies were conducted to characterize the compression flow of the underfill, estimate required chip placement forces, evaluate the effect of underfill geometry, and assess the potential formation of voids. Three types of voids were identified and experimentally observed that have the potential of forming during compression flow chip placement: compression voids, capture voids, and outgasing voids. With careful control of the underfill geometry, underfill rheology, placement parameters, and reflow process, void formation can be eliminated in the low-cost next-generation FC assembly process. Both analytical modeling and flow simulation results indicated that the placement forces required for compression flow chip placement are well within the ranges capable of standard placement systems. In addition, the results yield design guidelines that gave insight into process parameters such as the limits on underfill deposition geometry, potential of gravity flow, and underfill viscosity. The design guidelines provide a preliminary process window for compression flow chip placement.

1.6 COMPRESSION FLOW PLACEMENT MODEL AND ANALYSIS

The compression flow model is used to design the chip placement process and ensure that the chip makes contact with the substrate during placement.[19,38–40] The model uses a system of two equations that are quasi-steady-state approximations for the squeeze flow of a Newtonian fluid of viscosity μ between two parallel disks of radius R.[3,17,18] The force the underfill exerts on the chip can be approximated with the Stefan equation, Eq. (1.19). This is the quasi-steady-state approximation for the squeeze flow of a Newtonian fluid[3,17,18] between two parallel disks of radius R, a distance h apart, with a fluid of viscosity μ. Equation (1.19) can be used to estimate the force a placement system must exert to place a chip based on the compression flow process. Typically, placement machines move the gantry at high speed to the chip position. The chip is then moved at high speed toward the substrate to a programmed search height. Now the chip is placed onto the substrate at a constant programmed velocity $\dot{h}$ until the force sensor reaches a preprogrammed value $F_{placement}$. The placement machine then maintains the set force for a specified length of time, the dwell time. Using Eq. (1.19), F_{fluid} can be calculated at any height h_1 once the chip is fully wet by the underfill. The effective chip radius, R, can be approximated assuming a circular equivalent area of the rectangular chip given by Eq. (1.20). Equation (1.19) describes the motion of the placement machine until $F_{fluid} = F_{placement}$ at height h_1.

Based on computational fluid dynamics modeling of underfill compression flow during placement, there exists a theoretical limit to the placement velocity of a chip. The reason a limiting speed exists is due to the presence of underfill between the chip and the substrate during placement. There are three forces acting on a chip during the placement process: the force exerted by the placement system on the chip, the weight of the chip, and the hydrodynamic force exerted by the flowing underfill on the chip.

The compression flow model is used to better understand the yield characteristics of the placement process. Specifically, the effects of placement force and placement velocity on interconnection yield were analyzed. In addition, the compression flow model can be used to design the chip placement process and ensure that the chip bumps make contact with the substrate pads during placement. The process parameters are shown in Fig. 1.83. The model uses a system of two equations that are quasi-steady-state approximations for the squeeze flow of a Newtonian fluid between a chip and substrate. The force the underfill exerts on the chip F_{fluid} can be approximated with the Stefan equation, Eq. (1.19) where the chip and substrate surfaces are parallel. In Eq. (1.19), the underfill viscosity is μ, the chip is modeled as having an effective radius of R and the chip and substrate are separated by a distance h_0. The effective chip radius R can be approximated assuming a circular equivalent area of the rectangular chip given by Eq. (1.20).

$$F_{fluid} = \frac{3\pi R^4 \, \mu \, (-\dot{h})}{h_0{}^3} \tag{1.19}$$

$$R = \sqrt{\frac{L^2}{\pi}} \tag{1.20}$$

Equation (1.19) can be used to estimate the force a placement system must exert to place a chip based on the compression flow process. Typically, placement machines move the gantry at high speed to the chip position determined by the vision system based on substrate fiducials. The chip is then moved at high speed toward the substrate to a programmed search height. Now the chip is moved toward the substrate at a constant programmed velocity $\dot{h}$ until the placement head force sensor reaches a preprogrammed value $F_{placement}$. The placement machine then maintains the set force for a specified length of time, the dwell time or bond time t. Using Eq. (1.19), F_{fluid} can be calculated at any height h_0 under conditions where the chip is fully wet by the underfill. Similarly, Eq. (1.19) can be used to determine the height h_0 where the underfill hydrodynamic force equals that of the preprogrammed placement force limit, i.e., $F_{fluid} = F_{placement}$.

Once $F_{fluid} = F_{placement}$, the chip moves with a constant force $F_{placement}$ (i.e., rather than a constant speed). The motion of two parallel disks under constant applied force is described by Eq. (1.21). Here τ estimates the time required for the chip to move toward the substrate from h_0 to h_f where h_f is equal to the chip bump height, i.e., $h_f = h_{bumps}$. Equation (1.21) is derived from Eq. (1.19) by solving the differential Eq. in h with an initial condition of $h(t = 0) = h_0$. The parameter τ is the time $F_{placement}$ is applied, h_0 is the initial separation of the chip and substrate at the point, where $F_{fluid} = F_{placement}$, and h_f is the final separation distance between the chip and substrate equal to the bump height.

$$\tau = \frac{3\pi R^4 \mu}{2F_{placement}} \left[\frac{1}{h_f^2} - \frac{1}{h_o^2} \right] \tag{1.21}$$

Equations (1.19) through (1.21) can be used to determine why variations in placement force and placement velocity had no statistical effect on the interconnection yield in experiments 1 through 3. Consider the following case studies based on typical process configurations used in experiments 1 through 3. The viscosity of the underfill is 1.5 Pa·s. The area of the PB8-4 chip is 103.2 mm² having a circular equivalent radius of 5.73 mm. h_0 can be estimated using Eq. 1 given a programmed placement force, $F_{placement} = F_{fluid}$. The largest value for h_0 based on the process parameters in experiments 1–3, occurs at a velocity of 5 mm/s and a placement force of 400 g, (i.e., $F_{placement} = F_{fluid} = 3.92$ N), yielding $h_0 = 269$ mm. The maximum height of the solder bumps on the PB8-4 chips is $h_{bumps} = h_f = 98$ μm. Based on h_0, h_f and $F_{placement}$, the bond time needed to press the bumps into contact with the substrate can be estimated using $h_0 = 270$ mm, $h = 98$ μm, and $F_{placement} = 3.92$ N in Eq. (1.21). This yields a bond time, τ, of 0.176 s or 176 ms. The bond time for the assemblies in experiments 1 through 3 was 0.3 s. Therefore, there is adequate time to drive the bumps into contact with the substrate pads.

The other extreme case is at the minimum velocity and maximum force used in experiments 1 through 3. In this case, the placement force is 1600 g (i.e., $F_{placement} = F_{fluid} = 15.7$ N), and the placement velocity is 0.1 mm/s. Solving Eq. (1.19) gives the height above the substrate where the placement force is reached h_0 = 46 mm, where the remaining parameters are the same as above. This height is less than the bump height, implying that the chip bumps should make contact with the substrate pads prior to hydrodynamic force of the underfill reaching the placement force limit. In this case, the dwell or bond time is not needed because the placement force is large enough to force contact between the chip bumps and substrate pads during constant velocity chip placement.

1.7 MODELING AND ANALYSIS OF CHIP FLOATING

Under certain process conditions, chips can have a tendency to float during assembly if excess underfill is applied to the substrate.[19,38–40] Such observations are apparent in experiments 1 through 3. In such instances, x-ray inspection shows the chips to be properly aligned with the substrate bond pads before and after reflow, but no interconnections form during reflow. In addition, the chips are observed to easily slide over the substrate when a large amount of underfill is present on the substrate. This yield defect is attributed to a phenomenon called chip *floating*. For a chip to *float,* forces acting on the chip have to be sufficient to lift it off the substrate. Looking at a free body diagram of a chip after placement, Fig. 1.84, there are three forces acting on the chip: the weight of the chip and bumps W, the surface tension force F_{st} acting on the perimeter of the chip, and the force due to the pressure difference between the underfill below the chip and bumps P_u and the atmosphere P_a. This force balance forms the basis of the chip floating model and analysis.

Determining the weight of the chip and bumps can be accomplished by weighing the bumped chips prior to assembly. The surface-tension force can be estimated using Eq. (1.22), where L is the length of one side of the chip, and γ is the surface tension of the underfill, and α is the fillet contact angle at the top of the chip.

$$F_{st} = 4L\gamma \cos \alpha \tag{1.22}$$

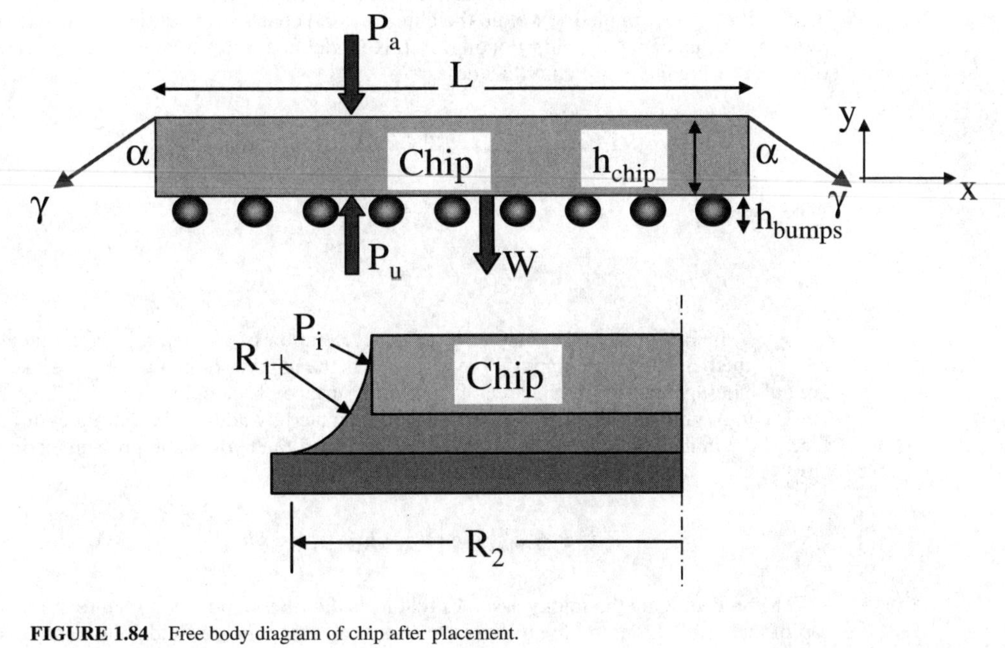

FIGURE 1.84 Free body diagram of chip after placement.

To determine the force caused by the pressure difference $P_a - P_u$, P_u must be broken down into its contributing components. P_u is comprised of two components; the pressure due to the surface energy of the underfill meniscus and the hydrostatic pressure on the active chip surface due to the underfill wetting up the side of the chip. In general, the hydrostatic pressure results from the column of fluid above the active chip surface. The hydrostatic pressure is referenced off the pressure difference due to the surface energy of the underfill meniscus. The Young-Laplace equation (Eq. 1.23) can be used to estimate the pressure difference caused by surface energy and the fillet meniscus. Eq. (1.23) can be used to calculate the magnitude and the sign of the pressure difference at the liquid-air interface based on the principal radii of curvature R_1 and R_2 and the surface energy γ of the underfill. As defined, R_2 is always negative, corresponding to the effective radius of the underfill wetted area on the substrate as shown in Fig. 1.85 (notice the sign convention that the critical radii are positive when measured outside the fluid and negative when measured through the fluid). R_1 can change sign depending on whether the fillet is concave ($R_1 > 0$) or convex ($R_1 < 0$) as shown in Fig. 1.85. P_i is the pressure inside the liquid meniscus at the top of the chip, as shown in the figure.

$$P_a - P_i = \gamma\left(\frac{1}{R_1} + \frac{1}{R_2}\right) \tag{1.23}$$

The hydrostatic pressure contribution can be estimated based on the height of the fillet above the active chip surface and the pressure component exerted by the respective fluid column. Equation (1.24) presents a conservative estimate of the hydrostatic pressure contribution due to the column of underfill. There, a conservative estimate of the pressure acting on the bumps is used, and the underfill fillet is assumed to wet the full height of the chip representing the maximum hydrostatic pressure tending to float the chip. Depending on the chip geometry and bump configuration, the hydrostatic pressure force ranges from 47 to 58 percent of the chip weight based on the test vehicles used in this study. This indicates that the chip geometry and underfill can create a condition where the hydrostatic force is important. In a typical configuration, the top of the chip is 800 μm above the substrate, whereas the gap between the bottom of the chip and substrate is approximately 100 μm. Therefore, a relatively large column of fluid is present where the fillet meets the chip, causing the relatively large hydrostatic pressure. It should also be noted that under process conditions where the chips have partial fillets, not completely wet up the chip side, the contribution of the hydrostatic force is lower, reducing the tendency for chip floating. In this model, a conservative estimate is used to model the fillet as wetting the entire chip thickness.

$$P_{hydrostatic} = \rho_{underfill}\, g\left[h_{chip}\,(1 - A_B) + (h_{chip} + h_{bumps})\,A_B\right] \tag{1.24}$$

where

$$A_B = \frac{nA_{bump}}{A_{chip}} \tag{1.25}$$

$\rho_{underfill}$ is the density of the underfill, g is acceleration of gravity, h_{chip} is the height of the chip, h_{bumps} is the height of the bumps before reflow, n is the number of bumps, A_{bump} is the cross-sectional area of a bump, and A_{chip} is the area of the active surface of the chip.

The total pressure under the chip P_u can be estimated by adding the fluid pressure at the top surface of the underfill resulting from the meniscus P_i and the hydrostatic pressure acting on the active chip surface and solder bumps, given by Eq. (1.26).

$$P_u = P_i + \rho_{underfill}\, g\left[h_{chip}\left(1 - A_B\right) + \left(h_{chip}\right) + h_{bumps})\,A_B\right] \tag{1.26}$$

The preceding methodology assumes that R_1 is the fillet radius of curvature in the region near the top of the chip. Therefore, the hydrostatic pressure caused by the fluid column along the side of the chip is added to P_i to obtain the pressure of the underfill below the chip and below the individual interconnection bumps. In general, R_1 is not constant along the length of the fillet. The fillet radius of curvature changes to accommodate the increase in pressure due to the hydrostatic pressure at varying depths in the underfill. Summing the forces in the y-direction gives the complete force balance on the chip after placement:

$$F_y = (P_a - P_u) A_{chip} + W_{chip} + W_{bumps} + F_{st} = 0 \tag{1.27}$$

Equation (1.27) can be rearranged in the form of a dimensionless parameter Π_f that can be used to predict conditions that promote chip floating. Specifically, Π_f, called for *float parameter,* is the ratio of the "buoyancy forces" acting on the chip to the "adhesion forces" acting on the chip. If Π_f is greater than 1, the chip will have a tendency to float. If Π_f is less than 1, the chip is under adhesion forces tending to hold the chip firmly to the substrate.

$$\Pi_f = \frac{(P_u - P_a) A_{chip}}{W_{chip} + W_{bumps} + 4L\gamma \cos \alpha} \tag{1.28}$$

Substituting for P_u and P_i yields:

$$\Pi_f = \frac{\left(\rho_{underfill} g [h_{chip} (1 - A_B) + (h_{chip} + h_{bumps}) A_B] - \gamma \left(\frac{1}{R_1} + \frac{1}{R_2}\right)\right) A_{chip}}{W_{chip} + W_{bumps} + 4L\gamma cos\alpha} \tag{1.29}$$

where

$$\Pi_f < 1 \quad \text{Adhesion Condition}$$

$$\Pi_f > 1 \quad \text{Floating Condition}$$

The key to predicting floating conditions based on the low-cost, high-throughput FC process and assessing chip floating with respect to the yield experiments presented earlier, centers on developing a relationship between the critical radii R_1 and R_2 and the applied no-flow underfill volume. In general, as the mass of underfill increases, the fillet radius increases until it reaches a point where it will flip from concave to convex, i.e., $R_1 \rightarrow \infty$ (see Fig. 1.85) changing sign. At this "critical" underfill mass, $(P_a - P_i)$ changes sign from positive to negative, which can change the sign of the numerator in Eq. (1.29) from negative to positive. This change of sign will cause Π_f to become greater than 0, and for sufficiently large underfill volumes Π_f becomes greater then 1, indicating conditions are favorable for chip floating. Lower underfill volumes ensure concave fillets typically leading to $\Pi_f < 1$ and an adhesion condition between the chip and substrate. It should also be mentioned that the chip floating model assumes the no-flow underfill behaves like a simple fluid, exhibiting negligible viscoelasticity when applied to the substrate prior to reflow curing. The model becomes considerably more complex under conditions where the underfill exhibits appreciable viscoelasticity after deposition on the substrate.

The underfill mass can be estimated using geometric approximations for the underfill fillets. In general, the underfill mass is equal to the mass of underfill under the chip plus the mass of underfill in the fillets less the mass of underfill occupied by the bump interconnections. As a first-order approximation, the fillet is modeled as a circular arc, as shown in Fig. 1.86. The underfill mass associated with the fillet is approximated as symmetric on the four sides of the chip. Based on the fillet geometry of Fig. 1.86 and a chip standoff gap of $(h - h_{chip})$ the underfill mass can be approximated by Eq. (1.30).

$$m_{underfill} = \left(h - h_{chip}\right) A_{chip} + 2L\rho_{underfill} \left(kh^2 \pm 2R_1^{\,2} \sin^{-1}\left(\sqrt{\frac{h^2 (k^2 + 1)}{2R_1}}\right) \pm h^2 (k^2 + 1)\right.$$

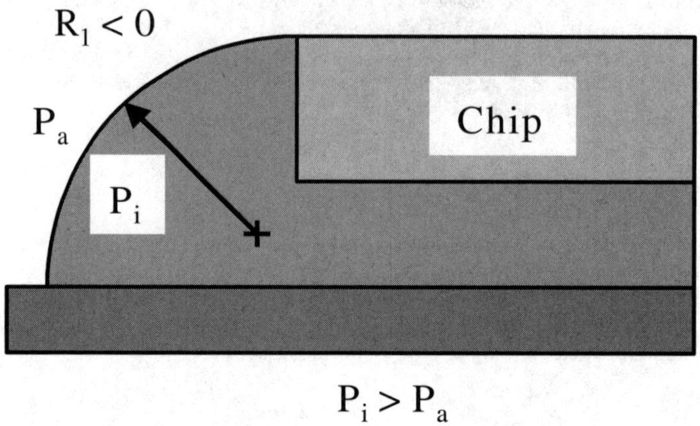

FIGURE 1.85 Influence of fillet shape on pressure difference.

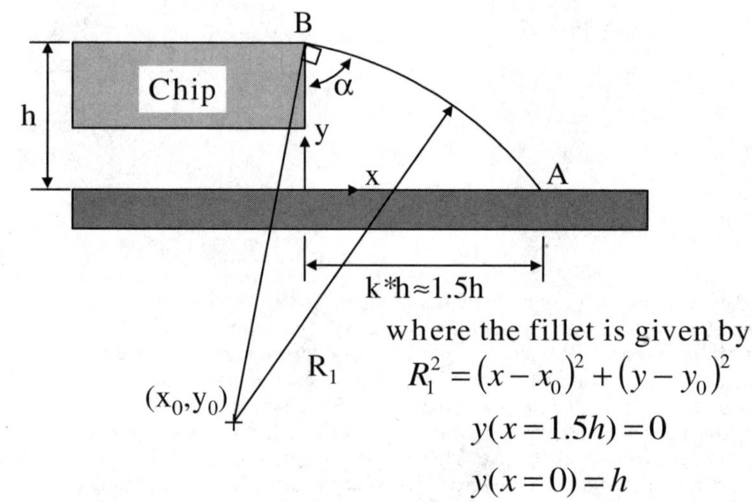

$$R_1^2 = (x - x_0)^2 + (y - y_0)^2$$
$$y(x = 1.5h) = 0$$
$$y(x = 0) = h$$

FIGURE 1.86 Approximation of fillet geometry for no-flow underfill mass estimation.

$$\left(R_1^{\,2} - \sqrt{\frac{h^2\,(k^2+1)}{4}}\right)\Bigg) - nV_{bump} \tag{1.30}$$

where k is the underfill wetting factor for the substrate, h is the total standoff gap height from the top of the chip, and V_{bump} is the average volume of a solder bump. Based on experimental observations for the no-flow underfill and substrate-solder mask combinations used, k is typically 1.5.

Assuming a constant fillet radius from the top of the chip to the bottom and using basic geometric relationships, the mass of underfill corresponding to specific values of R_1 and R_2 and hence Π_f can be

TABLE 1.26 Parameters for Chip Floating Model

Parameter	Value or range
γ	20 to 50 mN/m
k	1.5
A_{chip}	See Table 17.5
h	800 μm
$W_{chip} + W_{bumps}$	1.33 g (PB8-4)
	0.33 g (PB8-2)
	1.62 (FA10-4)
	0.36 (FA10-2)
$\rho_{runderfill}$	1.1 g/cm^3
h_{bumps}	98 μm (PB8)
	110 μm (FA10)
h_{chip}	700 μm
R_2	−4 mm
n	See Table 17.5

estimated. This enables prediction of the acceptable underfill volumes or process window for a given chip and underfill system with respect to chip floating. Table 1.26 presents the model parameters used. Figure 1.87 shows a graph of Π_f as a function of underfill mass for various underfill surface energies and FC test chips. The chip parameters (weight, dimensions, etc.) for this graph are based on the commercial FC test chips used in the yield analysis experiments presented earlier, namely, the PB8-2, PB8-4, FA10-2, and FA10-4 test chips. The plot (1.87*d*) shows that the PB8-4 chips should begin to float at an underfill mass of approximately 34.5 mg for an underfill surface energy of 30 mN/m. Experimental observations indicate that the test chip tended to float at an underfill mass of 36 ± 2 mg. Similar predictive capabilities are found for the other FC test vehicles studied that have both area array and perimeter array bump configurations. Specific results are presented in Table 1.27.

While underfill surface energy is a relatively difficult parameter to quantify,[41] notice that the critical underfill mass leading to floating is a relatively weak function of the underfill surface energy. Therefore, even with a poor estimate of the underfill surface energy, the predicted critical underfill amounts do not differ significantly. In contrast, the substrate wetting parameter k has a significant impact on the chip floating model predictions. A sensitivity analysis of k is presented in Fig. 1.88. It is clear from Fig. 1.88 that wetted length of the fillet beyond the edge of the chip can have a significant impact on the critical underfill mass inducing floating, i.e., where $\Pi_f = 1$. As the fillet wets and spreads a greater distance along the substrate, the critical underfill mass inducing floating increases. Therefore, under conditions where the density of components is high, tighter control limits must be held for the deposited underfill volume. However if the keep out distance between components is large, looser underfill volume tolerances can be accommodated without inducing chip floating. A typical k value for the no-flow underfill and substrate-solder mask systems used in this study is 1.5, which was used in the analysis. Careful experiments should be run to characterize k for the particular underfill and substrate-solder mask system used before using the chip floating model to estimate the process window for acceptable no-flow deposition volumes.

The chip floating model serves two purposes. First, it is a process design model enabling rapid estimation of the process window for the no-flow underfill mass to ensure high yield processing. Table 1.27 presents a summary of the results found for the FC test vehicles studied. Second, it helps explain the dramatic yield losses first experienced in experiment 1 prior to the underfill mass range being redesigned. Namely, early trials in experiment 1 showed significant yield loss when the underfill mass exceeded 37 g. The chip floating model predicts a limiting underfill mass of 35 g. In the final experimental matrix, the underfill was varied over a range of 15 to 35 g, minimizing chip floating and the associated yield loss.

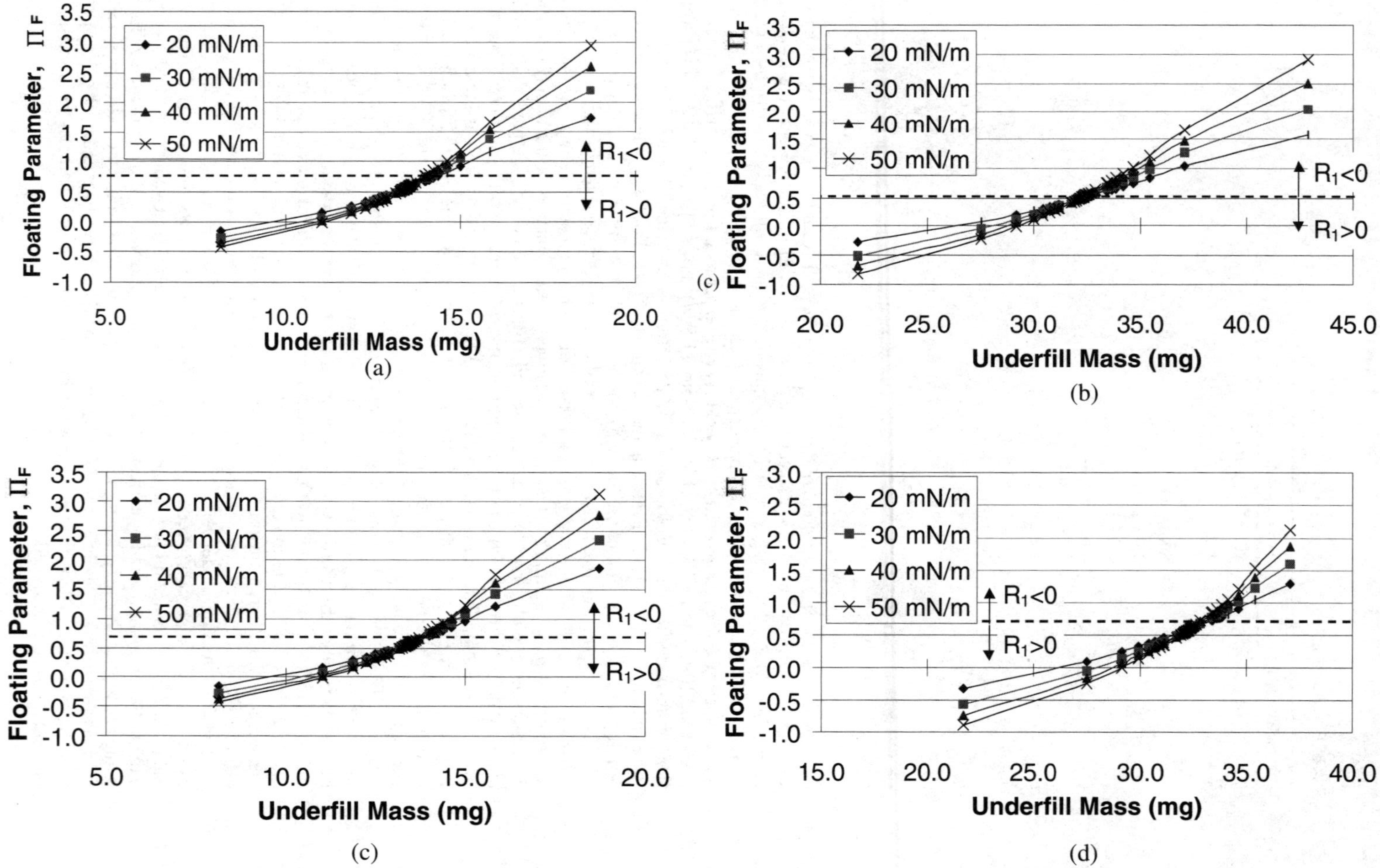

FIGURE 1.87 Floating parameter, Π_f for the FC test chips as a function of underfill mass. (*a*) FA10-2, (*b*) FA10-4, (*c*) PB8-2, and (*d*) PB8-4.

TABLE 17.27 Comparison of Predicted and Measured Underfill Amounts Resulting in Chip Floating

Test Vehicle	Chip size (mm)	I/O count	Predicted underfill mass inducing floating (mg)	Experimentally observed underfill mass inducting floating (mg)
PB8-2	5.08	88	15.0 to 15.8	14 ± 2
PB8-4	10.2	352	34.6 to 35.4	35 ± 2
FA10-2	5.08	317	15.0 to 15.7	16 ± 2
FA10-4	10.2	1268	35.5 to 37.1	37 ± 2

1.8 FC INTERCONNECTION YIELD ANALYSIS DURING THE REFLOW PROCESS

FC solder bumps are produced using a number of different manufacturing processes. One process is based on stencil printing solder paste, after which the solder is reflowed to form a spherical bump. The final heights of the solder bumps are inconsistent (Fig. 1.89) and "normally" distributed. Most of the bumps are in close proximity to the targeted or "mean" size, but depending on the standard, some bumps can be significantly larger or smaller than the targeted size.

Given that the solder bumps on the FC are of varied sizes, the chip must "collapse" during reflow in order for all bumps to make contact with their respective bond pads. In a conventional FC assembly process, the weight of the chip is sufficient to drive the collapse process. In the next-generation

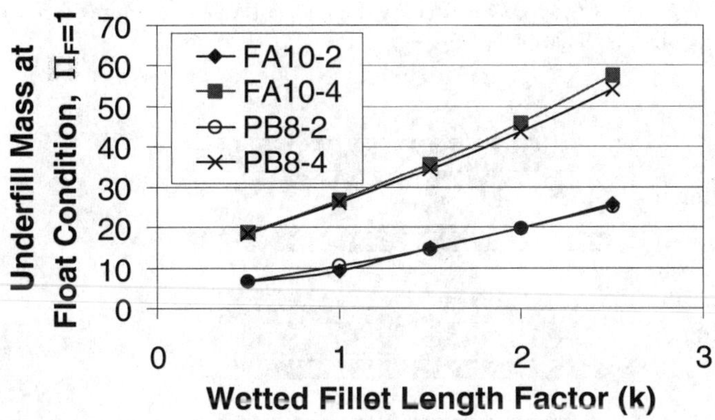

FIGURE 1.88 Underfill mass at the floating condition $\Pi_f = 1$ as a function of wetting length.

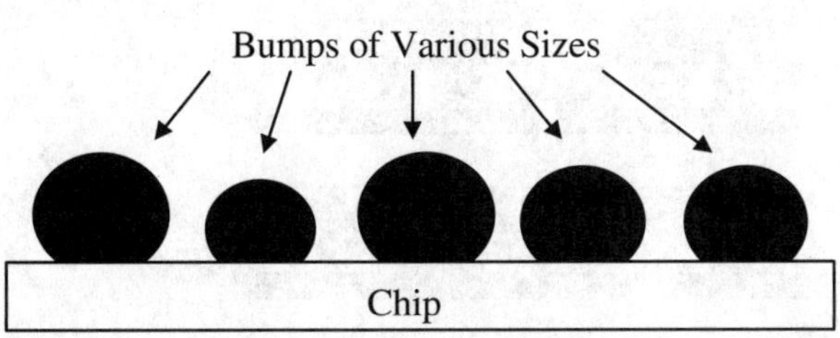

FIGURE 1.89 Chip with varied bump heights.

FC assembly process, a layer of viscous underfill material exerts a force during collapse that resists the collapsing motion.

No-flow underfill materials are formulated such that the underfill does not begin to cure until a latent catalyst has been activated. Ideally, the latent catalyst facilitates underfill curing at a temperature well above the melting point of eutectic solder. This ensures that the chip interconnections reflow, the chip collapses, and all bumps make contact with their respective bond pads prior to underfill hardening. Two potential problems can arise if the underfill begins to harden too early in the reflow process. The underfill viscosity can increase to a point that does not allow the chip to completely collapse. This incomplete collapse results in the shortest bumps not making contact with their respective bond pad. Second, the thin underfill layer between the smallest bumps and their respective bond pads generates a large enough compression flow force to cause the molten solder bumps to deform rather than push through the underfill. To address these issues, a process model was developed that simulates the chip collapse process assuming that the molten solder bumps do not deform as they move through the liquid underfill.

1.8.1 FC Interconnection Yield Analysis

The rigid solder model was developed to determine the time it takes for a complete chip collapse scenario to occur, assuming the mean and standard deviation of the bump height distribution is known. For a successful chip collapse, every bump must make contact with its respective bond pad. Figure 1.90 shows a schematic of the chip collapse process. As mentioned earlier, the heights of the solders bumps are inconsistent. The model assumes that prior to reflow, the chip rests stably on its three tallest bumps.

After chip placement, the assembly is placed in a reflow oven. Figure 1.91 is a schematic of the forces acting on the chip, once the solder has reached its melting point of 183°C.

F_{ss} is the total force due to steady state components: chip weight, underfill surface tension, and adhesion between the chip and board. F_{solder} is the force produced by the surface tension of the wetted solder joints. $F_{underfill}$ is the force exerted by the underfill due to compression flow. The next three sections give detailed explanations of F_{ss}, F_{solder}, and $F_{underfill}$.

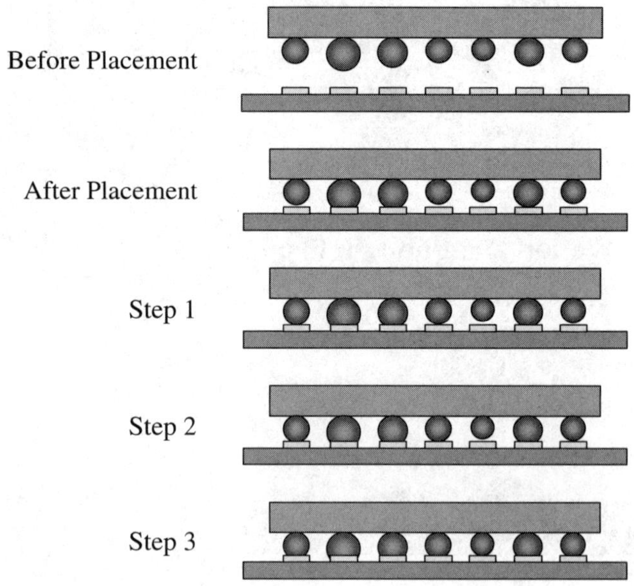

FIGURE 1.90 Chip collapse schematic.

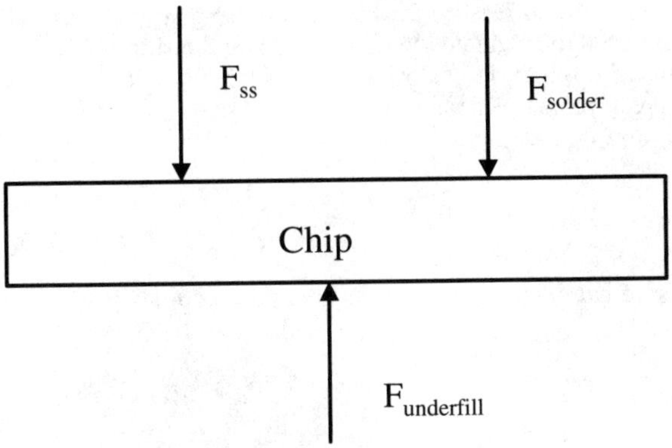

FIGURE 1.91 Free body diagram of collapsing chip.

1.8.2 Steady State Forces

The steady state force acting on a chip after placement is determined using methodology presented by Thorpe et al.[38–40]

Figure 1.92 is a free body diagram of the chip after placement. There are three forces acting on the chip:

1. The weight of the chip
2. Surface tension force F_{st} acting on the perimeter of the die
3. Force due to the pressure difference between the underfill below the die and solder bumps P_u, and the atmosphere P_a.

Determining the weight of the chip is straightforward. The surface tension force can be calculated using Eq. (1.31), where L is the length of one side of the chip, gunderfill is the surface tension of the $\gamma_{underfill}$, and α_c is the fillet contact angle at the top of the chip.

$$F_{st} = 4L\gamma_{underfill} \cos\alpha_c \tag{1.31}$$

To determine the force caused by the pressure difference $P_a - P_u$, P_u must be analyzed. P_u is comprised of two components; pressure due to the surface energy of the underfill, and hydrostatic pressure due to the underfill wetting up the side of the chip. The Young-Laplace equation (Eq. 1.32) can

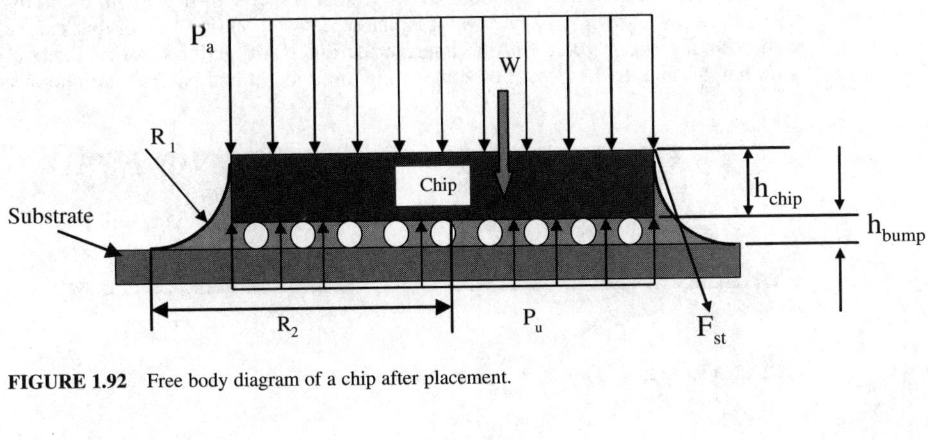

FIGURE 1.92 Free body diagram of a chip after placement.

be used to determine the pressure difference caused by surface energy and the fillet meniscus. This equation and Fig. 1.93 are used to calculate the magnitude and the sign of the pressure difference at the liquid-air interface based on the principal radii of curvature R_1 and R_2 and the surface energy γ of the underfill. P_i is the pressure inside the liquid at the fillet liquid-air interface at the top of the chip and R_1 and R_2 are shown on Fig. 1.93.

$$P_a - P_i = \gamma\left(\frac{1}{R_1} + \frac{1}{R_2}\right) \tag{1.32}$$

The hydrostatic pressure can then be added to P_i to obtain the total P_u, given by Eqs. (1.33) and (1.34).

$$P_u = P_i + \rho_{underfill}g\,[h_{chip}\,(1 - A_F) + (h_{chip} + h_{bumps})\,A_F] \tag{1.33}$$

$$A_F = \frac{(\text{Cross Sectional Area of Solder Interconnect}) \times (\text{\# of interconnects})}{\text{Area of Die}} \tag{1.34}$$

The preceding methodology assumes that R_1 is the radius of curvature of the fillet in the region near the top of the chip. Therefore, the hydrostatic pressure caused by the column of fluid along the side of the chip is added to P_i to obtain the pressure of the underfill below the chip and below the individual interconnection bumps. R_1 is not constant along the length of the fillet. It changes radius of curvature to accommodate the increase in pressure due to the hydrostatic pressure at varying depths in the underfill. Summing the forces in the y-direction gives the complete force balance:

$$\Sigma F_y = (P_a - P_u)\,A_{die} + W + F_{st} \tag{1.35}$$

For the rigid solder model, the steady state force F_{ss} is equal to ΣF_y.

1.8.3 Solder Force

When a solder ball contacts and wets a bond pad, the equilibrium shape of the interconnection is dictated by surface energy forces. The interconnection attempts to minimize its total surface energy. In doing so, the solder joint exerts either a pushing or pulling force on the chip. Figure 1.94 shows examples of both scenarios. If the interconnection is shorter than the equilibrium height, then it will produce a force that pushes up on the chip. If the interconnection is taller than the equilibrium height, then it will produce a force that pulls down on the chip.

Extensive research has been done in the area of solder self-alignment, including the normal reaction forces just shown.[12] They developed an explicit regression model that can calculate a normalized normal reaction force for one interconnection as a function of normalized chip misalignment, joint height, and the aspect ratio between the joint height and pad radius. The regression model has

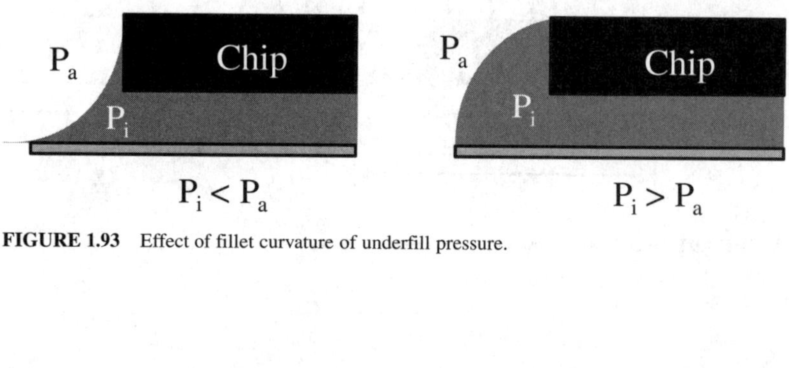

FIGURE 1.93 Effect of fillet curvature of underfill pressure.

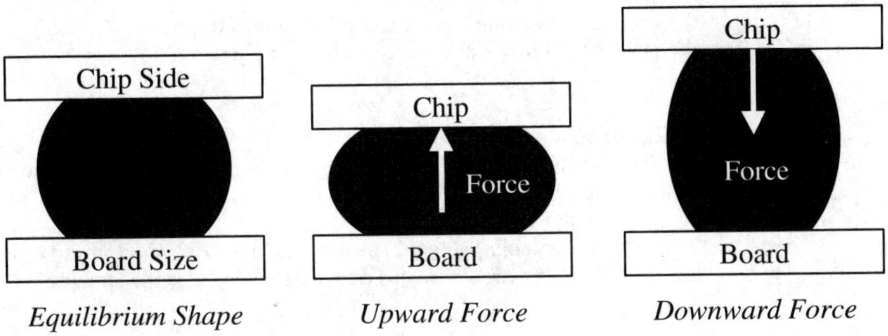

FIGURE 1.94 Force exerted by molten solder interconnections.

three nondimensional input parameters, N, P^*, H^* and one nondimensional output, F_N^*. Physically, N is a measure of the aspect ratio of the joint, P^* is the misalignment between the solder bump and the substrate pad, and H^* is the instantaneous joint height.

N is calculated using Eq. (1.36), P^* is assumed to be zero for simplicity, and H^* is calculated using Eqs 1.37 and 1.38.

$$N = \frac{V}{\pi R_p^3} \tag{1.36}$$

V is the volume of the solder bump, and R_p is the chip pad and substrate bond pad radius. These two pad radii must be equal to use the regression model.

$$H^* = \frac{H}{H_c} \tag{1.37}$$

$$H_c = \frac{V}{\pi R_p^2} \tag{1.38}$$

The coefficients for the regression model are a function of N and are presented in Eqs. 1.39 through 1.42.

$$a_0 = -907.38 + 2441.73N - 2823.98N^2 + 1439.13N^3 - 266.43N^4 \tag{1.39}$$

$$a_1 = 1881.44 - 5357.26N + 6525.01N^2 - 3431.96N^3 + 648.69N^4 \tag{1.40}$$

$$a_2 = -1192.24 + 3685.92N - 4799.78N^2 + 2622.64N^3 - 507.91N^4 \tag{1.41}$$

$$a_3 = 224.22 - 786.56N + 1122.26N^2 - 642.96N^3 + 128.19N^4 \tag{1.42}$$

F_N^* is calculated using Eq. (1.43), and the dimensioned F_N (N) is calculated using Eq. (1.44).

$$F_N^* = a_0 + a_1H^* + a_2H^{*2} + a_3H^{*3} \tag{1.43}$$

$$F_N = F_N^* = F_N^*\gamma R_P \tag{1.44}$$

F_N is the force that one interconnection exerts on the chip. F_{solder} is calculated by summing the F_N contribution from all bumps that have contacted and wetted a substrate bond pad (Eq. 1.45).

$$F_{solder} = \sum_i^k F_{Ni} \tag{1.45}$$

where F_{Ni} = the normal force contribution from interconnection '*i*'
k = the number of bumps in contact with a bond pad

1.8.4 Underfill Force

Hydrostatic squeeze film theory provides an approximation for the force that the underfill exerts on a collapsing chip. Basic viscous squeeze film theory assumes that a circular disk of radius R is approaching another circular disk of equal radius (Fig. 1.95).

Because the chip is not circular, an equivalent radius approximation must be made. Eq. (1.46) is used to calculate the equivalent circular radius of a square chip. L is the length of one side of the chip.

$$R = \sqrt{\frac{L^2}{\pi}} \tag{1.46}$$

The force that the underfill exerts on the upper disk (chip) is given by Eq. (1.47).

$$F_{underfill} = \frac{3\pi\mu\, hR^4}{h^3} \tag{1.47}$$

where μ = the viscosity of the underfill
h = the instantaneous separation distance between the disks
h^* = the velocity of disk 2 (chip)
R = the equivalent chip radius

To use Eq. (1.47), the collapse velocity of the chip is needed. This value is difficult to obtain. Solving the differential Eq. (1.47) for h results in Eq. (1.48), which gives the time of collapse from height h_2 to height h_1.

$$\tau = \frac{3\pi\mu R^4}{2F_{total}}\left[\frac{1}{h_{i+1}{}^2} - \frac{1}{h_i^2}\right] \tag{1.48}$$

where

$F_{total} = F_{ss} + F_{solder}$
h_i and h_{i+1} are subsequent bumps in the collapse process

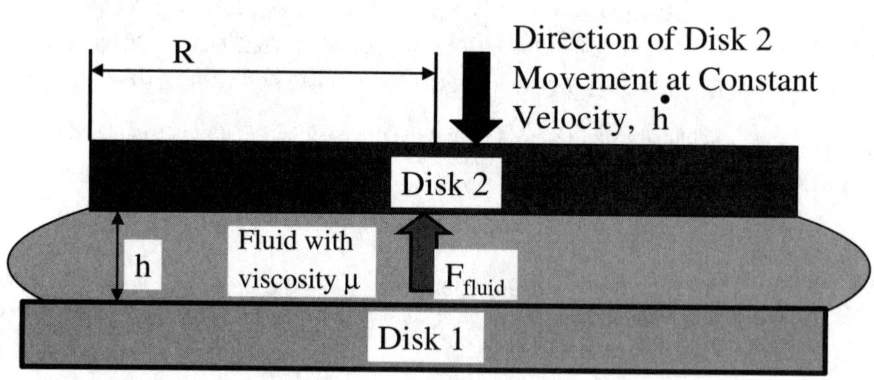

FIGURE 1.95 Compression flow with constant velocity.

1.8.5 Chip Collapse Time

The collapse time results are based on a numerical simulation of the collapse process. The simulation begins by sampling N bump heights from a "normal" distribution with mean μ and standard deviation σ. The bumps are numbered 1 to N, where h_i is the height of bump i, and $h_1 > h_2 > h_3 > \ldots > h_N$.

The collapse process is then simulated in an iterative stepwise manner. The distance steps are not constant. Each step is equal to the difference in height between bump i and bump $i+1$. For each step, F_{solder} is a unique value, based on the number of bumps in contact with bond pads, and the current shape of those interconnections. F_{solder} is added to the constant F_{ss} to find F_{total} for each step. Equation (1.48) is then used to determine the time of collapse from bump i to bump $i+1$. An example of steps 1 and 2 of the simulation is given below.

After placement, the chip is assumed to be resting on its three tallest bumps. The first distance step is the difference in height between bump 3 and bump 4 ($h_3 - h_4$) (Fig. 1.96). F_N for bumps 1, 2, and 3 is calculated based on the current height of h_3. F_{solder} is the summation of the F_N contributions from bumps 1 through 3. The collapse time from h_3 to h_4 is calculated using Eq. (1.48) where h_i is equal to h_3 and h_{i+1} is equal to h_4. At the conclusion of the first distance step, bumps 1 through 4 are wetted to their respective bond pads. The second distance step is the difference in height between bumps 4 and 5. F_{solder} for step two is the summation of the F_N contributions from bumps 1 through 4. The collapse time for step two is again given by Eq. (1.48), where h_i is equal to h_4 and h_{i+1} is equal to h_5.

This iterative step process continues until the Nth bump has made contact with a bond pad. A summation of the collapse times for each step gives the total collapse time.

1.8.6 Underfill Viscosity Effect

The most important results from the model examine the effect of underfill viscosity on collapse time. These results determine an allowable viscosity such that a complete chip collapse can occur. Figure 1.97 is a plot of collapse time versus underfill viscosity. Each line in the plot represents a standard deviation (in microns) of the initial bump height distribution. The results are based on the following assumptions: a PB8-4 chip with total steady state forces of 3 mN, solder surface tension of 300 mN/m, an initial "normal" bump height distribution with a mean of 98 μm, and a constant underfill viscosity. A steady-state force of 3 mN was chosen based on earlier modeling done by McGovern and Baldwin.[19] McGovern and Baldwin determined that using an optimal volume of underfill for the PB8-4 chip produced a steady-state force of 3 mN.

Figure 1.97 indicates that for a given bump height standard deviation, a linear relationship exists between viscosity and collapse time. As expected, the lower the underfill viscosity, the faster the chip collapse will occur. The figure also shows that the slope of collapse time versus viscosity decreases

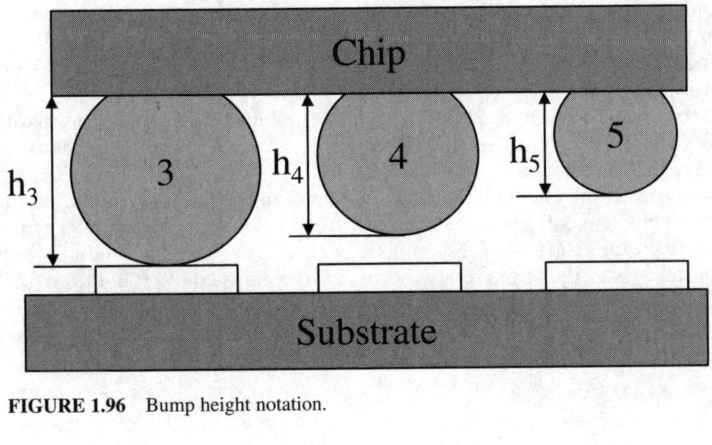

FIGURE 1.96 Bump height notation.

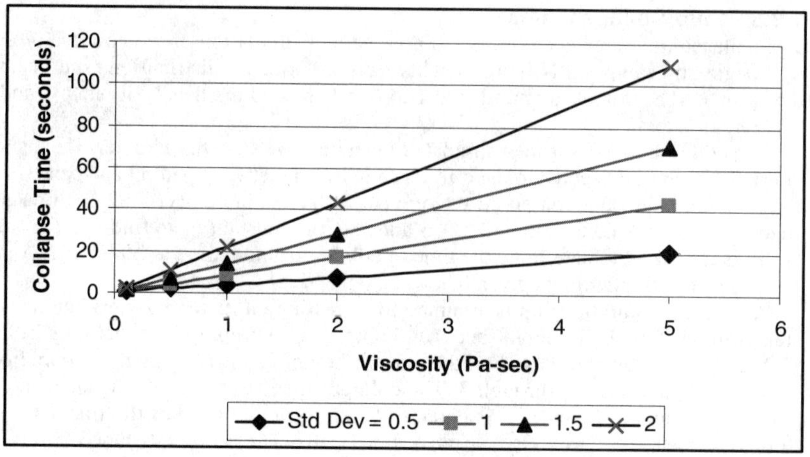

FIGURE 1.97 Collapse time versus underfill viscosity.

as standard deviation decreases. Therefore, a height distribution with a small standard deviation is less affected by increases in viscosity than a distribution with a larger standard deviation.

1.8.7 Processing Windows for Complete Chip Collapse

The results from the rigid solder model can be used to generate process windows that predict the conditions for which a complete chip collapse is likely. A process window graph allows a potential user of the high-throughput FC assembly process to understand the interactions of three important variables in the collapse process. The process windows presented in Fig. 1.98, relate underfill viscosity, maximum allowable collapse time, and initial bump height standard deviation. The shaded areas in the graph define combinations of underfill viscosity, maximum allowable collapse time, and bump height standard deviation that result in a successfully collapsed chip. Figure 1.98 shows that the maximum allowable collapse time parameter corresponds to the *x* axis, and the underfill viscosity corresponds to the *y* axis. The four shaded regions (AD) in the graph represent the bump height standard deviation. Regions A, B, C, and D define the processing window for a bump height standard deviation of 0.5 mm. Regions B, C, and D define the processing window for a standard deviation of 1.0 μm, regions C and D define the window for a standard deviation of 1.5 μm, and region D defines the process window for a standard deviation of 2.0 μm.

The solder bump manufacturer controls the bump height standard deviation, and in most cases, the user of the bumped FCs has little control over this parameter. The underfill viscosity and maximum allowable collapse time are the two parameters that users of the high-throughput process may have control over.

One approach to determining the maximum allowable collapse time is to find the length of time above 183°C that the underfill remains a constant viscosity. Typically, no-flow underfill materials remain at a relatively constant viscosity for x amount of time above 183°C, after which the viscosity quickly increases. If the chip collapse process takes longer than this time of constant viscosity, the likelihood of complete collapse is reduced. The time of constant viscosity depends on the underfill and the reflow profile. The underfill supplier may be able to adjust the chemistry to produce a desired time of constant viscosity, or the reflow profile can be adjusted to produce the desired time.

An alternative approach for estimating the maximum allowable collapse time is based strictly on the reflow profile. Reflow profiles are characterized by a number of different parameters, one of which is the time above 183°C. Knowing that the chip can only collapse while the solder interconnections

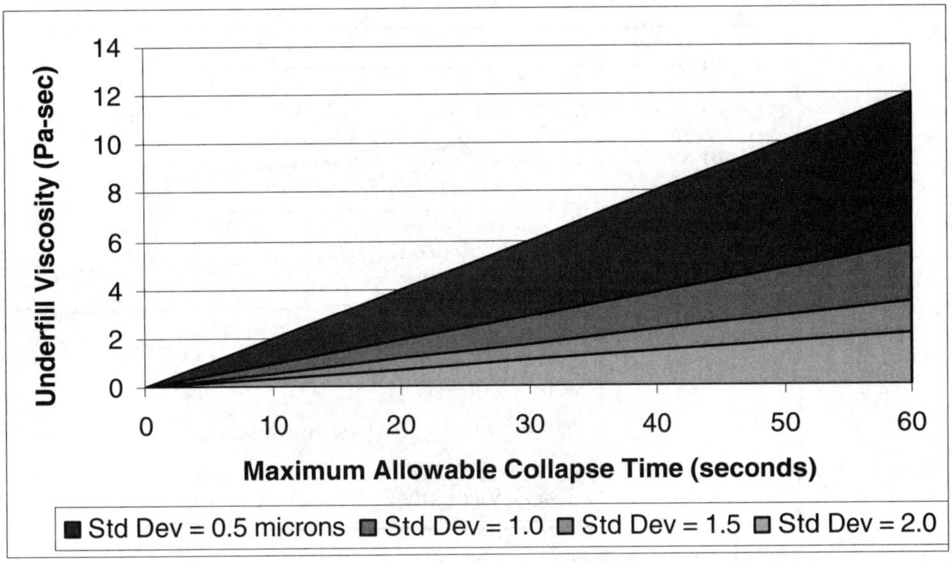

FIGURE 1.98 Process windows for successful chip collapse.

are molten, the time above 183°C reflow parameter can be used as the maximum allowable collapse time. It should be noted that in most cases, the underfill material will harden while the assembly is still experiencing reflow temperatures in excess of 183°C.

The process windows shown in Fig. 1.98 allow a user to specify a range of one parameter based on known ranges of the other two parameters. For example, if the bump height standard deviation is 0.5 μm, then any combination of allowable collapse time and underfill viscosity that lies in shaded regions AD will result in a completely collapsed chip. Alternatively, if known values of underfill viscosity and maximum allowable collapse time correspond to a point in region B, then the bump height standard deviation can be anywhere between 0 and 1.5 μm.

In general, low underfill viscosity, long allowable collapse times, and low bump height standard deviation increase the likelihood of a complete chip collapse.

1.8.8 Interconnection Density Effect

The rigid solder model can also be used to relate interconnection density and collapse time. Figure 1.99 shows collapse time versus number of bumps on a chip. The assumptions for this plot are a 10.16 × 10.16 mm chip, steady state force of 3 mN, bump height distribution with a mean of 98 μm, and a standard deviation of 2 μm. The data show that as the number of bumps per chip area increases, the collapse time decreases. Adding bumps increases the total amount of downward force produced by the solder interconnections, thus increased producing a faster collapse.

1.8.9 Bumps in Contact Due to Placement

The model results to this point have assumed that only the three tallest bumps are in contact with the substrate prior to reflow. During chip placement, it is possible to increase the placement force such that more than three solder balls come in contact with the substrate. Figure 1.100 shows the benefit of increasing placement force. If the tallest 10 percent of bumps are in contact with a bond pad prior to reflow, the collapse time is reduced by as much as 35 percent. If 50 percent of the bumps are in contact with bond pads, the collapse time is reduced by more than 50 percent.

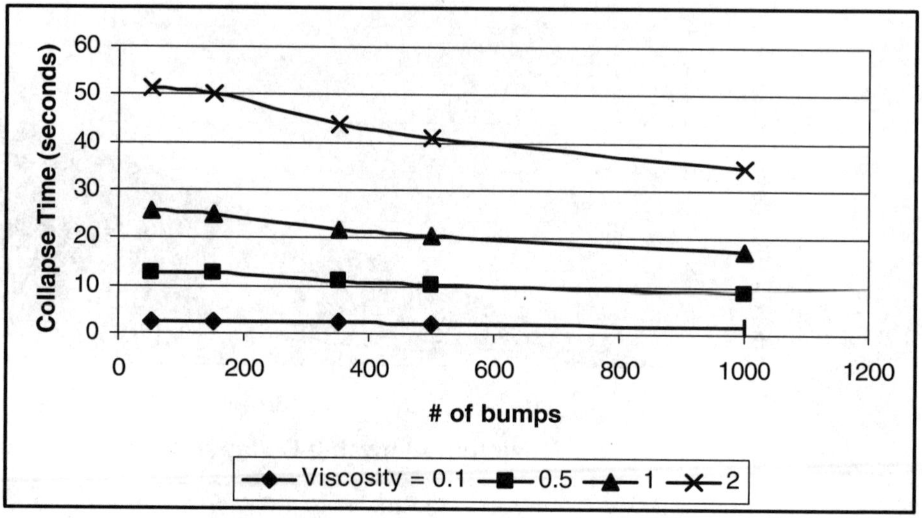

FIGURE 1.99 Collapse time versus number of bumps.

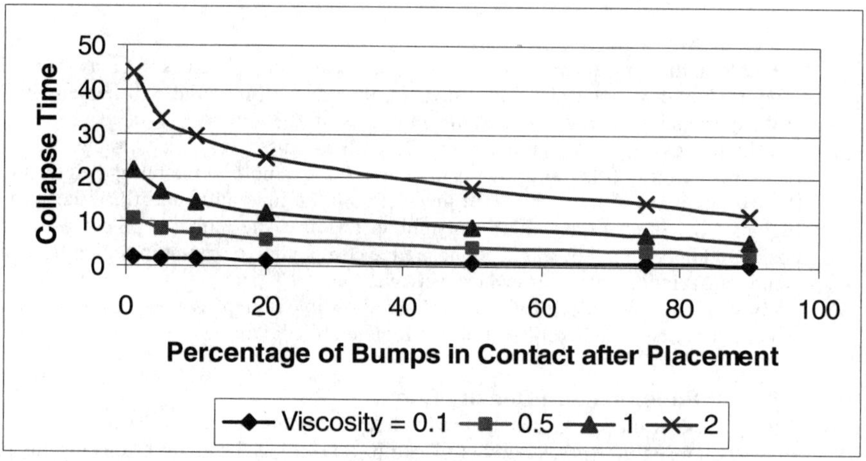

FIGURE 1.100 Collapse time versus percent of bumps in contact.

1.9 SUMMARY

FC technology represents a rapidly advancing area in commercial electronics. In order to ensure adequate reliability, FC assemblies undergo an underfill encapsulation process in which a polymer material is placed between the chip and the substrate. Conventional underfill processing is achieved through chip site to chip site dispensing and underfill flow via capillary action, making it a costly and time-consuming process, particularly as device sizes increase and standoff gaps decrease. To address the limitations associated with conventional FC processing, a low-cost, high-throughput FC process was developed. This process eliminates the need for time consuming capillary flow processing using a compression flow technique where the underfill is applied prior to chip placement. The innovative process integrates the chip placement and polymer underfill processes using a compression or squeeze flow technique. It results in significantly lower assembly costs and reduced cycle time.

The low-cost, high-throughput FC assembly process has converted the assembly process from a point-to-point process to an area-oriented process. This has resulted in decreased assembly time, allowing for increased throughput at lower cost. Parametric studies of FC assembly processes using the activity-based cost model highlighted the benefits of the low-cost, high-throughput assembly process. It has been shown that the assembly cost is reduced with increases in throughput. Furthermore, increased chip size showed minimal increases in cost due to the elimination of the time consuming capillary flow process, whereas the benchmark FC assembly process exhibits substantial cost increases. Cost reductions of 5 to 6 times were demonstrated. In the high-throughput process, underfill flow is no longer dependent on the chip size due to the compression flow chip placement process. Increased chips per panel showed little increase in cost for the next-generation process as a result of area-oriented processing. Cost reductions of 2 to 3 times were demonstrated.

In general, the compression flow of the underfill material governed assembly yield and reliability for the low-cost, high-throughput process using no-flow underfills. Both analytical modeling and flow simulation studies were performed based on the compression flow chip placement process. Simulation studies were conducted to characterize the compression flow of the underfill, estimate required chip placement forces, evaluate the effect of underfill geometry, and assess the potential formation of voids. Three types of voids were identified and experimentally observed that have the potential of forming during compression flow chip placement: compression voids, capture voids, and outgasing voids. With careful control of the underfill geometry, underfill rheology, placement parameters, and reflow process, void formation can be eliminated in the low-cost, high-throughput FC assembly process. Both analytical modeling and flow simulation results indicated that the placement forces required for compression flow chip placement are well within the ranges capable of standard placement systems. In addition, the results yield design guidelines that gave insight into process parameters such as the limits on underfill deposition geometry, potential of gravity flow, and underfill viscosity. The design guidelines provided an preliminary process window for compression flow chip placement.

The capability to assemble FCs using the low-cost, high-throughput assembly process incorporating no-flow underfill materials was demonstrated. The ability to implement the process on standard SMT assembly equipment, including placement machines and reflow ovens, was shown experimentally. The placement process was modeled based on parallel plate squeeze flow analysis both for constant placement speed and constant placement force conditions. Since placement machines have maximum placement forces, these models can be used to estimate whether a placement machine is compatible with the high-throughput assembly process for a given chip size and no-flow underfill material.

The placement process prefers a low to moderate viscosity, no-flow underfill material. Based on the compression flow model, for a given placement force, as underfill viscosity decreases, the allowable placement velocity increases, improving throughput. The die "floating" model was based on a steady-state force balance on the die after placement. The model determines an acceptable range of underfill mass such that the die remains in contact with the substrate after placement. Theoretical and actual data show that there exists a wide processing window for dispensed/printed underfill mass. Acceptable values for no-flow underfill viscosity and surface tension can be determined using the placement process models and the die floating analysis. For a given placement force, the allowable placement velocity increases as underfill viscosity decreases improving throughput. The low-cost, high-throughput assembly process also prefers an underfill with a low surface tension. From the die floating model, the higher the underfill surface tension, the more likely a die will float. It was also determined experimentally that the placement process does not cause voids in the underfill provided that appropriate deposition geometry is used.

Excellent reliability can be achieved using the low-cost, high-throughput FC process using no-flow underfill materials for standard FC on laminate applications. Liquid-to-liquid thermal shock reliability testing results indicate that no-flow materials have the ability to survive in excess of 1000 air-to-air and liquid-to-liquid cycles (−55 to 125°C). In general, the reliability data indicate that the high-throughput assembly process using no-flow underfills can satisfy the reliability requirements for many electronics applications.

1.10 REFERENCES

1. Baldwin, D., and Pascarella, N., "Manufacturability of Underfill Processing for Low-cost Flip Chip," *Structural Analysis in Microelectronics and Fiber Optics,* EEP-Vol. 21, ASME, New York, 1997, pp. 21–31.
2. Baldwin, D. F., and Beerensson, J. T., "Thermal Management in Direct Chip Attach and Chip on Board Assemblies," *Journal of Electronic Packaging,* Vol. 121 (4), ASME, New York, 1999, pp. 222–230.
3. Bird, R. B., Armstrong, R. C., and Hassager, O., *Dynamics of Polymeric Liquids, Vol. 1: Fluid Mechanics,* John Wiley & Sons, New York, 1977.
4. Chen, R., and Baldwin, D. F., "Smart Tooling for Assembly of Thin Flexible Systems," Proceedings of the 5th International Symposium and Exhibition on Advanced Packaging Materials, Processes, Properties, and Interfaces, Braselton, GA, IMAPS, New York, March, 1999, pp. 268–274.
5. Doot, R. K., "Motorola's First DCA Product: The Gold Line Pin Pager," Proceedings of the Electronic Components and Technology Conference, Orlando, FL, IEEE, New York, 1996, pp. 535–539.
6. Engelmaier, W., "Fatigue Life of Leadless Chip Carrier Solder Joints During Power Cycling," *IEEE Transactions on Components, Hybrids, and Manufacturing Technology,* Vol. CHMT-6 (3), 1983, pp. 232–237
7. Gamota, D., and Melton, C., "Materials to Integrate the Solder Reflow and Underfill Encapsulation Processes for Flip Chip on Board Assembly," *IEEE Transactions on Components, Packaging, and Manufacturing Technology,* Part C, Vol. 21 (1), January, 1998.
8. Gamota, D., and Melton, C., "Reflowable Material Systems to Integrate the Reflow and Encapsulant Dispensing Process for Flip Chip on Board Assemblies," International Conference Electronics Assembly Materials Process Challenges, Atlanta, GA, IPC-TP-1098, IPC, Chicago, IL, 1996.
9. Gamota, D., and Melton, C., "The development of reflowable materials systems to integrate the reflow and underfill dispensing processes for DCA/FCOB assembly," *IEEE Transactions on Components, Packaging & Manufacturing Technology,* Part C Manufacturing, Vol. 20 (3), 1997, pp. 183–187.
10. Gamota, D., and Melton, C., "Materials to Integrate the Solder Reflow and Underfill Encapsulation Processes for Flip Chip on Board Assembly," *IEEE Transactions on Components, Packaging, and Manufacturing Technology,* Part C, Vol. 21 (1), January 1998.
11. Gilleo, K., and Blumel, D., "The Great Underfill Race," in 1998 International Symposium on Microelectronics, Philadelphia, PA, IMAPS, New York, 1996, pp. 701–706.
12. Goldman, L. S., "Self-Alignment Capability of Controlled-Collapse Chip Bonding," in Proceedings of the Electronic Components Conference, Orlando, FL, IEEE, New York, 1972, pp. 332–339.
13. Han, S., Wang, K., and Cho, S., "Experimental and Analytical Study on the Flow of Encapsulant During Underfill Encapsulation of Flip-Chips," Proceedings of the 46th Electronic Components and Technology Conference, Orlando, FL, May, IEEE, New York, 1996, pp. 327–334.
14. Houston, P. N., Baldwin, D. F., Crane, L. N., and Konarski, M., "Low Cost Flip Chip Processing and Reliability of Fast Flow, Snap Cure Underfills," Proceedings of the Electronic Components and Technology Conference, San Diego, CA, IEEE, New York, May 1999, p. 61.
15. Johnson C. D., and Baldwin, D. F., "Wafer Scale Packaging Based On Underfill Applied At The Wafer Level For Low-Cost Flip Chip Processing," Proceedings of the Electronic Components and Technology Conference, San Diego, CA, IEEE, New York, May 1999, p. 950.
16. Lasky, R. C., and Baldwin, D. F., "Throughput: The Critical Cost Variable in DCA Assembly," Proceedings of the SMTA 3rd Annual National Symposium, Research Triangle Park, NC, 1996.
17. Leider, P. J., and Bird, R. D., "Squeezing Flow Between Parallel Disks. I. Theoretical Analysis," *Industrial Enginnering Chemical Fundamentals,* Vol. 13, 1974, pp. 336–341.
18. Leider, P. J., "Squeezing Flow Between Parallel Disks. II. Experimental Results," *Industrial Engineering Chemical Fundamentals,* Vol. 14, 1974, pp. 342–346.
19. McGovern, L. P., and Baldwin, D. F., "High-throughput Low-cost Flip Chip on Board Assembly Processing," *Electronics Packaging and Production,* Vol. 38 (2), February 1998, pp. 68–76.
20. Mizutani, M., "A Study of a New Flip Chip Packaging Process for Diversified Bump and Land Combination," Proceedings of the Electronic Components and Technology Conference, Seattle WA, IEEE, New York, 1998, pp. 316–319.

21. Montgomery, D. C., *Design and Analysis of Experiments,* John Wiley & Sons, New York, 1997.
22. Palaniappan, P., and Baldwin, D. F., "In Process Stress Analysis of Flip Chip Assemblies During Underfill Cure," *Applications of Experimental Mechanics to Electronic Packaging,* EEP-Vol. 22/AMD-Vol. 226, ASME, New York, 1997, pp. 7–14.
23. Palaniappan, P., Selman, P., Baldwin, D. F., Wu, J., and Wong, C. P., "Correlation Of Flip Chip Underfill Process Parameters And Material Properties With In-Process Stress Generation," *IEEE Transactions on Electronics Packaging Manufacturing,* Vol. 22 (1), IEEE, New York, January 1999, pp. 53–62.
24. Pascarella, N., and Baldwin, D. F., "Compression Flow Modeling of Underfill Encapsulants for Low-cost Flip Chip Assembly," Proceedings of the 48th Electronic Components and Technology Conference, Seattle, WA, May, IEEE, New York, 1998, pp. 463–470.
25. Pascarella, N. W., and Baldwin, D. F. , "Manufacturing Analysis of Underfill Processing for Low-cost Flip Chip," *Journal of Electronics Manufacturing,* Vol. 8 (1), ASME, New York, October 1998, pp. 39–50.
26. Pascarella, N. W., and Baldwin, D. F., "Advanced Encapsulation Processing for Low-cost Flip Chip Assembly—A Cost Analysis," 3rd International Symposium on Advanced Packaging Materials Processing Properties and Interconnection, Braselton, GA, IMAPS, New York, March 1997.
27. Pascarella, N. W., and Baldwin, D. F., "Advanced Encapsulation Processing for Low Cost Electronics Assembly—A Cost Analysis," *Advances in Electronic Packaging 1997,* EEP-Vol. 19–1, ASME, New York, 1997, pp. 359–363.
28. Pascarella, N. W., and Baldwin, D. F., "Compression Flow Modeling of Underfill Encapsulants for Low-cost Flip Chip Assembly," *IEEE Transactions on Components, Packaging, and Manufacturing Technology—Part C,* Vol. 21 (4), IEEE, New York, October 1998, pp. 325–335.
29. Pascarella, N. W., and Baldwin, D. F., "Cost Analysis of Low Cost High Throughput Next Generation Flip Chip Assembly," *International Journal of Microcircuits and Electronic Packaging,* Vol. 20, IMAPS, New York, 1997, pp. 571–577.
30. Pascarella, N. W., and Baldwin, D. F., "Cost Analysis of Low-cost High-throughput Next-generation Flip Chip Assembly," *International Journal of Microcircuits and Electronic Packaging,* Vol. 20, 1997, pp. 571–577.
31. Rodriguez, G., and Baldwin, D. F., "Analysis of Solder Paste Release in Fine Pitch Stencil Printing Processes," *Journal of Electronic Packaging,* Vol. 121, September, ASME, New York, 1999, pp. 1–11.
32. Rodriguez, G., and Baldwin, D. F., "Analysis of Solder Paste Release in Fine Pitch Stencil Printing Processes," *Advances in Electronic Packaging 1999,* EEP-Vol. 26-1 ASME, New York, 1999, pp. 649–660.
33. Schubert, A., Dudek, R., Leutenbauer, R., Doring, R., Kloeser, J., Oppermann, H., Mechel, B., Reichl, H., Baldwin, D. F., Qu, J., Sitaraman, S., Swaminathan, M., Wong, C. P., and Tummala, R., "Do Chip Size Limits Exist for DCA?," Proceedings of the 5th International Symposium and Exhibition on Advanced Packaging Materials, Processes, Properties, and Interfaces, Braselton, GA, IMAPS, New York, March 1999, pp. 150–162.
34. Schwiebert, M., and Leong, W., "Underfill Flow as Viscous Flow Between Parallel Plates Driven by Capillary Action," *IEEE Transactions on Components Packaging and Manufacturing Technology—Part C,* Vol. 19, 1996, pp. 133–137.
35. Sitaraman, S., Hanna, C., Mechaelides, S., Palaniappan, P., and Baldwin, D. F., "Numerical and Experimental Study of the Evolution of Stresses in Flip Chip Assemblies During Assembly and Thermal Cycling," Proceedings of the Electronic Components and Technology Conference, May 1999.
36. Smith, B. A., Houston, P. N., Baldwin, D. F., and Thorpe, R., "A Reliability Analysis of No Flow Underfill Materials," Proceedings of the Electronic Components and Technology Conference, Las Vegas, NV, IEEE, New York, May 2000, pp. 1719–1730.
37. Suhir, E., "Calculated Thermal Induced Stresses in Adhesively Bonded and Soldered Assemblies," Proceedings of the 1986 International Symposium on Microelectronics, Chicago, IL, IEEE, New York, pp. 383–392.
38. Thorpe, R., Baldwin, D. F., and McGovern, L. P., "High Throughput Flip Chip Processing and Reliability Analysis Using No-Flow Underfills," Proceedings of the Electronic Components and Technology Conference, San Diego, CA, IEEE, New York, May 1999, p. 419.
39. Thorpe, R., Baldwin, D. F., and McGovern, L. P., "Yield Analysis and Process Modeling of Low Cost, High Throughput Flip Chip Assembly Based on No-Flow Underfills," IEEE Transactions on Electronics Packaging Manufacturing, Vol. 24 (2), April 2001, pp. 123–135.
40. Thorpe, R., McGovern, L. P., and Baldwin, D. F., "Analysis Of Process Yield In Low-cost Flip Chip On Board Assembly Processes," Thermo-Mechanical Characterization of Evolving Packaging Materials and Structures, ASME, New York, EEP—Vol. 24, 1998, pp. 27–33.
41. Van Krevelen, D. W., *Properties of Polymers,* Elsevier, New York, 1972.
42. Wong, C. P., and Baldwin, D. F., "Novel No-flow Underfills for Low-Cost Flip-Chip Applications: Materials and Processes," *Future Circuits International,* Issue 3, London, 1998, pp. 67–70.

43. Wong, C. P., and Shi, S., "Study of the Fluxing Effects on the Properties of No-Flow Underfill Materials for Flip Chip Applications," Proceedings of the 48th Electronic Components and Technology Conference, Seattle, WA, IEEE, New York, 1998.
44. Wong, C. P., Shi, S., and Jefferson, G., "High Performance Low-cost Underfills for Flip-Chip Applications," Proceedings of the 47th Electronic Components and Technology Conference, San Diego, CA, IEEE, New York, 1997.

CHAPTER 2
FLIP CHIP ASSEMBLY AND UNDERFILLING

R. Wayne Johnson
Auburn University

2.0 INTRODUCTION

IBM first introduced flip chip (FC) technology in 1964 in the solid logic technology (SLT) hybrid modules in the System 360 mainframe.[1] The technology was developed by IBM to replace wire bonding as a means of interconnecting semiconductor dies to thick-film metallization on alumina. The dies were three-terminal transistors with Au/Ni-plated Cu balls embedded in a Pb/Sn solder bump on the three input-output (I/O) pads of the transistor. A Cr/Cu/Au interface layer was deposited between the Al transistor bond pads and the solder bump. The device was assembled to the hybrid substrate by inverting (flipping) the dies and reflow soldering the copper balls to corresponding metal pads on the substrate. The copper balls maintained a constant standoff between the dies and the substrate after reflow.

As the I/O count of the dies increased, the copper spheres were replaced with high-lead tin-lead alloy solder bumps.[2] The solder balls collapse somewhat during the reflow soldering process, balancing the weight of the chip and the surface tension forces of the molten solder. This phenomenon gave rise to the IBM terminology *controlled collapse chip connection* (C4). The flow of the solder (collapse of the chip-substrate spacing) was controlled during reflow by controlling the solder volume and the wettable metal exposed on the dies and the substrate.

The advantages of the FC assembly process for high-volume manufacturing were recognized by the automotive industry. FC-on-ceramic (FCOC) was adopted by the automotive industry in the late 1970s as a high-volume surface-mount assembly technology for applications such as ignition modules. The dies typically were small (<3 mm) with a low I/O count (<20). The solder bumps primarily were high-lead alloys, although some eutectic bumps were used. The large pitch of the bumps allowed solder paste to be printed on the thick-film metal pads. The FC die was then placed and reflowed along with other surface-mount components. With the high-lead bumps, a high-lead solder paste was used. The thick-film ceramic substrate allowed reflow at temperatures in excess of 325°C. The FCOC assembly process paralleled the development of surface-mount technology (SMT) on laminate with plastic leaded packages such as plastic leaded chip carriers (PLCCs), quad flatpacks (QFPs), and small outline transistors (SOTs).

FC technology offers other advantages as semiconductor device speed, size, and I/O count increase. For high-speed applications, the small solder balls add minimal parasitics (resistance, capacitance, and inductance) and propagation delays to the electrical signal path. In addition, the elimination of the die package permits the dies to be placed close together on the substrate, further reducing interconnection parasitics and propagation delay. Elimination of the die package and close spacing of dies are also an advantage in portable electronics, where size and weight are critical. Finally, as the die I/O count continues to increase, perimeter electrical interconnection by wire bonding becomes limited. In FC technology, I/Os can be arrayed over the entire area of the die, dramatically increasing the number of I/Os.

While there is a coefficient of thermal expansion (CTE) mismatch between alumina (6.5 ppm/°C) and silicon (3 ppm/°C), millions of reliable, under-the-hood modules were built using small dies on ceramic without the need for underfill. Millions of FC solder joints also were used in mainframe computers without failure. Increasing either the size of the die or the CTE of the substrate (switching from ceramic to laminate at approximately 16 ppm/°C) increases the strain on the solder joints during thermal cycling or thermal shock (Fig. 2.1). To improve reliability, underfill was added to the FC assembly process.[3–5] Underfill is a polymer material (typically a filled epoxy) that fills the space between the die and the substrate. The underfill bonds the die and the substrate together, forming a trilayer structure. The net effect is to reduce the strain on the individual solder joints by creating a structure that warps as the temperature changes (Fig. 2.2).

The objective of this chapter is to discuss the materials and processes for FC assembly with underfill.

2.1. PROCESS OVERVIEW

The application and cure of the underfill material are the primary assembly steps that have an impact on FC assembly process flow. The underfill can be applied after the FC is placed and reflowed (capillary flow) or prior to die placement. Fluxing underfills are applied to the substrate just prior to die placement, whereas wafer-applied underfills are coated onto the wafer prior to singulating the wafer into individual dies. These three options are introduced below and will be detailed in later sections of this chapter.

2.1.1 Capillary Flow Underfills

Liquid capillary flow underfill materials are dispensed along the edge of the die after the die has been reflow soldered to the substrate. Capillary action pulls the liquid under the die, as illustrated in Fig. 2.3. After the liquid underfills the die, a second dispense may be required to form a fillet around the edge of the die. Self-filleting is observed with some materials, eliminating this step. Following the dispense step, the underfill must be cured. Cure times and temperatures vary from 5 to 90 min at 150 to 165°C depending on the underfill chemistry. Capillary flow underfills are available commercially from a number of suppliers and are used most commonly today.

2.1.2 Fluxing Underfills

Fluxing underfills include the fluxing chemistry for the soldering step in the underfill. The liquid is applied to the substrate, typically by dispensing. The die is then placed and reflowed. The underfill may cure during the reflow cycle, or a postreflow cure may be required depending on the material.

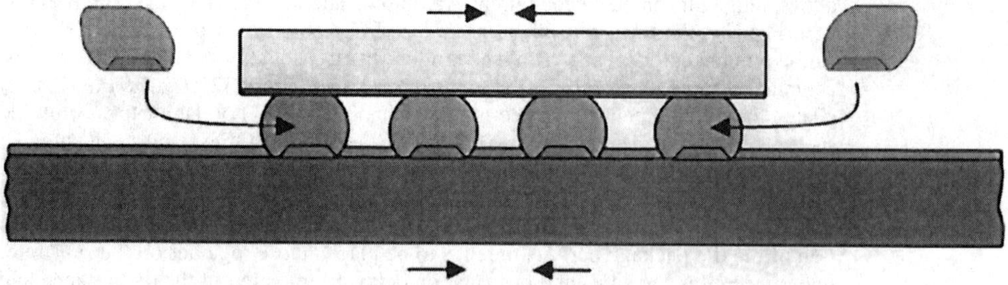

FIGURE 2.1 Stress on solder joints due to CTE mismatch when assembly is cooled from solder solidification temperature to room temperature. The substrate contracts more than the silicon die during cool-down.

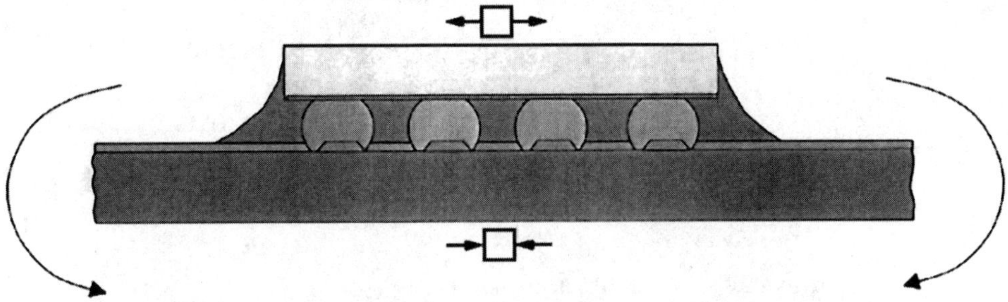

FIGURE 2.2 Warpage when underfilled structure is cooled from underfill cure temperature to room temperature.

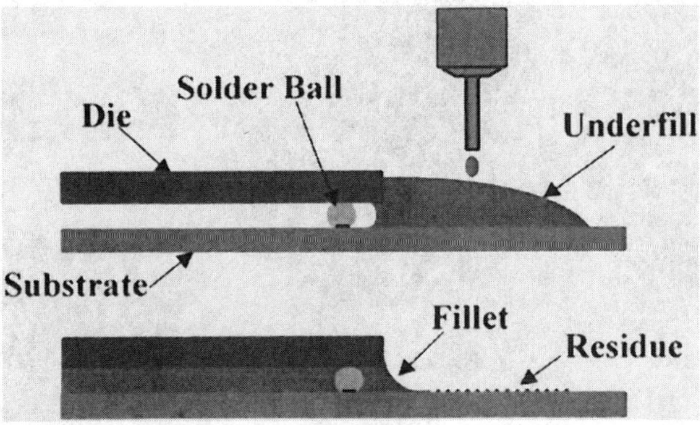

FIGURE 2.3 Capillary underfill process.

The process is illustrated in Fig. 2.4. Fluxing underfills are available commercially from a number of suppliers.

2.1.3 Wafer-Applied Underfills

Rather than applying the underfill one die at a time during the assembly operation, wafer-applied underfills are coated onto all dies on a wafer at one time. After coating, the underfill must be a dry (nontacky) film for handling. During the reflow cycle, the underfill liquefies, providing fluxing action. The underfill may cure during the reflow cycle or may require a postreflow cure step. The process is illustrated in Fig. 2.5. Wafer-applied underfills are in the developmental stage and not available commercially.

2.2 SUBSTRATE DESIGN

With packaged SMT components, the design process is simplified through industry standardized I/O configurations and recommended substrate pad designs. However, FC I/O configurations are custom to each die; therefore, substrate pad designs have not been standardized. The design will depend on the I/O pitch, the number of I/Os, the arrangement of the I/Os, and the substrate technology. Within this parameter space, design options exist and are discussed in the following section.

2.2.1 Layout

While FC bump patterns can be irregular, single-row perimeter, two-row perimeter, and full area array are most common. Each presents its own design and substrate fabrication challenges.

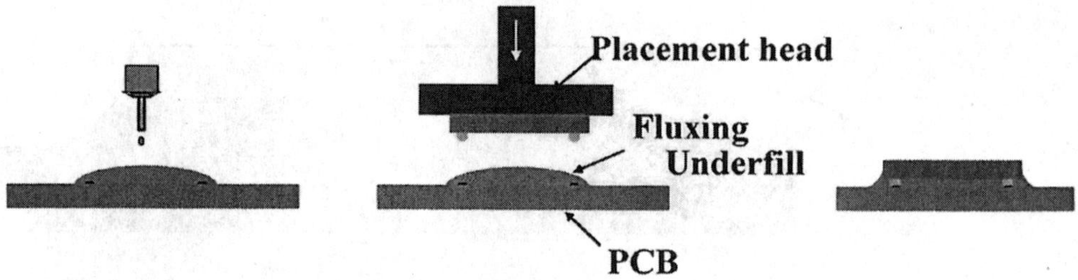

FIGURE 2.4 Assembly with fluxing underfill.

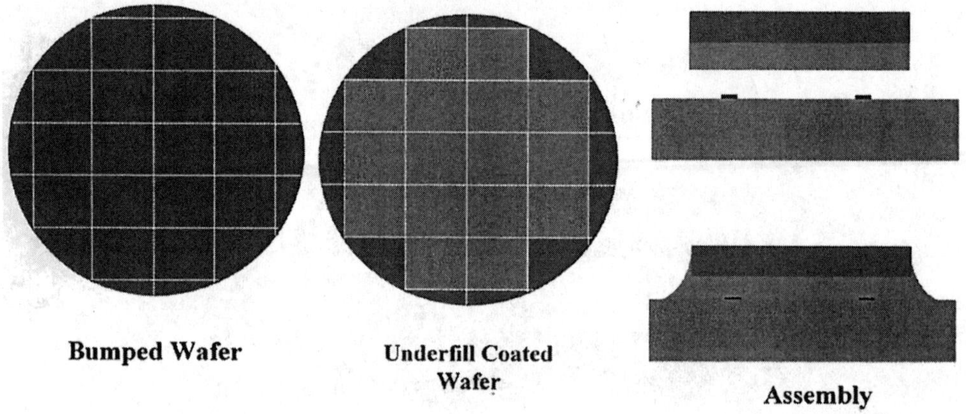

FIGURE 2.5 Wafer-applied underfill assembly process.

Single-Row Perimeter. FC dies with a single row of perimeter solder bumps are the simplest to route. The printed wiring board (PWB) technology must provide line and space pitch equal to the pitch of the solder bumps, as illustrated in Fig. 2.6. In this design, the copper trace width and the solder mask opening define the solderable copper area exposed to the solder ball. Increasing the solderable area decreases the gap height between the die face and the solder mask. In turn, decreasing this gap height increases the time required for capillary underfills to flow, adversely affecting the underfill process. If the gap becomes sufficiently small, capillary underfill may not flow into the gap.

For fine-pitched solder bumps, decreasing the solderable area increases the risk of shorting between bumps. If the solder mask opening (trench) becomes too narrow, the solder ball will rest on the edges of the solder mask and not make contact with the copper trace. An open solder joint after reflow can be the result.

PWB manufacturing variations must be considered. Variations in copper etching (trace width) and solder mask developing (trench width) from board to board will lead to variations in gap height. In turn, this will change the volume of material required to underfill the die. In production, the underfill dispense volume is not changed from board to board. Therefore, the underfill fillet must provide the reservoir to ensure that the die is completely underfilled. With gap variations there will be a corresponding variation in the amount of fillet formed. Consistent PWB fabrication is important.

A second issue in PWB fabrication is the solder mask registration. As illustrated in Fig. 2.7, if the solder mask is shifted to the right, the solder balls on the two sides can be aligned, but the solder balls on the top and bottom are misaligned. The placement in Fig. 2.7*a* assumes that copper-defined fiducials are used. If solder mask–defined fiducials are used, the placement result is shown in Fig. 2.7*b*. In both cases, the solder balls do not line up with the pads. At the top and bottom of the die, the solderable pads are defined by the copper trace in the x direction and the solder mask in

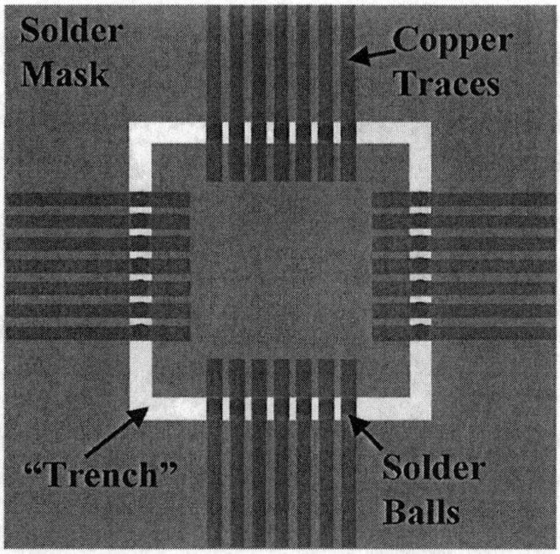

FIGURE 2.6 "Trench" design for single-row perimeter FC dies.

the y direction. On the other two sides of the die, the solderable pads are defined by the solder mask in the x direction and the copper trace in the y direction. Thus no placement shift can compensate for solder mask misregistration. The surface tension of the molten solder during the reflow process will tend to center the chip, creating approximately even solder joint distortion on all sides, as shown in Fig. 2.8. The impact of this distortion on reliability has not been documented. With fine-pitch solder bumps, solder mask-to-copper misregistration increases the likelihood of solder bridging.

An alternate design can be used to address misregistration issues (Fig. 2.9). In this design, the die placement is based on copper fiducials. The maximum solder mask misregistration must be considered in the determination of the length of the exposed copper fingers. Overetching of the copper is more critical in this design because overetching reduces not only the trace width but also the trace length. Overetching combined with a solder mask shift to the left can result in very small pads on the right-hand side. If the solder mask is misregistered as shown in Fig. 2.9*b,* the gap height will be less on the left side (more solderable copper area). This design approach decreases the potential for bridging of fine-pitch FC bumps. The die-to-PWB gap height is increased in this design by eliminating the solder mask under the die. Other designs for single-row perimeter dies are possible, but these two are the most common.

Two-Row Perimeter. The first determining factor for a two-row perimeter design is the FC I/O pitch relative to the pitch of the PWB. Can a PWB trace be routed between adjacent solder ball pads (Fig. 2.10)? The design in Fig. 2.10 is similar to Fig. 2.6. Isolated solder mask openings (as shown) or two trenches can be used depending on the I/O pitch and the solder mask resolution. The two-trench approach would impede underfill flow. Consistent control of copper etching, solder mask resolution, and solder mask-to-copper registration are again critical.

Figure 2.11 is a two-row perimeter design with characteristics similar to Fig. 2.9. The solder mask misregistration should be less than 50 percent of the line-to-line spacing on the PWB. Depending on pitch, this may be challenging.

If the PWB technology will not support one trace between solder bump pads, the second row of bumps must be routed toward the center of the die and down through plated through-holes or vias. If conventional PWB technology with plated through-holes for z-axis interconnections is to be used, the plated through-holes under the dies must be tented or plugged. This is necessary to prevent the underfill from flowing out through the hole during the underfill process. Plated through-holes typically are tented with dry-film solder mask. However, FC PWBs are fabricated almost exclusively with liquid photoimagable (LPI) solder mask to achieve thinner coatings and better resolution. LPI

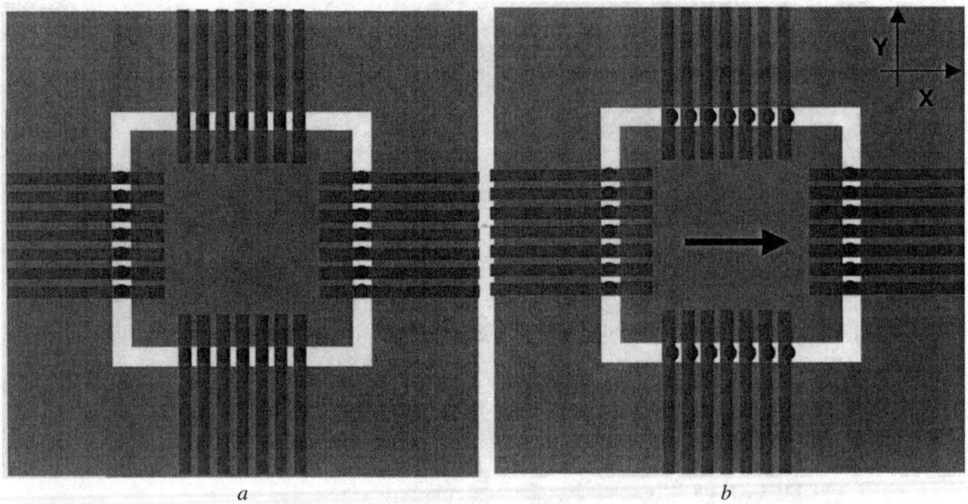

FIGURE 2.7 Solder mask misregistration to the right. (*a*) Placement based on copper-defined fiducials; (*b*) Placement based on solder mask–defined fiducials.

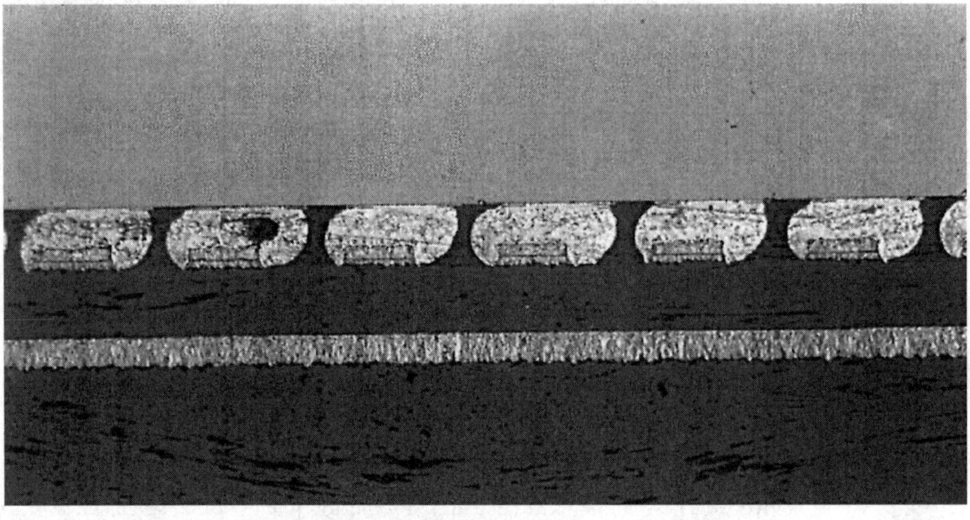

FIGURE 2.8 Solder joint distortion due to solder mask misregistration.

solder mask cannot tent plated through-holes. Thus plugged holes typically are used for FC PWBs. After plating, the holes are filled (plugged) with epoxy and cured. The LPI solder mask can then coat over the plugged hole. The design is illustrated in Fig. 2.12. Copper etching and solder mask-to-copper registration are critical to this design.

Area Array. With smaller dies or full area arrays, the pad diameters for the plated through-holes are too large to allow routing. High-density interconnect (HDI) substrates are required. The substrates are fabricated by buildup of sequential dielectric and copper layers processed on a laminate core. The HDI structure is illustrated in Fig. 2.13. This figure is of the IBM surface laminar circuitry

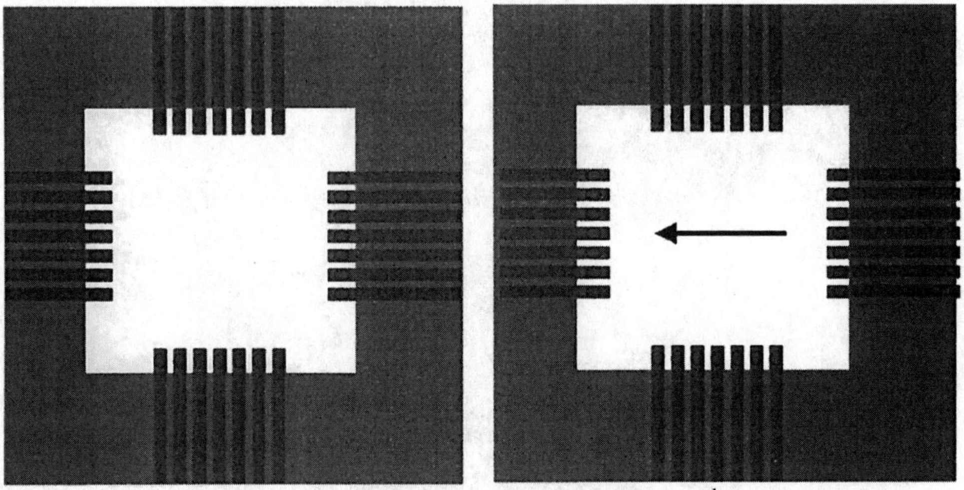

a *b*

FIGURE 2.9 Alternate design for single-row perimeter FC: (*a*) perfect copper-to-solder mask registration; (*b*) solder mask is misregistered to the left.

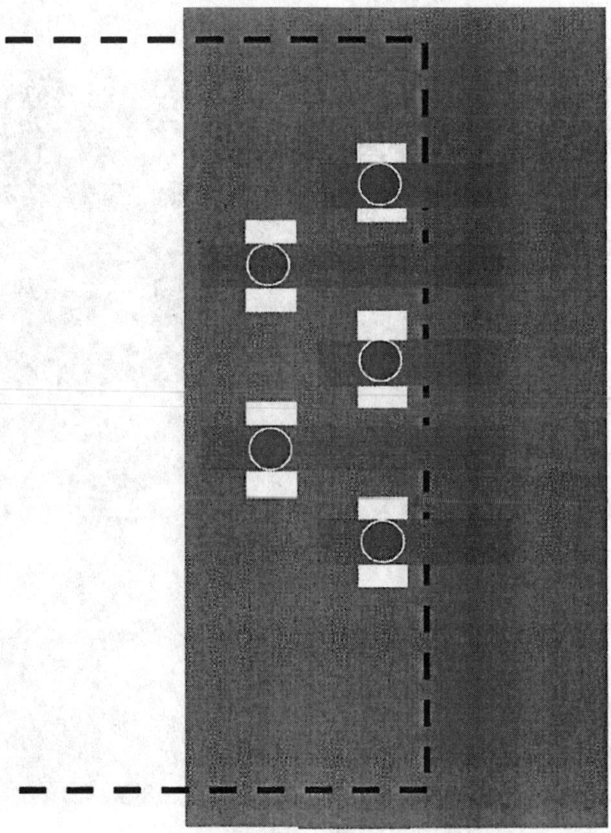

FIGURE 2.10 Two-row perimeter design.

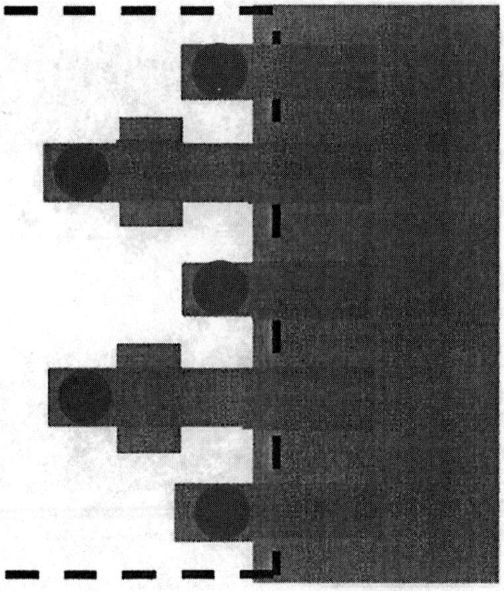

FIGURE 2.11 Alternate two-row perimeter design.

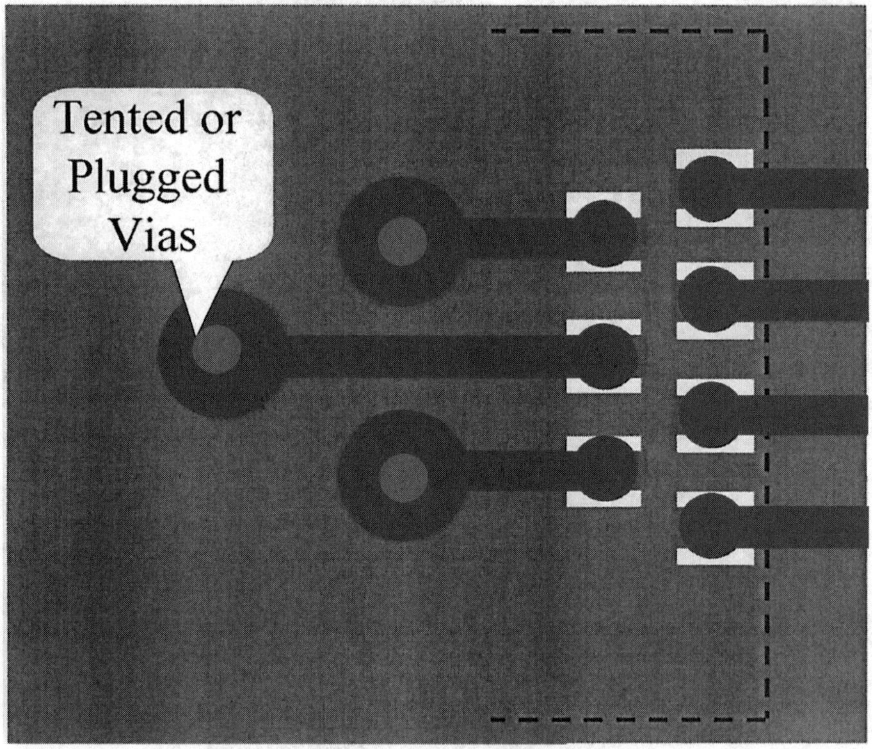

FIGURE 2.12 Two-row perimeter design with plugged plated through-holes.

(SLC) technology, one of the first HDI approaches. The SLC technology used a liquid, photoimageable dielectric layer with additive copper to fabricate the HDI layers.

A number of HDI approaches subsequently have been developed. The starting dielectric material may be either liquid (nonreinforced) or a dry film (reinforced or nonreinforced). The formation of vias is by photoimaging, laser, or plasma, as shown in Table 2.1. The vias are through only one layer of dielectric and are less than 125 μm in diameter (Fig. 2.14). Additive processing is used commonly for copper deposition/patterning. In some approaches, the vias are filled with a conductor-filled epoxy. This is an advantage in "pad in via" designs to be discussed below.

With the smaller pads and vias achievable with HDI technology, area array FC dies can be routed. The pitch of the vias must equal that of the solder bumps on the dies. If the via pad can fit in the interstitial space between solder bump pads, a "dog bone" design similar to ball grid array (BGA) and chip scale packaging (CSP) patterns can be used (Fig. 2.15).

The ultimate in density is achieved with a "pad in via" design (Fig. 2.16). One concern with pad in via design is the potential for voids trapped in the solder joint (Fig. 2.17). The impact of these voids on reliability has not been determined. The second concern is the increased solder wettable area, which decreases the standoff height. The vias may be either filled or plated solid to avoid entrapped voids during solder reflow. In addition to density, the absence of a solder mask and potential for misregistration is a benefit of this design.

2.2.2 Copper and Solder Mask Requirements

Voids (air pockets) can be trapped during underfill flow at steep topology features such as the edge of a thick copper trace. Thinner copper reduces the probability of voids in the underfill at the base of the copper trace. As the pitch of the FC device decreases, thinner copper is also required to fabricate finer lines and spaces. The thickness of the surface copper layer is the combination of starting foil thickness and the plating thickness required for reliable plated through-holes in conventional PWB fabrication. Foil thicknesses less than ¼ oz/sq. ft. (8.3 μm) are available but more expensive and difficult to process. Typical plated through-hole copper thicknesses range from ½ oz/sq. ft. (16.6 μm) for consumer/portable electronics applications to 1 oz/sq. ft. (33.2 μm) for automotive applications. The thickness of the plated through-hole copper required is a function of the total board thickness and the reliability (thermal cycle) requirements. In most HDI structures, no beginning foil is used, and the surface copper thickness is determined by the plating thickness.

Electroless nickel/immersion gold is the most commonly used surface finish for FC assembly. For long-term reliability, the gold volume within the solder joint should be less than 3 percent and preferably less than 1 percent to avoid embrittlement of the joint and Au/Sn intermetallic formation. As the solder ball volume decreases, the maximum gold thickness decreases. Other finish alternatives include silver and tin. Organic solderability preservative (OSP) coatings also are used commonly for FC boards. The number of reflow cycles and high-temperature process steps and the processing sequence should be reviewed if OSPs are to be used. The effectiveness of OSPs decreases with

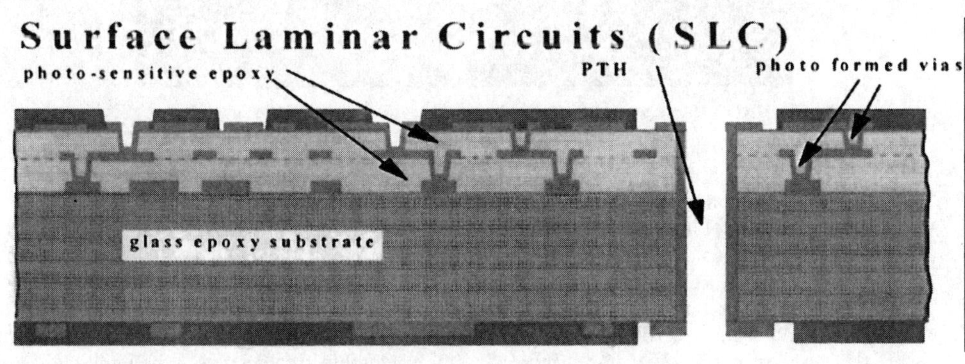

FIGURE 2.13 Example of high-density interconnect (HDI) substrate.

TABLE 2.1 HDI Dielectric Types and Via Processing

Dielectric type	Form	Via processing
Non reinforced	Liquid	Photoimage, laser
	Dry film	Photoimage, laser, plasma
Reinforced	Dry film	Laser

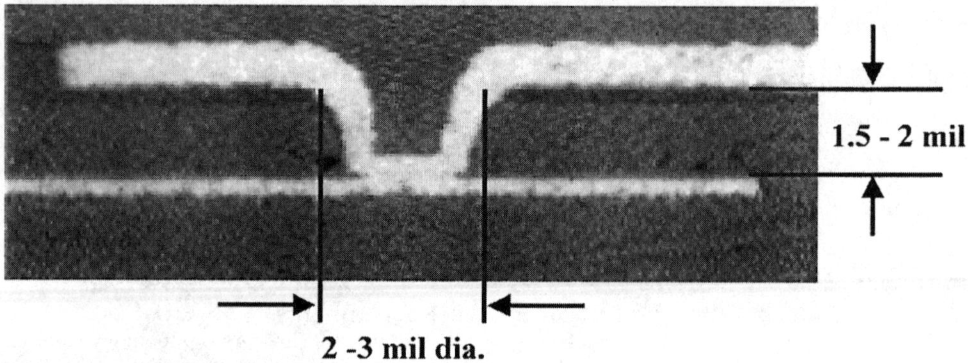

FIGURE 2.14 Example of HDI via.

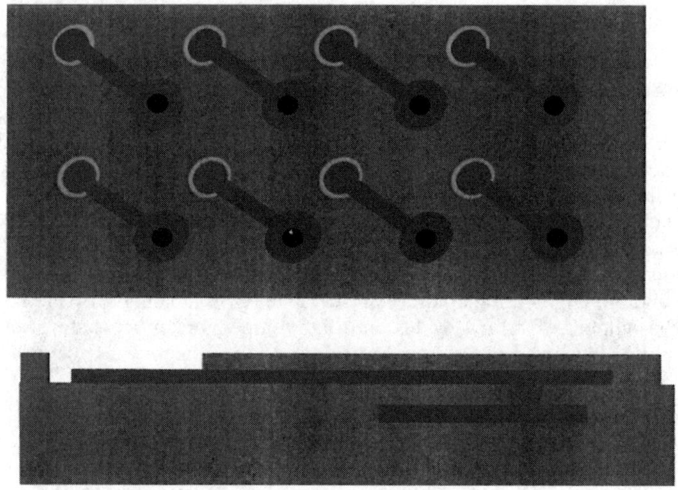

FIGURE 2.15 "Dog bone" design for area array FC using an HDI substrate.

multiple high-temperature exposures. Some OSP coatings are formulated to withstand more high-temperature exposures.

The solder mask has an impact on the assembly process and yield. As discussed in the preceding section, solder mask registration is critical. While the industry standard registration tolerance is 3 mils (75 μm), fine-pitch FCs may require a tolerance of less than 2 mils (50 μm). This is a challenge for the industry, particularly if processing 18 × 24-in panels for manufacturing efficiency. Solder mask resolution is also important. In many designs, the solderable area is partially defined by the solder mask opening. Increasing the solderable area decreases the gap height, making the underfill process more difficult. If the solder mask opening is too small, solder bridging can result with fine-pitch assemblies. In the extreme case, the solder mask opening may be too small for the solder ball

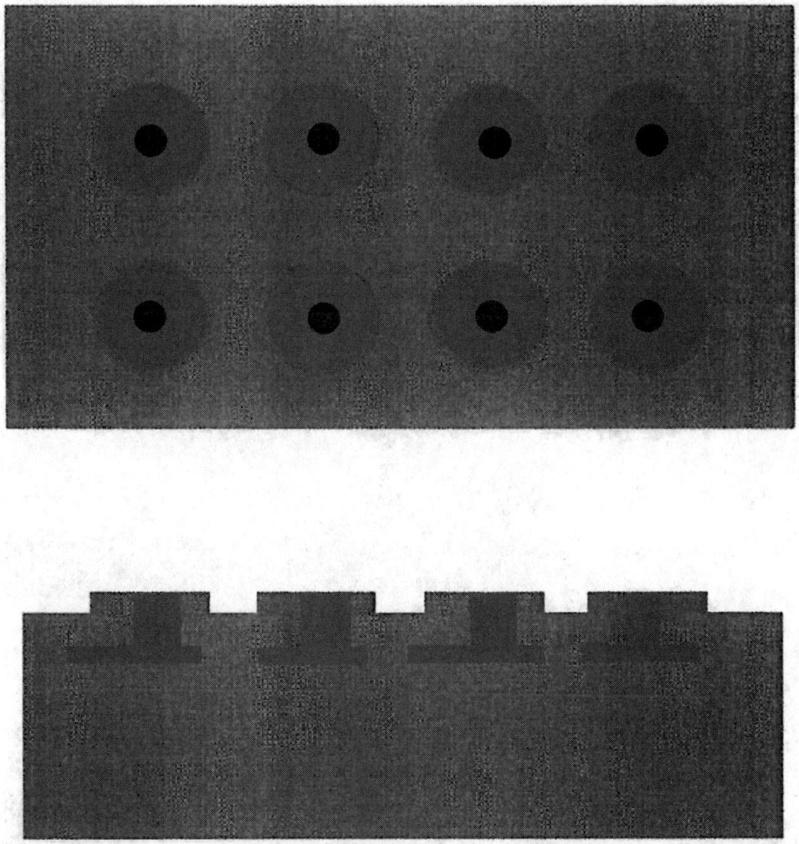

FIGURE 2.16 Pad-in-via design with HDI substrate.

to touch the copper pad, resulting in an open connection. The solder mask thickness has an impact on the gap height. The solder mask thickness over the copper traces should be less than 12 μm. It is a challenge to apply thin solder mask without having skips or exposed copper, particularly at the edge of traces.

Proper board design, materials selection, and fabrication are important to FC assembly yield and reliability. Each design is unique, and compromises must be considered.

2.3 ASSEMBLY WITH CAPILLARY FLOW UNDERFILL

The process flow for FC assembly with capillary flow underfill is illustrated in Fig. 2.18. This process assumes that the flux residue is not cleaned after reflow, which is typical of most FC assemblies. Each of the process steps and the associated materials will be discussed in the following subsections.

2.3.1 Die Presentation

Dies can be presented for pick and place either as sawn wafers, in waffle packs, or in tape and reel. By placing dies directly from the wafer, intermediate handing steps are eliminated. Since wafers are sawn bump side up, the dies must be picked from the bump side and inverted for placement. This adds an extra step to the pick-and-place operation. Care must be taken that ejector pins used to assist

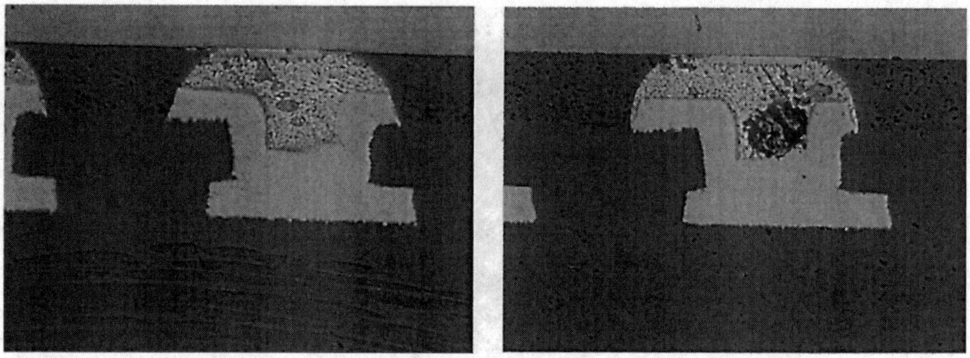

FIGURE 2.17 Example of pad-in-via design with and without entrapped void. (*Jabil Circuits.*)

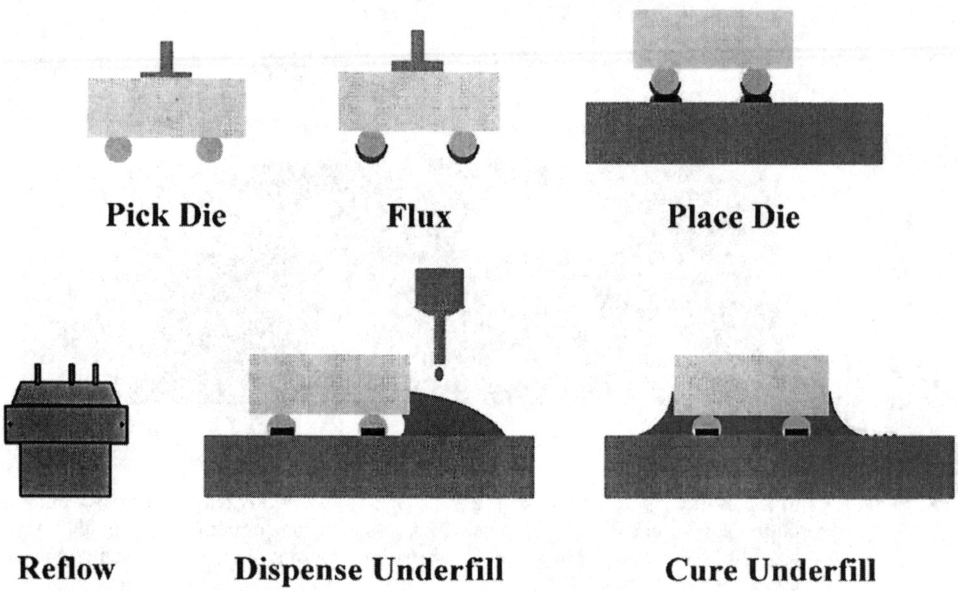

FIGURE 2.18 FC assembly with capillary flow underfill.

in the removal of dies from the expanded tape do not cause microcracks or flaws in the backside of the dies. Motorola has shown that ejector pin damage of the die backside can lead to die fracture during assembly or in the field.[6] This is an issue both for pick and place from wafers and for placing of dies into waffle packs or tape from the wafer.

Waffle packs are used commonly for FC dies, but are limited in the number of dies they contain. For high-volume assembly, tape and reel is preferred, since a single reel will hold hundreds of dies. In both cases, the dies are inverted, solder balls down for direct pick and place. The individual die cavities in the waffle pack or tape should be only slightly larger than the dies to avoid rotation and translation of the dies in the cavity.

2.3.2 Flux and Flux Application

In standard SMT assembly, solder paste is stencil printed, and the components are placed into the paste. Flux, not solder paste, generally is used in the assembly of FC dies. Flux application is required prior to placement of the FC. Fluxing options are listed in Table 2.2. Dip fluxing is illustrated in Fig. 2.19. The dipping operation is performed on the pick-and-place system and adds to the FC placement cycle time. The variables to control are flux depth, dip force, and hold time. Increasing these variables increases the amount of flux transferred to the solder balls. This, in turn, increases solder wetting and self-centering of dies placed off center. However, increasing the volume of flux increases the amount of residue, which may interfere with underfill flow and decrease reliability if a no-clean process is used. The viscosity and surface tension of the flux also affect flux transfer. Since the flux is exposed to the atmosphere as a thin layer, the evaporation rate for volatile components must be extremely low at room temperature.

Stencil or screen printing also has been used for flux application.[7] If both FC and SMT components are to be assembled to the board, the flux is printed first. A step stencil (Fig. 2.20) is then used to print the solder paste. Flux printing can apply flux to multiple sites on a board in one pass and eliminates the flux dip step from the pick-and-place operation, increasing throughput at pick and place. The cost analysis of print versus dip depends on the product and the assembly line.

Stencil/screen design, flux printing characteristics, flux slumping, and flux life on the stencil are important processing considerations when developing a flux printing process. The design of the stencil or screen and printing characteristics determine the flux volume. Slumping of the flux can cause solder bridging of fine-pitch FCs. Similar to dip fluxing, the evaporation rate for volatile components must be extremely low at room temperature. Otherwise, the print characteristics will change during the production run.

Spray, jetting, and flood dispense are used with no-residue (low) fluxes. These are low-viscosity, low-solids-content fluxes that are applied over the entire area of the chip site. The coverage for flood dispense depends on substrate design/topology and the wetting characteristics of the substrate surfaces. Spray dispense and jetting are more uniform and less dependent on topology. Some pick-and-place systems are equipped with flood dispense capability. Separate fluid dispense systems can be used for flood, jet, and spray dispense.

The wetting characteristics of the flux can be evaluated by cross-sectioning solder joints and examining the solder-pad interface (Fig. 2.21). The die shear strength and the failure mode also can be used to evaluate wetting. With good solder wetting, the failure point for die shear of FC-on-laminate (FCOL) should be either at the die–solder ball interface (Fig. 2.22) or at the copper pad–laminate interface for small pad (fine-pitch) designs. In Fig. 2.22, the wetting at some pad sites was minimal, and the failure was at the solder ball–substrate pad interface. Since some wetting did occur, electrical continuity was measured after assembly. Thus electrical testing alone is not sufficient for process development. With fine-pitch FC dies, the copper pad is small, and pad adhesion to the laminate is the limiting factor in shear strength.

With all fluxes and application methods, the flux must provide sufficient tack to hold the FC dies in place prior to melting of the solder in the reflow process. Mechanical shock and vibration in the conveyor systems, as well as airflow in the reflow oven, can cause chip movement.

No-clean fluxes are used most often in FC assembly because cleaning in the small FC gap is difficult. No-clean fluxes leave residue after the reflow process. Figure 2.23 is a thermal gravimetric analysis (TGA) plot of a typical no-clean flux used for FC assembly.[8] The TGA thermal profile approximates a reflow cycle. After the thermal cycle, approximately 35 percent by weight of the flux remains as residue. As mentioned previously, flux residue can impede the flow of underfill material. Chemical interaction between the underfill and the flux residue during underfill curing also can degrade adhesion and reliability. Fluxes and underfills must be evaluated as a materials system.

TABLE 2.2 Flux Application Options

Option	Flux	Comment
Dip	Medium viscosity, no-clean or solvent clean	Increases placement cycle time
Print	Medium viscosity, no-clean or solvent clean	High throughput for boards/panels with many FC dies
Spray or jetting	Low viscosity, no (low) residue	Flux applied over entire chip site including nonsoldering areas
Flood dispense	Low viscosity, no (low) residue	Flow of flux substrate topology-dependent

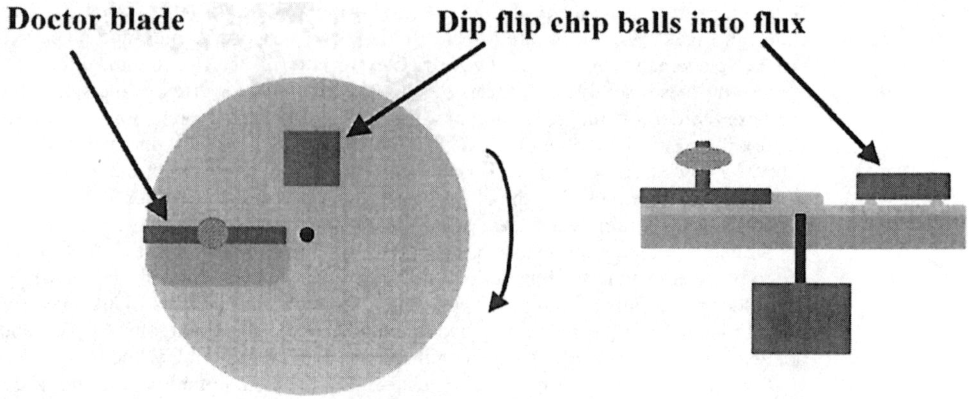

FIGURE 2.19 Dip fluxing.

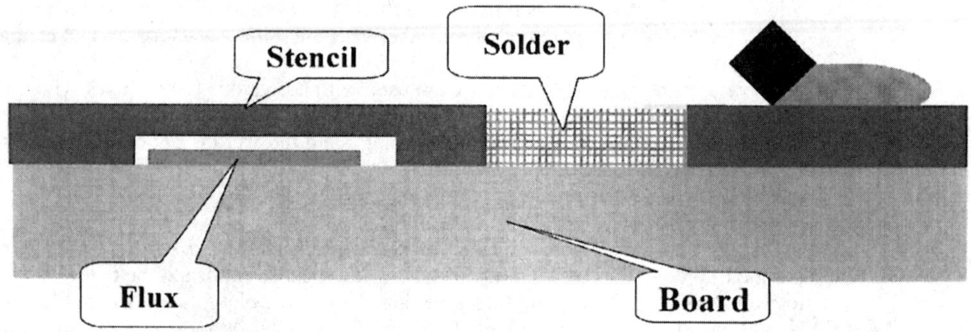

FIGURE 2.20 Example of flux and solder paste printing for mixed SMT assembly.

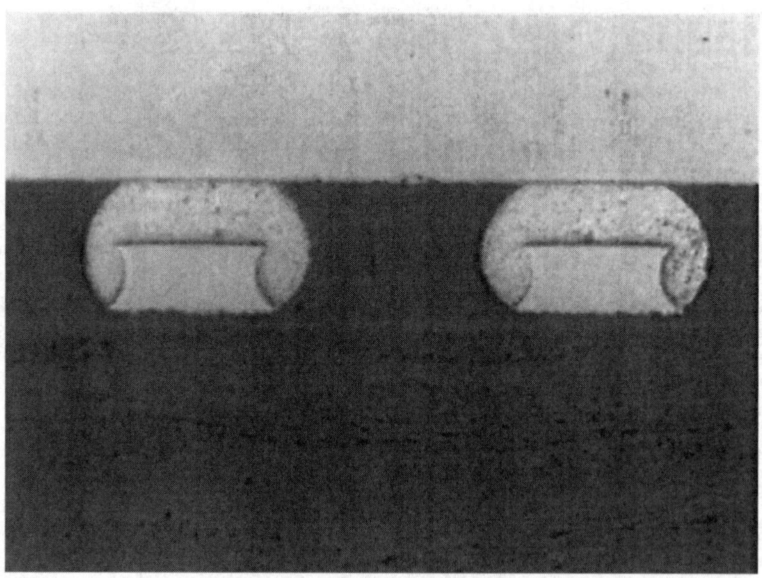

FIGURE 2.21 Cross section showing excellent solder wetting of copper pad.

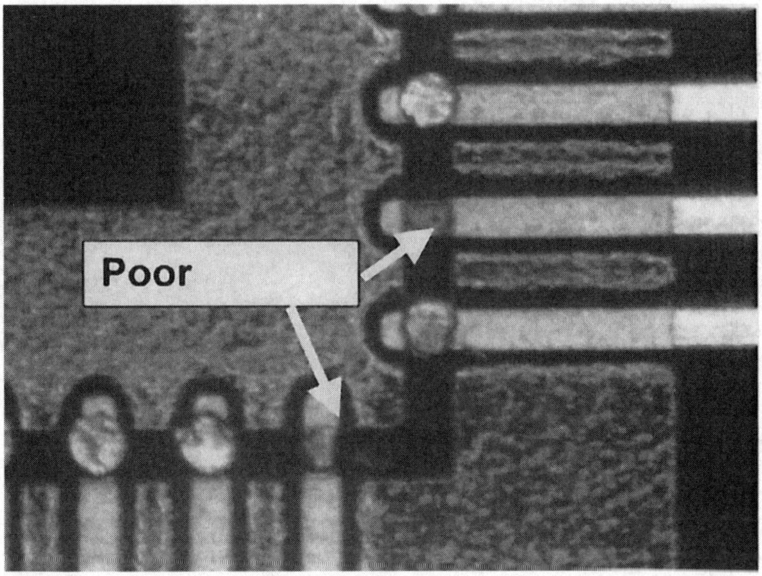

FIGURE 2.22 Example of poor solder wetting at two sites. The assembled FC die was sheared from the PWB. With good solder wetting, the solder ball remains on the PWB pad. The failure is at the die-to-solder ball inferface.

No-residue (low) fluxes are being evaluated to minimize these issues. These materials leave less than 2 percent by weight residue when tested using the same TGA temperature profile used in Fig. 2.23. Another no-clean alternative is epoxy flux. The chemistry of these fluxes incorporates epoxy groups. After the reflow cycle is completed, the flux "residue" is an adherent epoxy. The challenge in the formulation of epoxy fluxes is that the acids common to fluxes react with the epoxy groups to cross-link the material. Thus the fluxing activity and shelf-life must be balanced.

If cleaning is required, solvent cleaning typically is used. The surface tension of water is too high for cleaning in the small gap under the FC. Centrifuge-based cleaning systems have been shown to be effective at cleaning in small gaps. For under-the-hood automotive applications, Delphi-Delco has described the evaluation of cleaning processes and materials for FC-on-laminate using quartz dies and dyed fluxes.[9] If flux cleaning is used, a process-control method must be developed because inspection is not a production option.

2.3.3 Pick and Place

Pick and place for FC dies is similar to that for other fine-pitch SMT devices, with added emphasis on the vision system and placement accuracy. Standard wafer sawing tolerances do not allow FC placement based on visioning of the die outline. Exceptions are very course pitch FC dies or very tight sawing specifications. The CCD camera in the vision system must have sufficient resolution to recognize individual solder balls. Sufficient contrast between the solder balls and the die surface is also required. Vertical lighting (Fig. 2.24) typically leads to reflection from both the solder balls and the die surface and often cannot be used. Because of the curvature of the solder balls and the resulting angles of reflection, angled lighting can provide the required contrast. Transforms are used to further increase the contrast and image recognition. If dip fluxing is used, fluxing before or after visioning must be considered, as well as the impact of the flux on the image.

The required placement accuracy (maximum misalignmnet) is often specified at 20 percent, as defined in Fig. 2.25. Lewis has shown that high assembly yields can be achieved with higher levels of misplacement due to the self-centering nature of molten solder.[10] As noted previously, increasing the volume of flux increases self-centering and improves assembly yield. X-ray can be used to measure

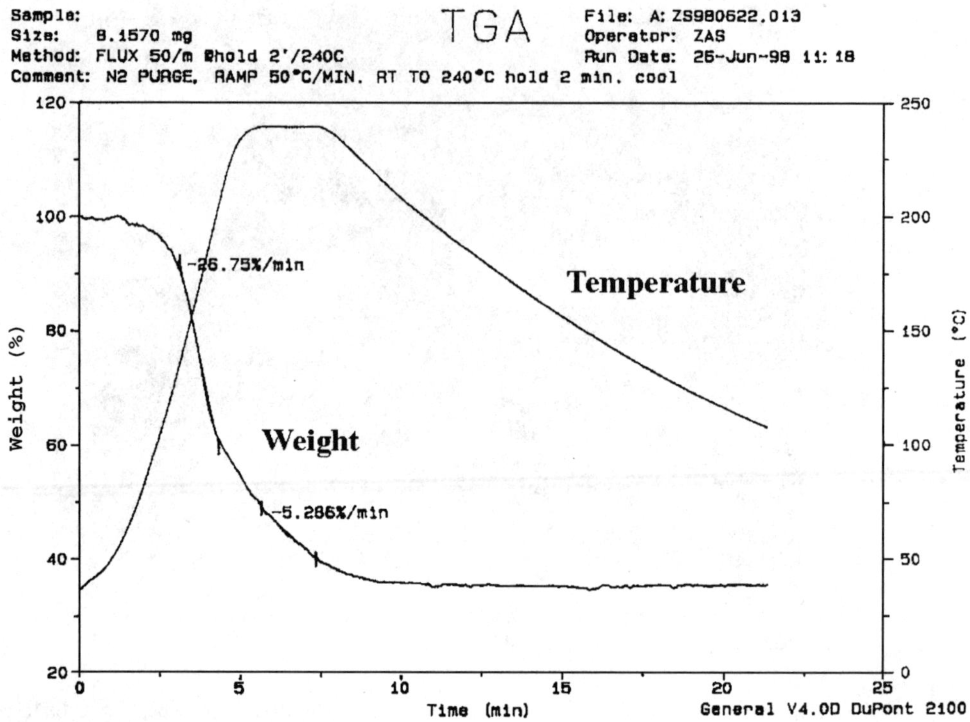

FIGURE 2.23 TGA plot of a no-clean flux.

the initial placement accuracy (to characterize the placement system capability) as well as to measure the alignment after reflow (to measure the limits of self-centering).

2.3.4 Reflow

The reflow temperature profile for FC assembly should follow the manufacturer's recommended profile for the flux used and is similar to standard SMT reflow profiles. Depending on the flux and board surface finish, air or nitrogen reflow can be used. In forced convection reflow ovens, the force of the gas flow on the dies must not be sufficient to move the dies. This will be a function of die size, the number of solder balls, the tackiness of the flux, and the change in tackiness during the reflow until the solder melts and wets the PWB metallization. As mentioned previously, vibration in conveyors or reflow ovens also can cause die movement and low assembly yields.

X-ray is often used to characterize the solder joints after reflow. Figure 2.26 is a typical x-ray image after reflow. Bridging between adjacent solder balls is easily detected by x-ray. X-ray (Fig. 2.27) also can detect voids in the solder joint. The voids may have existed in the solder balls as originally formed on the die or result from trapped flux vapors during the reflow profile. Void formation during reflow can be reduced or eliminated by flux selection and reflow profile modifications to minimize vapor generation once the solder ball liquefies. The impact of voids on FC reliability has not been quantified. Voids in solder balls are also common to BGAs and CSPs. An industry accepted limit of 20 percent or less void content generally is accepted for BGAs.

X-ray also can be used to measure self-alignment during the soldering process and to detect shorts between solder balls. Solder bridging can result from die misplacement or movement after placement prior to reflow. Insufficient wettable substrate pad area also can lead to bridging of fine-pitch solder balls. This is aggravated by solder mask misalignment, as has been mentioned previously.

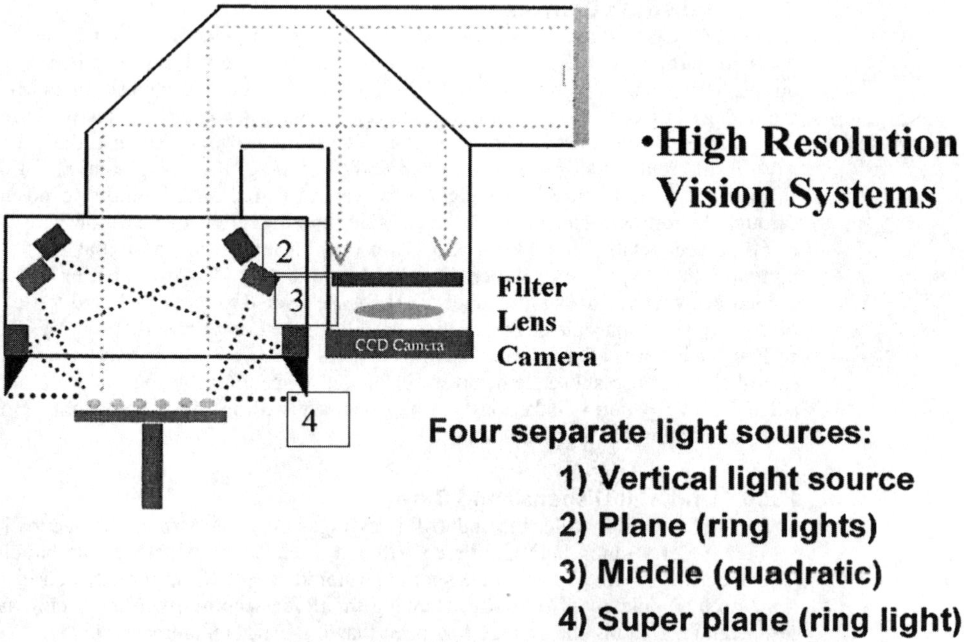

FIGURE 2.24 Example of vision system lighting options for FC dies. (*Siemens Energy and Automation.*)

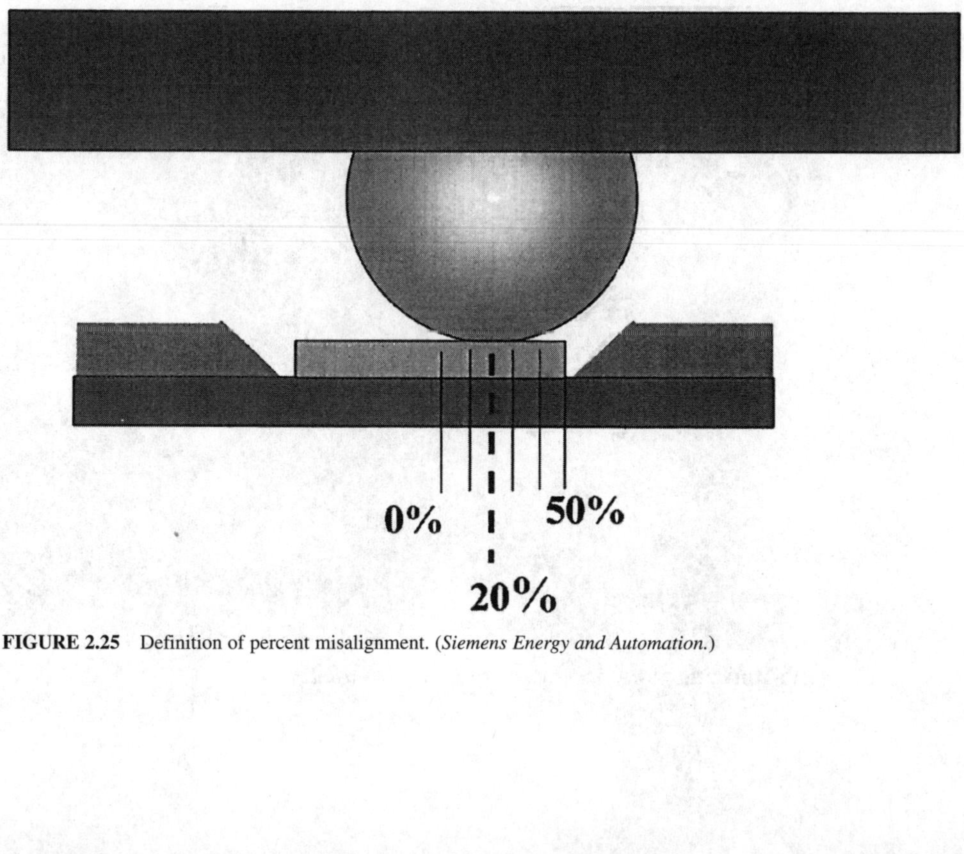

FIGURE 2.25 Definition of percent misalignment. (*Siemens Energy and Automation.*)

2.3.5 Substrate Dehydration

If the PWB has absorbed moisture from the environment, this moisture may be liberated and create moisture bubbles (voids) in the underfill during the underfill curing process. Moisture from the atmosphere is absorbed by both the PWB laminate and the solder mask. In an in-line SMT process, the underfill dispense follows the reflow cycle. This reflow cycle is effective in removing a significant amount of moisture absorbed by the PWB (Fig. 2.28). In this test, the PWBs (bare and with several different solder masks) were exposed to 85 percent relative humidity at 85°C for 168 h to fully saturate the boards. The weight gain was measured after humidity exposure, after one pass through the reflow cycle, and after one pass through the reflow cycle followed by humidity exposure at 60 percent relative humidity at 30°C for 6 h, all relative to the initial "dry" weight of the test coupon. The reexposure to 60 percent relative humidity at 30°C for 6 h after reflow was to simulate a delay between reflow and the underfill dispense related to possibly board testing prior to underfill or some production delay. As can be seen, little moisture is absorbed during this 6-h period. The reflow cycle drives off the "easy to remove" moisture. The lower-temperature underfill cure cycle is not likely to cause subsequent moisture release. Depending on the underfill and the solder mask selected, a dehydration bake may not be necessary with in-line FC assembly. However, individual material combinations should be tested.

2.3.6 Underfill Dispense and Cure

The manufacturer provides the underfill in syringes of various sizes. The underfill is loaded into the syringe under vacuum and may be centrifuged to ensure that there are no bubbles in the syringe. The underfill is shipped frozen and should be stored at −40°C until ready to use. The underfill must be thawed completely prior to use. Thawing should be accomplished by placing the syringe at room temperature. Heating the syringe for speed thawing is not recommended.

Automated dispense systems are used to dispense a controlled volume of underfill in a precise pattern. The volume of underfill dispensed is critical to ensure complete underfilling of the die and

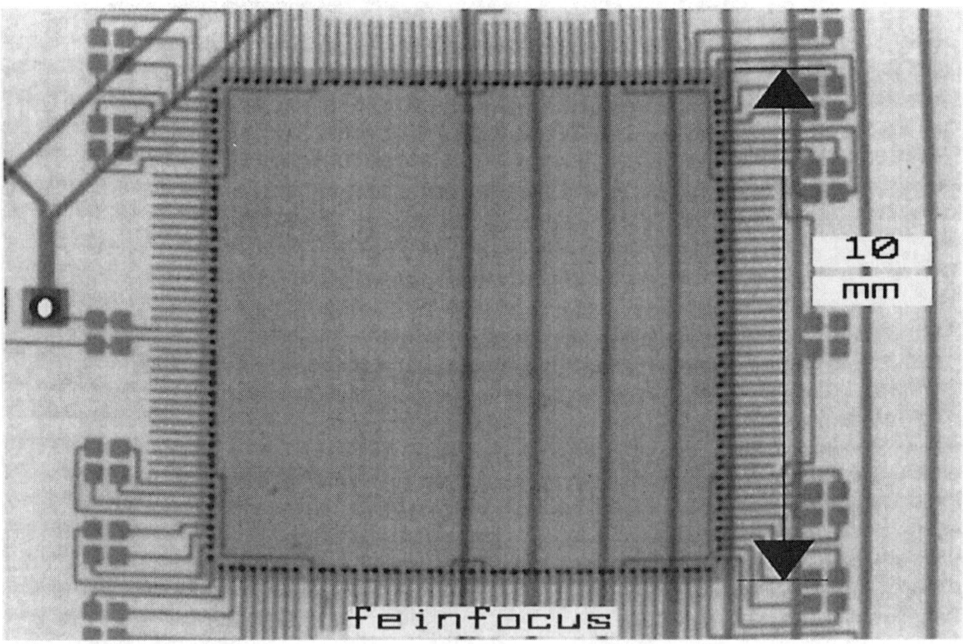

FIGURE 2.26 X-ray image of an FC assembly. (*FeinFocus.*)

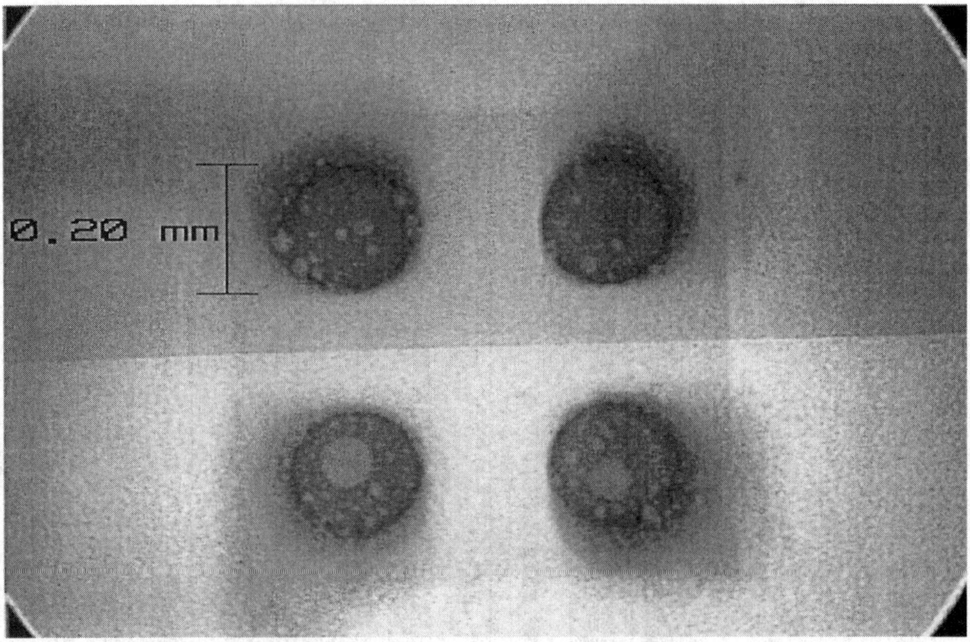

FIGURE 2.27 Example of voids in FC solder joints as detected by x-ray. (*FeinFocus.*)

fillet formation. In production, the gap height and hence the volume of underfill required to fill the space between the die and the board will vary somewhat from board to board. The fillet provides the reservoir to accommodate this variation. Volumetric control is achieved using a rotary valve or positive-displacement dispense valve. Conventional time-pressure dispense valves do not provide the required volume control and repeatability. Since the viscosity of the underfill may increase with time at room temperature, the dispense parameters may require minor adjustments over a period of time to maintain a constant volume. With current production dispense systems, the parameters can be adjusted automatically based on calibration using either volumetric measurement or weight-based systems. The density of the underfill is required in the weight-based system to calculate actual dispense volume. A baseline volume or weight of underfill is dispensed with a known set of dispense parameters. For a rotary-valve system, these parameters would be dispense time and valve rpms. At a later calibration, the same parameters would be used, and the corresponding volume or weight would be measured. If the amount of material dispensed was less than during the baseline case, the rpms would be increased to compensate.

The dispense pattern is critical to achieving complete underfilling without entrapping voids under the dies. The dispense pattern is a function of the underfill, the die size, the die bump pattern, and the PWB topology and is determined experimentally. Common dispense patterns are shown in Fig. 2.29.

The dispense needle must be placed close to the die so that material contacts the gap between the die and the PWB. Capillary forces then pull the material under the die. The needle travel speed, the valve rpms for a rotary-valve system, the needle gauge, and the travel length determine the volume of underfill dispensed. From a production point of view, needle travel speed should be maximized. However, if the underfill does not flow rapidly under the die, a large volume of material may accumulate beside the die. This material may get on top of the die or flow away from the die before it has an opportunity to flow under the die. Board heating typically is used to increase the flow rate under the die. PWB temperatures range from 60 to 100°C depending on the underfill material and the die size. Heating the underfill initially lowers its viscosity but also causes curing of the underfill, which increases its viscosity. For large dies, the underfill time is longer, and

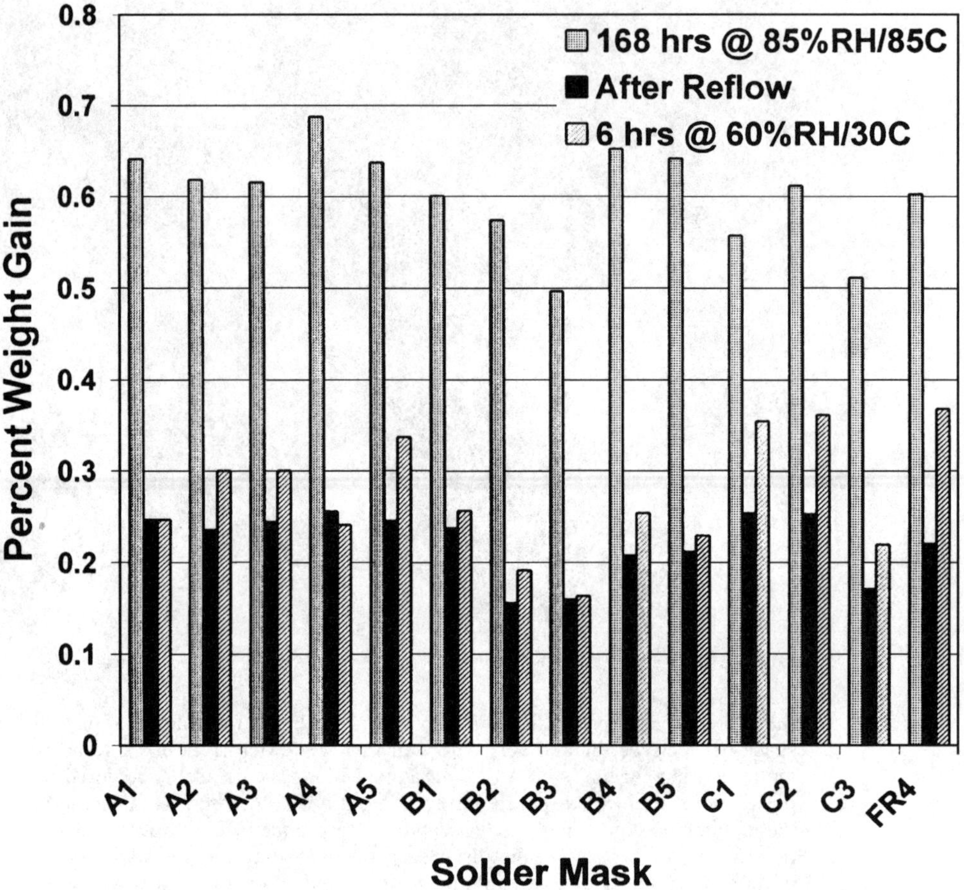

FIGURE 2.28 Percent weight gain after 168 h of exposure at 85 percent relative humidity at 85°C, after one reflow cycle, and after 6 h of exposure at 60 percent relative humidity at 30°C.

partial underfill curing can be a concern at the higher PWB temperatures. A design keep-out region must be defined around the FC die for underfill dispense. If the dispensed underfill contacts adjacent components, some of the underfill may be "pulled" to that component. The variation in the amount pulled would lead to a lack of underfill volume control for the FC die and possibly incomplete underfilling.

A particular challenge with perimeter bumped dies is that the solder bumps tend to accelerate the flow of the underfill around the perimeter of the die. If the flow around the perimeter exceeds the flow under the area of the die, a void will be trapped under the die as the two perimeter flow fronts meet. An example of this is shown in Fig. 2.30, which is a scanning acoustic microscopic image through the FC die. The light area is due to reflection of the ultrasonic wave at the discontinuity caused by a large void in the underfill. The gray area indicates complete underfill. The length of the line or the line segments of the L pattern must be adjusted to limit this "racing" of material around the die perimeter.

Following the initial dispense of the underfill, a fillet dispense may be required to form a fillet around the die on the sides of the die where underfill was not dispensed originally (Fig. 2.31). Time

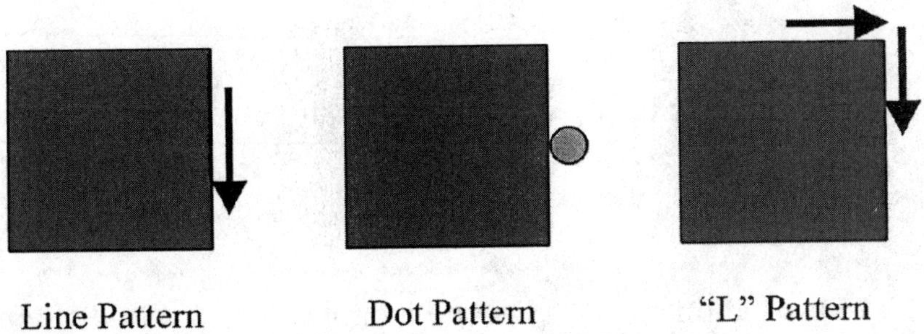

FIGURE 2.29 Line, dot, and L dispense patterns.

FIGURE 2.30 Scanning acoustic microscopic image of an underfilled FC die. The underfill "raced" around the perimeter solder bumps, trapping a void under the die.

must be allowed for the underfill initially dispensed to flow completely under the die to prevent trapping a void under the die with the fillet dispense. Many fast-flow underfills are self-filleting, eliminating the need for this step. The surface tension and wetting characteristics of the underfill continue to pull material under the die to form a fillet on the nondispensed sides of the die.

Underfill time is a critical issue in high-volume production. Newer-generation "fast flow" underfills have been introduced to increase production throughput. This is a particular issue as the die size increases. Equation (2.1) relates the flow time t for a viscous fluid between two parallel plates with a spacing of h. The flow time increases as the square of the distance traveled and linearly with decreasing gap height. Thus doubling the die size increases the flow time by a factor of 4. The trend to finer-pitch solder balls on the die (higher I/O counts) leads to a reduction in solder ball volume, resulting in lower gap heights, further increasing the flow time. The equation simplifies the actual FC structure by ignoring the presence of solder balls and their impact on increasing flow speed due to additional surface and capillary forces.

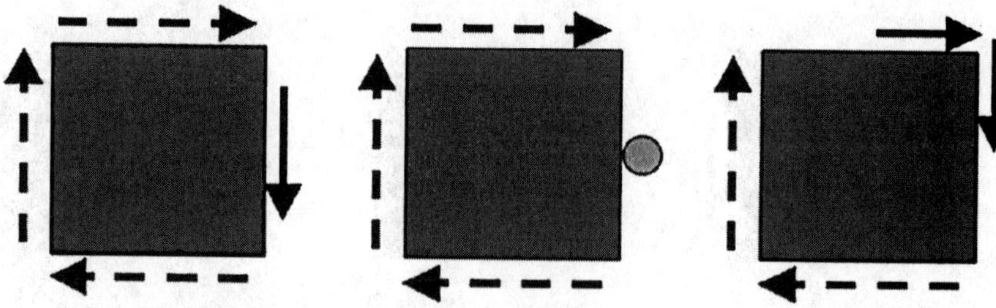

FIGURE 2.31 Underfill dispensing patterns. Dotted lines indicate fillet dispense pattern.

$$t = \frac{3\mu L^2}{h\gamma \cos \theta} \tag{2.1}$$

where t = flow time
μ = absolute viscosity
L = distance traveled
h = gap height
γ = surface tension
θ = wetting angle

Figure 2.32 illustrates a common method to measure underfill flow speeds. The underfill is dispensed between two glass plates with a known separation, and the time and distance traveled are measured. This test method is equivalent to the conditions (parallel plates, no solder balls) described by Eq. (2.1), and the data can be plotted as t versus L^2. The wetting angle for the surfaces used in the test setup may differ from the wetting angle of the surfaces in an actual FC assembly.

Sandia National Laboratories developed an alternate approach to measuring flow speed.[12] The test uses a V groove machined into an aluminum plate (Fig. 2.33). Markers are also machined into the aluminum to facilitate measurement of underfill flow distance. An equation was developed to relate flow distance to the flow time (Eq. 2.2). The underfill is dispensed into a reservoir, and the flow distance versus time is measured. The aluminum piece can be heated to investigate the effects of temperature on flow speed. An example is shown in Fig. 2.34.

$$X^2 = (\gamma/\mu)\, df(\alpha, \theta)\, t \tag{2.2}$$

where X = distance traveled
t = time
γ = surface tension of fluid
μ = viscosity of fluid
d = characteristic depth of groove
$f(\alpha, \theta)$ = function of groove and contact angle

After underfill dispense, the material must be cured. Cure conditions vary from supplier to supplier and range from 150 to 165°C for 5 to 90 min. "Snap" cure underfills with cure times of less than 10 min can be cured using in-line conveyor ovens, whereas the 90-min cures usually are performed in batch convection ovens.

Scanning acoustic microscopy (SAM) is used commonly to develop underfill processes and dispense patterns as well as in-process sampling and process requalification. High-frequency transducers (>100 MHz) are used to provide the necessary resolution to detect small defects. Figure 2.30 is a C-mode SAM image of a good underfill dispense with no voids, whereas Fig. 2.33 shows small voids in the underfill. Voids can lead to early delamination, which in turn leads

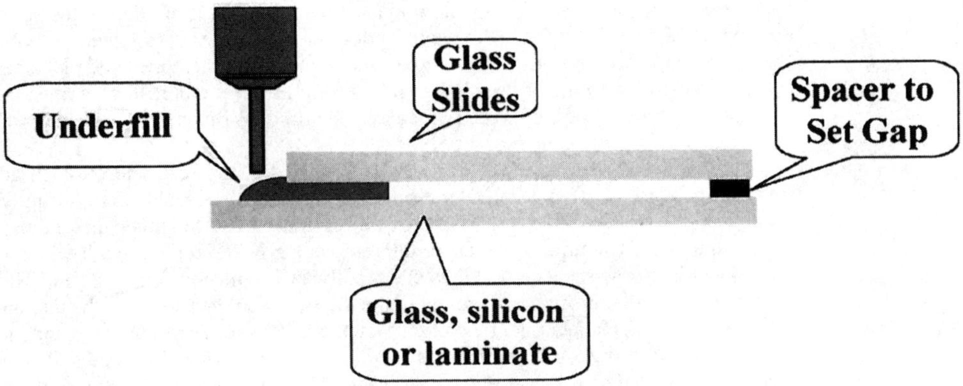

FIGURE 2.32 Test setup for measuring underfill flow rate.

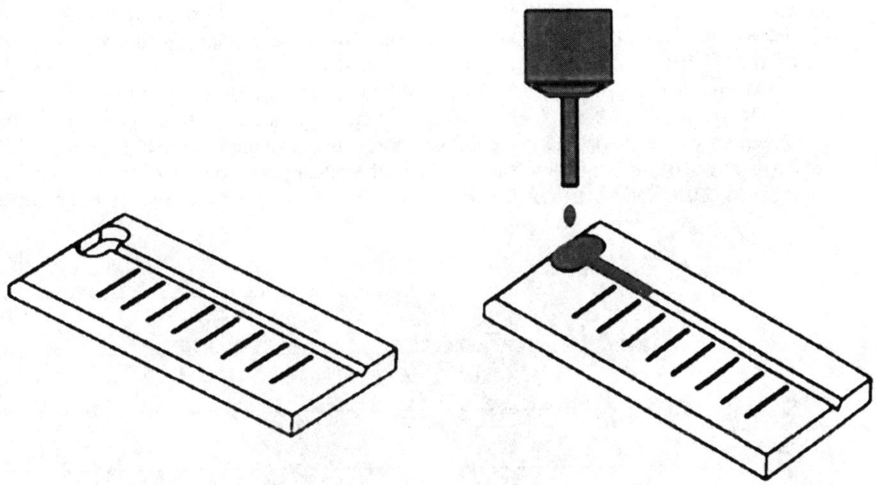

FIGURE 2.33 Sandia National Laboratories underfill flow test method.

to electrical failures. Solder extrusion from the solder balls into adjacent voids during thermal cycling also leads to early electrical failures. SAM also can be used to detect filler separation during the underfill flow process (see Fig. 2.33). Filler separation results in a nonuniform CTE within the underfill. In particular, the resin-rich region in contact with the silicon dies will have a higher CTE than the filler-rich area. Cross sections and flat sections (polishing either the PWB or the die away) are used to complement and verify SAM results. Figure 2.30 is a cross section illustrating filler separation.

2.3.7 Rework

The removal and replacement of a defective FC is relatively straightforward prior to underfill. The die can be heated using a hot-air rework station until the solder balls melt and the die can be removed. With continued heating, the excess liquid solder remaining on the pads after die removal can be vacuumed or wicked away. To attach a replacement die, flux is applied to the solder balls on the die by dipping, the die is aligned using split optics, and the solder is reflowed by hot air. The process is similar to BGA or CSP removal and replacement. However, the optics and placement resolution must be greater for FC rework. If boards are to be tested prior to underfill, care must be used in handling and probing. Flexing of the PWB, particularly the thin PWBs used in portable products, can fracture the delicate FC solder joints.

In many cases, testing is not done prior to underfill dispense and cure. In addition, rework of field-return units is often a consideration. The purpose of underfill is to strongly bond the FC die to the PWB. However, if rework is required after underfill, some mechanism must exist for decreasing the underfill structural strength for die removal and site preparation. Two approaches have been taken to address rework. The first is to chemically attach the underfill material. This process requires application of the appropriate chemical to the die site and penetration of the chemical into the underfill under the die. The chemical must not attach the PWB or surrounding components.

The second approach is thermally reworkable underfills.[13,14] The underfill is designed to break down when heated to solder reflow (rework) temperature. This breakdown of the thermoset network is a result of the incorporation of a monomer that has a special linkage designed to break apart on heating. The breakdown of the network reduces the adhesion to the die, allowing easy removal of the die. The underfill is also now ready for easy removal from the PWB for site preparation. Using a conventional hot-air rework station, the die is heated to typical reflow temperatures of 220°C, and the die is removed with a slight twisting motion. After die removal, significant underfill material remains on the PWB and must be removed prior to placement of the new die (Fig. 2.35).

Mechanical brushing is used to clean the die site. A custom cleaning machine has been designed and fabricated (Fig. 2.36). The machine provides *x, y,* and *z* control. Several different brush styles and materials were tested. The flat-end horsehair brush provided the best cleaning with the least amount of damage to the board and the brush. A flat-end nylon brush wore away before one die site was cleaned. Disk-style brushes are not recommended because the disk is more difficult to control in

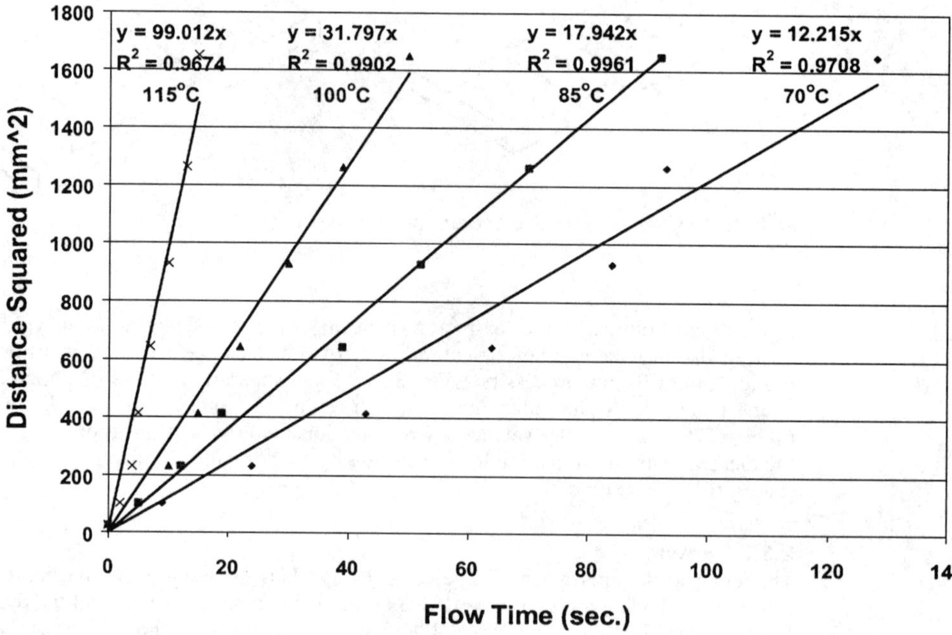

FIGURE 2.34 Distance squared versus flow time as a function of temperature for a commercial "fast flow" underfill.

the small work area and because its hard core can cause damage to the pads. Figure 2.37 shows the site after preparation by brushing. The die site is then ready for die placement, reflow, and underfill.

Since the thermally reworkable underfill is triggered at the rework (reflow) temperature, dies that have been underfilled with this material cannot be subjected to subsequent reflow cycles in the manufacturing process. Reworkable underfills with higher trigger temperatures are being developed to allow reflow after underfill. These materials will require a higher rework temperature.

2.4 ASSEMBLY WITH FLUXING UNDERFILLS

To reduce the number of process steps associated with FC assembly, Pennisi[15] patented an adhesive and encapsulant material with fluxing properties. The idea was to incorporate fluxing activity into the underfill material and then dispense sufficient quantity of the fluxing underfill on the die site prior to die placement to serve as both the flux for soldering and, when cured, the underfill. The assembly concept is illustrated in Fig. 2.38. The flux application step is replaced by the fluxing underfill dispense. The postreflow underfill dispense and associated capillary flow time are eliminated. The fluxing underfill may cure during the reflow cycle, eliminating another process step, or it may require a postreflow cure depending on the material.

The formulation of a fluxing underfill is challenging. Acids typically are found in solder fluxes to provide the fluxing action of removing metal oxides to facilitate metallurgical wetting. However, acids initiate the cross-linking of epoxies. By increasing the acid content, the fluxing activity increases, but the shelf life decreases. Shelf life, fluxing activity level, and lot-to-lot consistency are issues being addressed by the materials suppliers, and newer material systems are being introduced.[16]

Fluxing underfills have no or low filler content. If filler particles are trapped between the solder ball and the PWB pad during placement, they may prevent the solder ball from touching the PWB pad during reflow, and wetting will not occur. Addition of filler to the fluxing underfill can decrease assembly yield. Without filler, the CTE of the fluxing underfill is significantly higher than that of traditional capillary underfills. Reliability studies have shown that even with the higher CTE, fluxing underfills meet the thermal cycling requirements for many applications.[17]

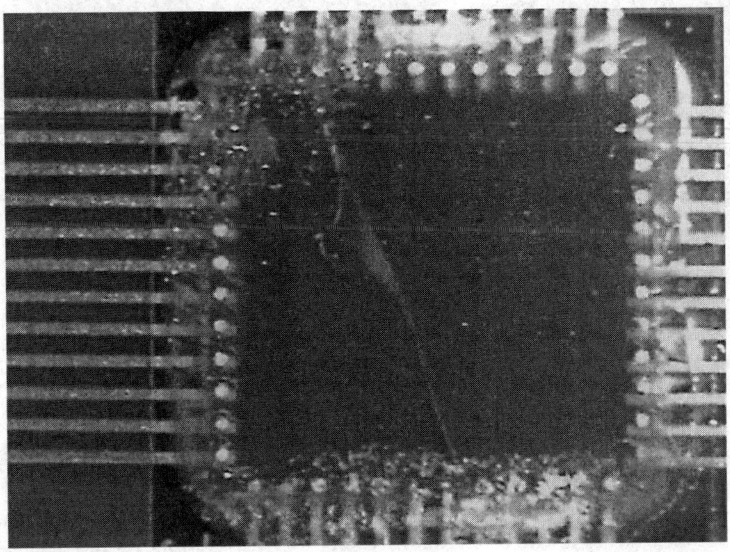

FIGURE 2.35 Die site after die removal.

FIGURE 2.36 Automated brushing machine for site preparation.

FIGURE 2.37 Die site after brush cleaning.

2.4.1 Substrate Dehydration

As discussed previously, PWB laminate and solder mask absorb moisture from the atmosphere. In the capillary underfill approach, the PWB has passed through a reflow cycle just prior to underfill dispense, and much of the absorbed moisture has been removed, potentially eliminating the need for a dehydration bake. With fluxing underfills, the reflow cycle follows the dispense of the fluxing underfill, and the potential exists for voids to form due to moisture release from the PWB or solder mask during the reflow cycle. A second source of gas-generated voids in the fluxing underfill is

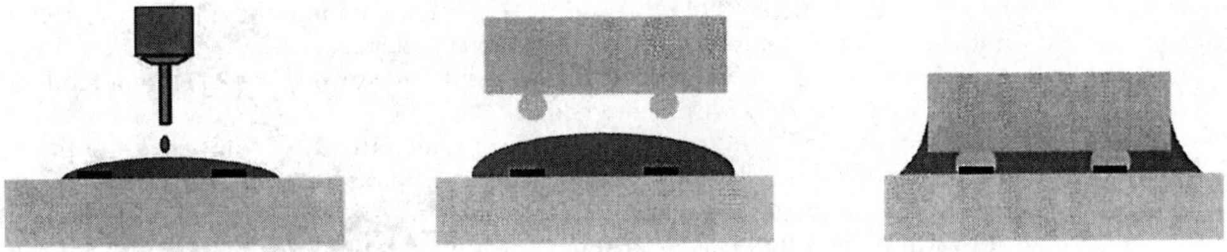

FIGURE 2.38 Fluxing underfill assembly process.

outgassing from incompletely cured solder mask. A dehydration bake is recommended prior to fluxing underfill dispense to remove moisture and to ensure that the solder mask is cured completely. In a two-sided SMT assembly process with the FC assembly on side 2, dehydration may not be necessary if the board goes immediately from side 1 reflow to side 2 assembly. If a dehydration step is required, this may be an added process step compared with the capillary flow underfill process and should be included in cost models.

2.4.2 Application of Fluxing Underfill

The fluxing underfill is applied to the substrate by dispensing. Stencil printing has been proposed as a higher-speed application method, but this approach does not produce the smooth, dome-shaped deposit that is required for void-free die placement.[18] During placement, the fluxing underfill should contact the center of the die first and then spread radially as the die continues to move toward the PWB. Other dispense patterns have been used.[16] Volume control is important in dispensing fluxing underfill. Insufficient volume will result in nonwetting by some solder balls and incomplete underfilling, whereas excess fluxing underfill can cause the die to "float." If the die floats and the solder balls do not contact the pads on the PWB during reflow, opens will occur. Shifting of the die with respect to the PWB pads also can result from die floating.

2.4.3 Die Placement

During placement of an FC die in fluxing underfill, the underfill must flow outward as the die nears the PWB in a "squeeze flow" process. As the underfill is squeezed between the die and the PWB, it generates a force opposing the placement force. To understand the interaction of variables in the squeeze-flow process, the placement of a flat, circular disk on a substrate with a newtonian fluid in between can be considered analogous to the placement of an FC die in fluxing underfill. Assuming that the disk and substrate are parallel and that the volume between the disk and substrate is filled with underfill, the Stefan equation (Eq. 2.3) can be used to approximate the force.[19,20]

$$F_{\text{underfill}} = \frac{3\pi R^4 \mu (-v)}{h_0^3} \tag{2.3}$$

where $F_{\text{underfill}}$ = force exerted by underfill
R = radius of the disk
μ = viscosity of underfill
v = placement velocity
h_0 = height between disk and substrate

From this equation, the force increases with increasing placement velocity, underfill viscosity, and disk (die) size. The force also increases as the disk approaches the substrate (h_0 decreases). If $F_{\text{underfill}}$ equals the placement force before the solder balls contact the PWB pads ($h_0 >$ solder ball height), then sufficient hold time must be programmed into the placement program to allow the die under constant force to traverse the remaining distance. For practical underfill viscosities, die sizes, and placement velocities, excessive placement force or hold times are not required to place FC dies into fluxing underfills with high yields.

As mentioned previously, the fluxing underfill should have a smooth, domed shape, and the center of the die should first make contact with the underfill. As the die proceeds downward, the underfill is squeezed outward, and the underfill contact circle with the die surface will expand outward, dispelling air as it advances. If the outward movement is too fast, there is a potential to form voids behind the solder balls as the underfill flows outward. Placement experiments should be conducted to optimize the placement process for high electrical interconnection yield and no voiding.

2.4.4 Reflow

The reflow profile for a fluxing underfill is critical to high assembly yields. Figure 2.39 is an example of a good solder joint using fluxing underfill. The viscosity of the fluxing underfill must remain low until the solder balls melt to allow collapse of the die as the solder wets the PWB pads. While the viscosity of the underfill decreases with temperature, the thermal energy input to the fluxing underfill during the early stages of the reflow profile accelerates cross-linking of the polymer, increasing the viscosity. Figure 2.40 is a cross section illustrating the effect if the underfill viscosity increases before the solder melts.

Three approaches can be taken to address the profile issue. The first approach is to limit the cross-linking rate of the underfill by selection of the catalyst and the amount used. These underfills can be reflowed in a conventional SMT reflow profile (Fig. 2.41). With a slower cross-linking rate, complete curing of the underfill is not achieved during the reflow cycle, and a postreflow cure is required. This reduces the sensitivity to the reflow profile but adds a cure step similar to the capillary underfills. In fact, due to the slow cross-linking rate, the postreflow cure time may be longer than the cure time for "snap" cure capillary flow underfills, reducing the process advantage of fluxing underfills.

The second approach is the use of a "volcano" reflow profile, as shown in Fig. 2.42. This profile rapidly heats the assembly to the soldering temperature, minimizing the thermal exposure of the underfill prior to melting of the solder. In a mixed assembly with SMT components and solder paste to be reflowed simultaneously with the fluxing underfill, the compatibility of the SMT components and solder paste with this fast-temperature-rise profile must be established.

The third approach is to use a catalyst that does not initiate cross-linking until temperatures rise slightly above the melting point of the solder. Work is ongoing to develop and refine these materials.

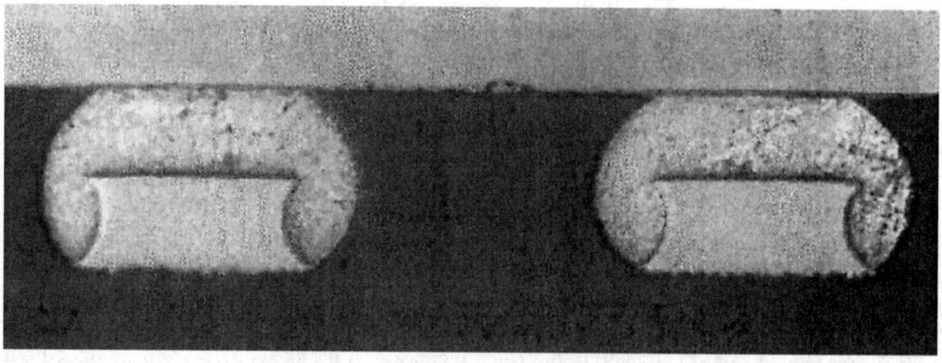

FIGURE 2.39 Cross section illustrating good solder joint formation when the fluxing underfill.

FIGURE 2.40 Cross section illustrating poor solder joint formation when the fluxing underfill viscosity increases too much in the reflow profile prior to melting of the solder.

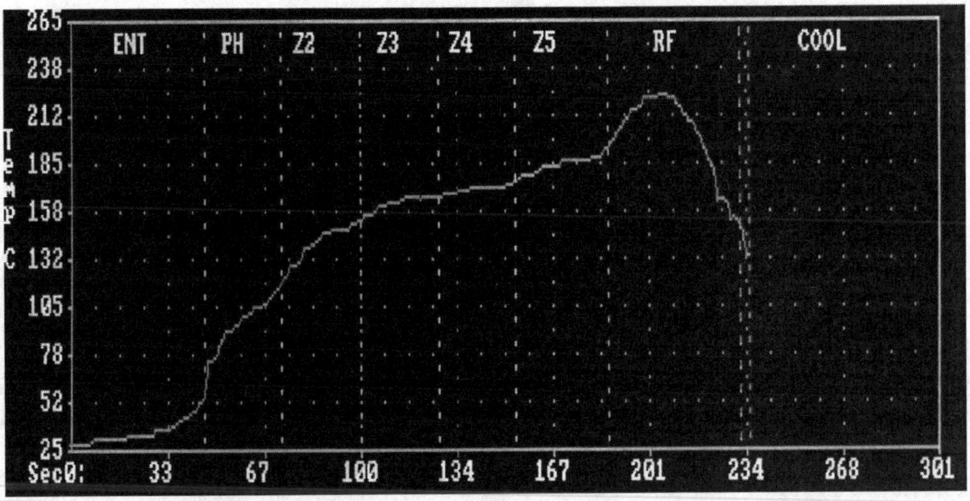

FIGURE 2.41 Typical SMT reflow profile.

New materials are being introduced into the market. With this approach, the underfill is much less sensitive to the temperature and duration of the soak in a conventional SMT reflow profile. These underfills typically cure during the reflow cycle. With mixed SMT assembly, the challenge will be getting large SMT components to temperature without the FC dies and underfill soaking at too high a temperature prior to solder melting. Advanced profiling tools are available to assist in profile development.

2.4.5 Rework

Reworkable, fluxing underfills have not been commercialized but are under development. It has been noted that the fluxing underfills that do not fully cure during reflow can be "reworked" prior to the postreflow cure. Since the underfill has not cured fully, it has not achieved its full adhesion and mechanical strength, and the die may be removed. Further work is required to develop truly reworkable fluxing underfills.

2.5 WAFER-APPLIED UNDERFILLS

To achieve an FC manufacturing process that is transparent to a standard SMT assembly line, the fluxing underfill should be applied at the wafer level and should cure during the reflow cycle.[21] The capillary, fluxing, and wafer-applied assembly approaches are compared in Fig. 2.43. Applying the fluxing underfill to the wafer simultaneously coats hundreds of dies, eliminating the need for underfill dispense for each die individually at the assembly stage. The materials and coating and assembly processes are being developed currently in the industry. While significant progress has been made, wafer-applied underfills are not available commercially at the time of this writing.

2.5.1 Application to Wafer

Finding a consistent and accurate coating method by which the material may be deposited onto the wafer is crucial to the successful implementation of wafer-applied underfill. Several techniques potentially could be used for coating a bumped wafer: (1) screen/stencil printing, (2) pad printing, and (3) extrusion coating.

Screen/Stencil Printing. The stencil and screen-printing processes are used widely in the electronics industry for the fabrication of PWBs, flexible circuits, and other related items. The challenge in this process is the irregular surface topology presented by the solder bumps. A uniform coating thickness that does not coat the tops of the solder balls is required.

Pad Printing. Pad printing is an offset gravure printing method, where the gravure or cliché is patterned with the artwork to be printed, and ink is transferred from the cliché to the part using a silicon pad. First, artwork is transferred to a stainless steel cliché plate using a photoetch process. Ink is spread onto the cliché, and a silicone pad is pressed against the cliché and then lifted, with the

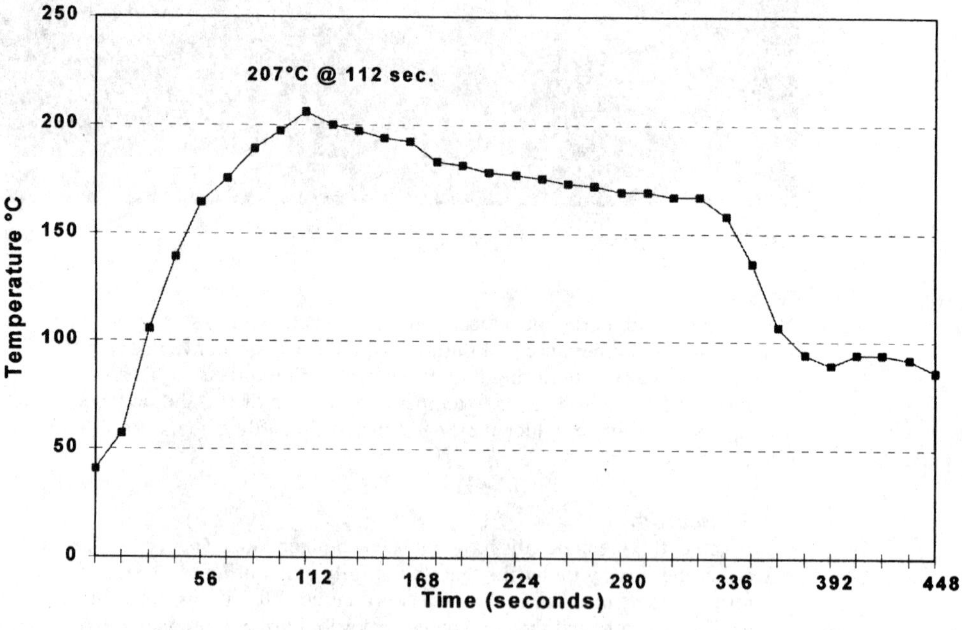

FIGURE 2.42 Example of "volcano" reflow profile. (*Emerson & Cumming.*)

ink adhering to the silicone pad. The part to be printed is then aligned under the pad, and the pad is pressed against the part and lifted off, with the ink transferring to the part. A thin film of ink is deposited on the part, with a fine resolution and a locational accuracy of ±25.4 μm.

Pad printing has the ability to print over irregular surfaces, with the silicon pad conforming to the shape of the substrate. It is this characteristic that gives it great potential for coating a bumped wafer.

Extrusion Coating. Extrusion coating is the direct application of process fluids to a substrate via a patented, specially designed fluid pumping and delivery system. This creates a highly uniform coating over the entire substrate with virtually no waste. The extrusion coating process has been used extensively for thick-film application on semiconductors wafers up to 300 mm in diameter. This unique technique holds great potential for coating bumped wafers.

Experimental Wafer Coating. In recent work, the wafer-applied underfill was applied successfully on a 101.6-mm wafer. A thin uniform coating layer with clean solder balls and clean saw streets was produced (Fig. 2.44).

After coating, the underfill is B-staged to produce a tack-free surface compatible with die placement in tape and reel for shipping, storage, and automated pick and place. The shelf life and required

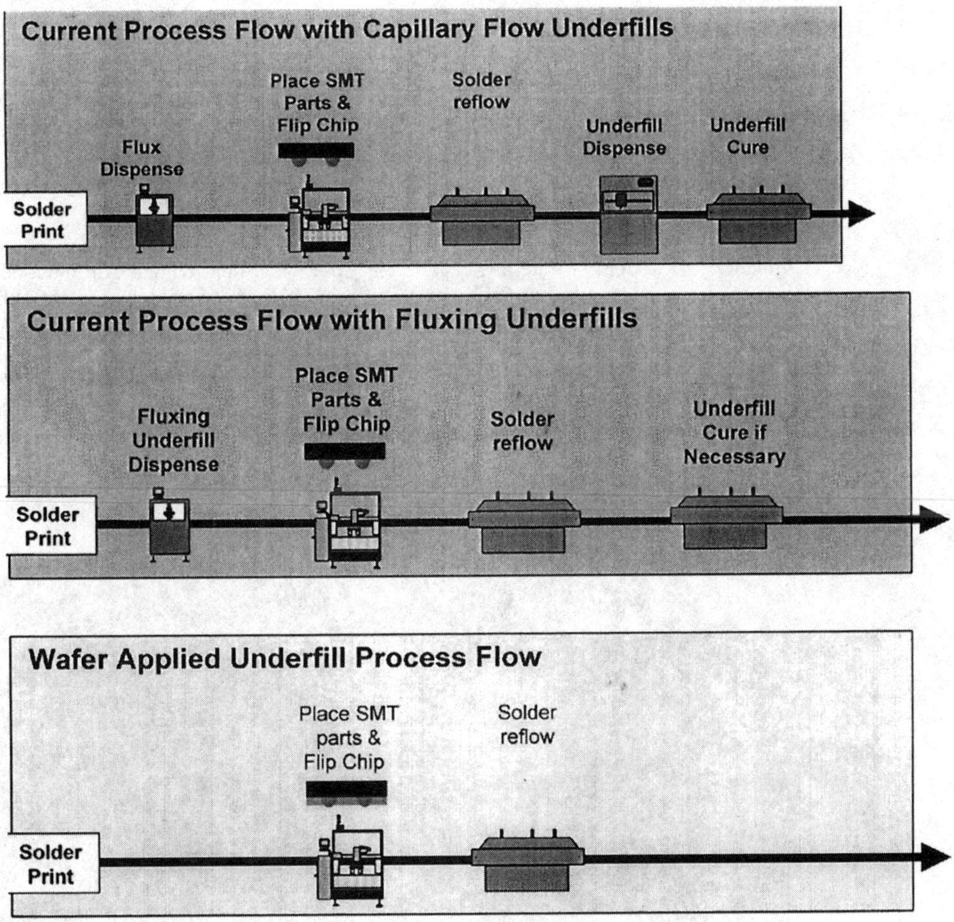

FIGURE 2.43 Comparison of capillary, fluxing, and wafer-applied underfill assembly processes.

storage conditions for the B-staged underfill are still being determined but must be compatible with transportation and inventory management.

One additional issue that is being addressed is the sawing of the wafer. In a typical wafer-sawing operation, water is used to remove sawing debris and cool the saw blade. The compatibility of the B-staged underfill with water must be understood. If the coating process coats underfill into the saw streets, the saw characteristics of the underfill must be examined, particularly any tendency to "gum" the saw blade.

2.5.2 Placement

The placement issues with a wafer-applied underfill are vision recognition and holding the die in place after placement. Vision studies have been performed using standard pick-and-place equipment used for traditional FC placement. In evaluating underfill color, black was found to be the best (Fig. 2.45). As the materials and coating process mature, additional vision studies will be required, but the results to date are encouraging.

Tack is required to hold the die in place during the time between placement and melting of the solder. In FC assembly with capillary underfill, a tacky flux is used, whereas the fluxing underfill provides the tack in that assembly process. The wafer-applied underfill is tack-free after B-staging. However, it can be softened by heating to produce a tacky surface. It is proposed that a focused heat source or soft-beam laser be used to slightly heat the die just prior to placement. The necessary pick-and-place equipment modifications are being investigated.

2.5.3 Reflow

The reflow cycle for wafer-applied underfill is a standard SMT profile. The underfill cures during the reflow cycle, eliminating the need for a postreflow cure step. Figure 2.46 shows a cross section of a solder joint reflowed with wafer-applied underfill. Good solder wetting is observed.

2.5.4 Rework

The ultimate wafer-applied underfill will incorporate reworkability, but it is too early in the development cycle to demonstrate this capability. If the wafer-applied underfill must be exposed to multiple reflow cycles in the assembly process flow, a higher rework temperature will be required if a thermally activated rework chemistry/process is used.

2.6 RELIABILITY

The reliability requirements of FC assemblies fall into two broad categories: (1) component level and (2) board or product level. If FC technology is used to assemble a die into a package that is to be

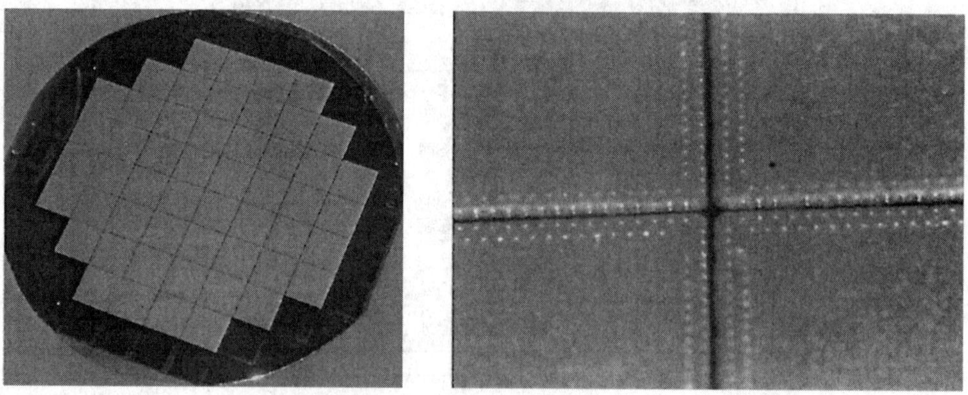

FIGURE 2.44 Photographs of wafer-applied underfill.

assembled subsequently onto a board, component-level testing is required. An example would be an FC reflowed onto a laminate carrier to create an FC BGA package (Fig. 2.47). As a component, the FC must first survive the board-level assembly process and then still meet the reliability requirements of the final product. The primary concerns are the effects of moisture and reflow cycles on reliability. This will be discussed in the next subsection.

In a board-level assembly, the FC must meet the reliability requirements of the final product. The environmental exposure and expected product life vary significantly for mainframe computers, pagers, and automotive engine controllers. The potential field failure modes for the product must be determined and appropriate laboratory test requirements defined. Since the actual field-use conditions cannot be employed for reliability testing due to the test time required, accelerated laboratory conditions are used. It is important in defining laboratory testing that new failure modes that do not occur in the field are not introduced while accelerating the test. Potential FC failure modes and test methods are shown in Table 2.3.

In field use, electronic products are not exposed to a single test condition, such as high temperature, but see combined environmental stresses. Laboratory tests are typically a single test condition (thermal cycling, high temperature, humidity, etc.). There has been some effort to define tests that expose the assembly to multiple tests in a prescribed sequence. However, with ever-decreasing time-to-market pressures, sequential testing often is not practical. Qualification testing of new materials and processes is time-consuming and expensive. Sometimes, potentially superior new materials are not used because of the time and cost associated with their qualification.

Effort has been devoted to developing "physics of failure" models to evaluate new product designs in an attempt to reduce test time and cost. This requires complete, accurate material properties, which often are not available. A combination of modeling and stress testing is used commonly in the design for reliability and qualification processes.

2.6.1 Component-Level Testing

As mentioned previously, component-level testing adds assembly-related exposure of the component to the product-level reliability testing requirements. The primary issue is the exposure to multiple

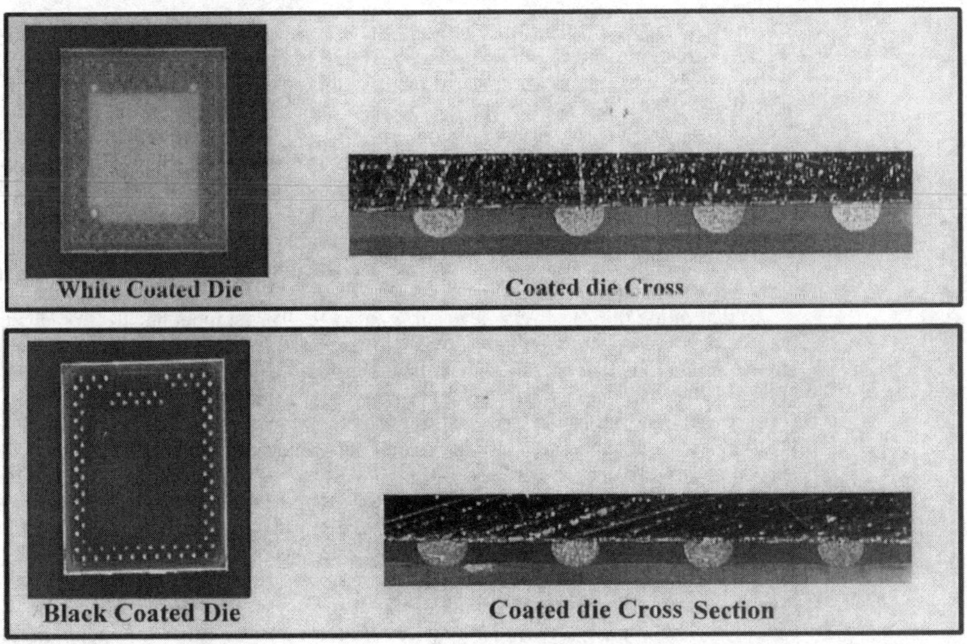

FIGURE 2.45 Comparison of visual contrast with white and black underfill.

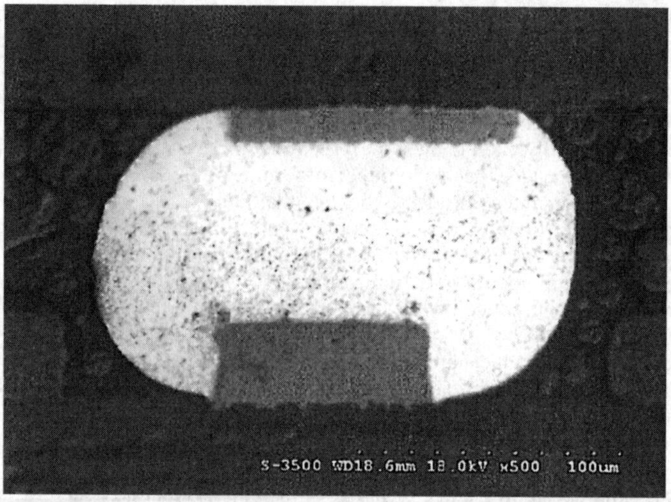

FIGURE 2.46 Cross section of a solder joint assembled with wafer-applied underfill.

reflow cycles. If the component package absorbs moisture from the environment, the rapid heating during the reflow cycle can liberate this absorbed moisture in the form of pressurized steam that can delaminate or crack the package. This assembly failure mode is known as *popcorning* because the mechanism is the same for popping popcorn. In an FC package, the underfill can absorb moisture. Moisture exposure can reduce the adhesion of the underfill. If a laminate substrate is used in the package, the laminate and the solder mask also may absorb moisture. During reflow, delamination at the die-underfill and the underfill-laminate interfaces may occur.

Moisture Sensitivity. The JEDEC Solid State Technology Association (*www.jedec.org*) has developed test methods and an industry-accepted definition of moisture sensitivity (J-STD-020A). The moisture exposure test conditions and factory storage conditions are listed in Table 2.4. If a component passes level 1 test conditions, it is deemed moisture-insensitive and can be stored indefinitely at 30°C or less at 90 percent relative humidity. At level 3, the component may be exposed to 30°C or less at 60 percent relative humidity for only 72 h. Moisture-sensitive components are dehydration baked to remove moisture and sealed in dry bags by the manufacturer for shipment and storage. Once the dry bag is opened, the component must be used in assembly within the time limit indicated by its moisture sensitivity level. The moisture sensitivity level is marked on the dry bag. If the component is exposed to the atmosphere longer than the time specified, it must be dehydrated prior to use to avoid popcorning during reflow.

To determine the sensitivity level of a package, the components are first electrically tested and inspected. The inspection includes a visual inspection for cracks and scanning acoustic imaging for any delamination. Then the components are exposed to the time, temperature, and humidity conditions specified for the sensitivity level of interest in the right columns (soak requirements) of Table 2.4. If the sensitivity level is not known, tests must be preformed at multiple levels to determine the lowest level the package can pass. After the temperature and humidity exposure, the components are subjected to three reflow cycles defined in the J-STD-020A. The reflow cycles are to follow no sooner than 15 min after removal from the humidity chamber and not longer than 4 h after removal. Following the three reflows, the components are electrically tested and inspected. Electrical failure, cracking, and delamination are criteria for failure at a given sensitivity level.

Preconditioning. For environmental testing of components, moisture preconditioning is often specified. The objective of moisture preconditioning is to induce any potential degradation of the

TABLE 2.3 Potential FC Failure Mechanisms and Test Methods

Failure mode	Test methods
Underfill delamination	Thermal cycling, thermal shock, humidity, highly accelerated stress testing (HAST)
Solder fatigue	Thermal cycling, thermal shock, power cycling, flexing
Creep	High-temperature storage, thermal cycling, thermal shock
Corrosion	Humidity, HAST
Electromigration	High-temperature storage with current flow
Excess intermetallic formation	High-temperature storage
Die fracture	Thermal cycling, thermal shock, flexing
Silicon cratering	Thermal cycling, thermal shock, flexing
Alpha particles	Electrical testing

TABLE 2.4 JEDEC J-STD-020A Definition of Moisture Sensitivity Level

	Assembly floor life		Soak requirements	
Level	Time	Conditions	Time (h)	Conditions
1	Unlimited	≤30°C/85%RH	168	85°C/85%RH
2	1 y	≤30°C/60%RH	168	85°C/60%RH
2a	4 wks	≤30°C/60%RH	696*	30°C/60%RH
3	168 h	≤30°C/60%RH	192*	30°C/60%RH
4	72 h	≤30°C/60%RH	96*	30°C/60%RH
5	48 h	≤30°C/60%RH	72*	30°C/60%RH
5a	24 h	≤30°C/60%RH	48*	30°C/60%RH
6	Time on label	≤30°C/60%RH	Time on label	30°C/60%RH

*Standard soak time includes a default value for semiconductor manufacturer's exposure time between bake and bag plus the maximum time allowed out of the bag at the distributor's facility of 24 h.

component due to assembly to simulate the condition of the component on an actual assembly prior to environmental testing. To moisture precondition the component, it is exposed to the time, temperature, and humidity test level (right columns of Table 2.4) consistent with the moisture sensitivity level rating of the component. Following exposure, the component is exposed to three reflow cycles (240°C peak) within 0.5 to 4 h. The component is then tested (electrical, visual, and acoustic microscopy) and ready for subsequent environmental testing (thermal shock, thermal cycling, etc.).

Multiple melting of the solder ball during reflow cycles can lead to excess intermetallic formation. Depending on the FC underbump metallurgy, the intermetallic will be either Sn/Cu or Sn/Ni. The growth rate for Sn/Cu intermetallics is greater than for Sn/Ni intermetallics. Excess intermetallic formation can lead to failure by delamination, dewetting, or embrittlement. To evaluate intermetallic formation, solder ball shear strength and cross sections after multiple reflow cycles can be examined. Figure 2.48 plots the solder ball shear strength as a function of reflow cycles. In this case, no degradation in strength was observed.

2.6.2 Environmental and Board-Level Testing

Thermal Cycling and Thermal Shock. Electronic products normally are exposed to variations in environmental temperatures in actual use. The mismatch in CTEs of the materials used to construct the product results in mechanical stresses as the temperature varies. Solder is susceptible to failure under stress by two modes. Fatigue failure results from cyclic stress, whereas creep is a failure under constant stress or load. The creep failure rate increases with increasing temperature. Figure 2.49 shows creep failure of FC solder joints held at constant force at 165°C.

FIGURE 2.47 Example of an FC BGA.

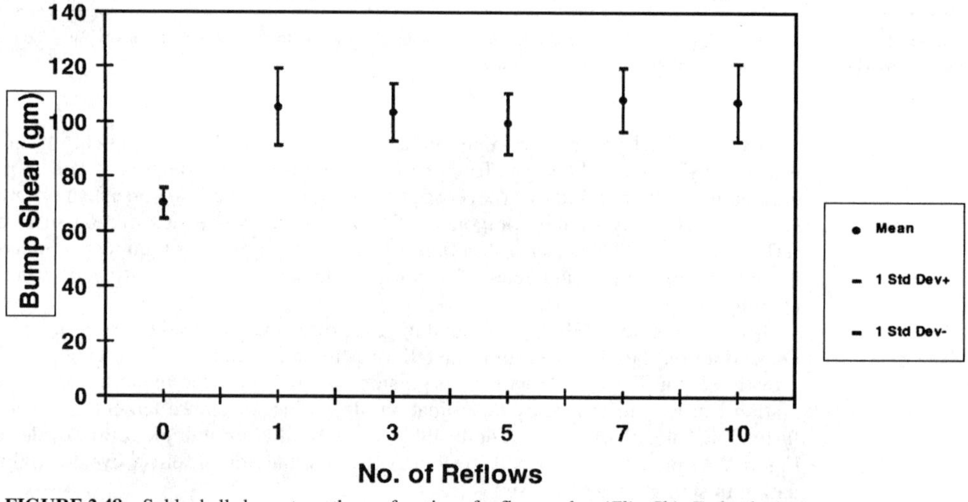

FIGURE 2.48 Solder ball shear strength as a function of reflow cycles. (*Flip Chip Technologies.*)

Failure of FC solder joints in thermal cycling and thermal shock testing results from a combination of fatigue and creep. The role of each failure mechanism depends on the materials, the temperature range, the temperature ramp rate, and the hold times (particularly the high-temperature hold time). In laboratory testing, the ramp rate and the temperature range are increased, whereas the hold times are decreased to accelerate the test and reduce test time. Direct correlation of laboratory test results with actual field life is complex.

The difference between thermal cycling and the thermal shock is the *ramp rate*. In thermal shock, the transition time from hot to cold and back is very rapid. In liquid-to-liquid thermal shock, the test boards are moved from one liquid held at high temperature to a second liquid held at low temperature. The transfer of heat by liquid contact is very efficient. For air-to-air thermal shock, the test boards are automatically shuttled between two chambers. The air in the two chambers is maintained at the two test temperatures. While the rapid transfer between chambers is comparable between liquid-to-liquid and air-to-air testing, the heat transfer rate is less for the air-to-air system. Since the thermal mass of the test boards and any fixtures or test cables has an impact on the amount of heat transferred, thermocouples are required to measure the actual temperature ramp rate and to ensure that all parts reach the temperature extremes.

Thermal cycling often uses a single chamber that is alternately heated and cooled. The temperature ramp rate is significantly slower. While this may better represent the actual temperature cycle rate in the field, the cycle time is longer, extending the test time. There is increasing use of thermal shock testing in the industry to accelerate testing to meet product development cycle requirements.

As mentioned in Sec. 2.1, underfill is used to increase the thermal cycle/thermal shock life of FC solder joints. During thermal cycling/thermal shock, there is also stress on the underfill globally and locally at the underfill-die and underfill-laminate interfaces. A typical failure mode is delamination of the underfill (most commonly at the die-underfill interface) and then rapid failure of the nearby solder joint. An example is shown in Fig. 2.50. With high adhesion underfills that do not match the CTE of the solder, solder joint fatigue failures ultimately can occur without delamination (Fig. 2.51).

As discussed in the underfill dispense section, voids in the underfill are a processing defect. During thermal cycling, extrusion of solder into voids between adjacent solder balls can lead to shorts, as shown in Fig. 2.52.

If the CTE of the underfill is higher than that of the solder, the solder is compressed by the underfill on the cold side of the cycle and is in tension in the z direction on the hot side of the cycle. Either the solder joint ultimately will fail open or the underfill may crack and solder will extrude into the crack, forming a short (Fig. 2.53).

Cracking of the underfill fillet and cracking of the PWB also are observed in thermal cycle and thermal shock testing. Propagation of these cracks can lead to either delamination or cracking of large silicon dies. Solder extrusion has been observed in PWB cracks at the base of solder balls. This extruded solder could lead to shorts within the PWB, particularly in HDI boards.

In situ resistance monitoring is required for accurate thermal shock or thermal cycle testing of underfilled FC dies. Failures (high resistance/open circuits) have been observed with daisy-chain test dies in the hot bath hundreds of cycles before failure is observed at room temperature or at the cold extreme. This can be explained using the illustration in Fig. 2.54. From finite-element modeling, the maximum stress on the solder joint occurs at the low-temperature extreme of the cycle because the "stress-free" temperature should be near the cure temperature of the underfill. However, at the cold temperature extreme, the contraction of the underfill (the CTE of the solder is lower than that

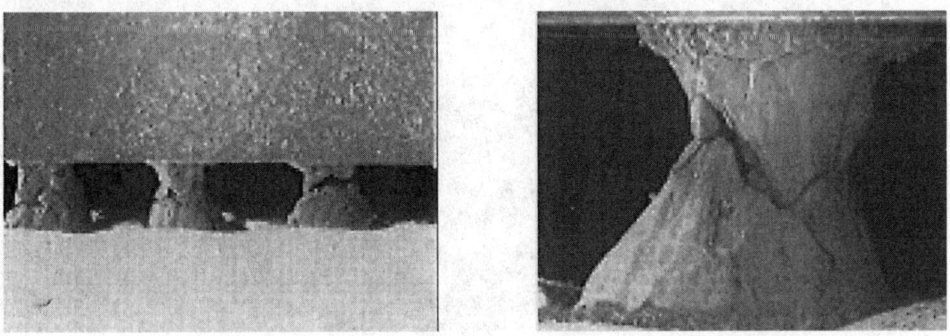

FIGURE 2.49 Creep failure of FC solder joints with a constant force at 165°C. (*Flip Chip Technologies.*)

FIGURE 2.50 C-SAM image of underfill delamination due to thermal shock cycling.

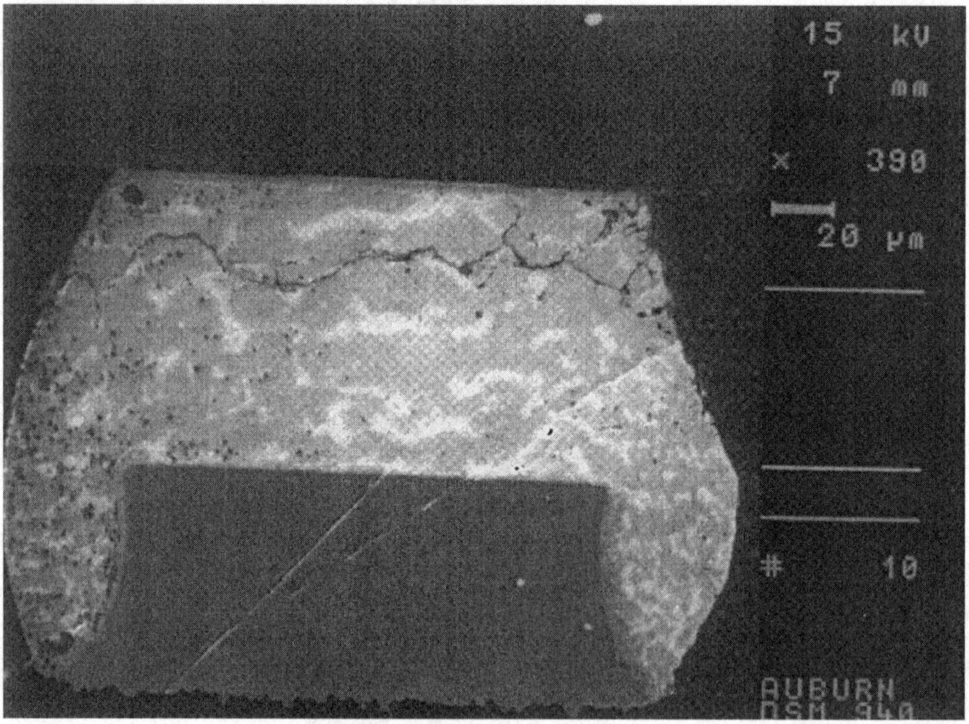

FIGURE 2.51 Solder joint fatigue failure. There was no delamination of the underfill.

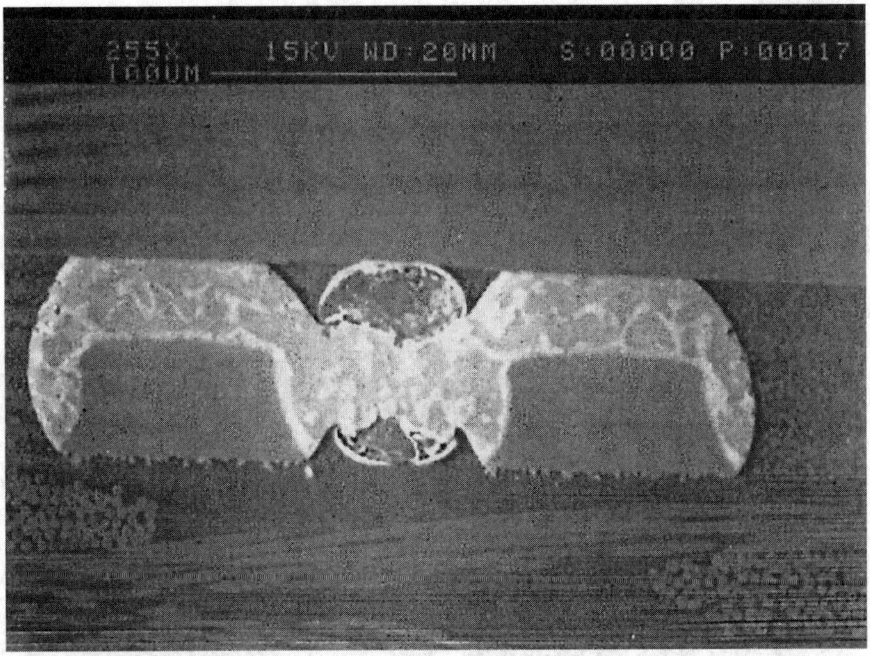

FIGURE 2.52 Cross section of a short due to extrusion of solder into the void between adjacent solder balls.

of the underfill) will hold the cracked solder joint together, maintaining electrical continuity. On heating to the high-temperature extreme, the expansion of the underfill will open the crack, and a failure can be measured. Thus periodic measurements for continuity at room temperature will not discover high-temperature intermittent opens and will overestimate the reliability of the connections.

Thermal cycling and thermal shock also can lead to die fracture. The underfill couples the silicon die to the PWB. As the structure is cycled, it bends or warps. Since the PWB contracts most, the backside of the silicon die is placed in tension during the cold side of the cycle. Any defects in the backside of the die such as microcracks from ejector pins can serve as stress concentrators, and a crack can propagate through the silicon. An example of die cracking is shown in Fig. 2.55.

Power Cycling. When an electronic product is turned on, the semiconductor devices dissipate power and get hot first. Thus the silicon die starts to expand before the underfill, solder, and PWB do. Power cycling can result in stress on the assembly and failure. Power cycling tests should be conducted for dies that dissipate significant power and that are subject to power cycling in the product application.

Flex Testing. Many portable products have integrated keypads. Pressing of the keypad flexes the thin PWB. FCs on the opposite side of the PWB would experience cyclic mechanical stresses. Underfill significantly increases the reliability of these applications.

Temperature-Humidity-Bias Aging. Electronic products are exposed to atmospheric moisture. Temperature-humidity-bias aging is used to evaluate the product reliability. The potential failure mode is corrosion in the presence of humidity and electrical bias. Temperature increases the moisture level and is an accelerating factor in corrosion. The epoxy underfills are formulated with very low extractable chlorine levels to minimize the potential for corrosion. No-clean flux residues are

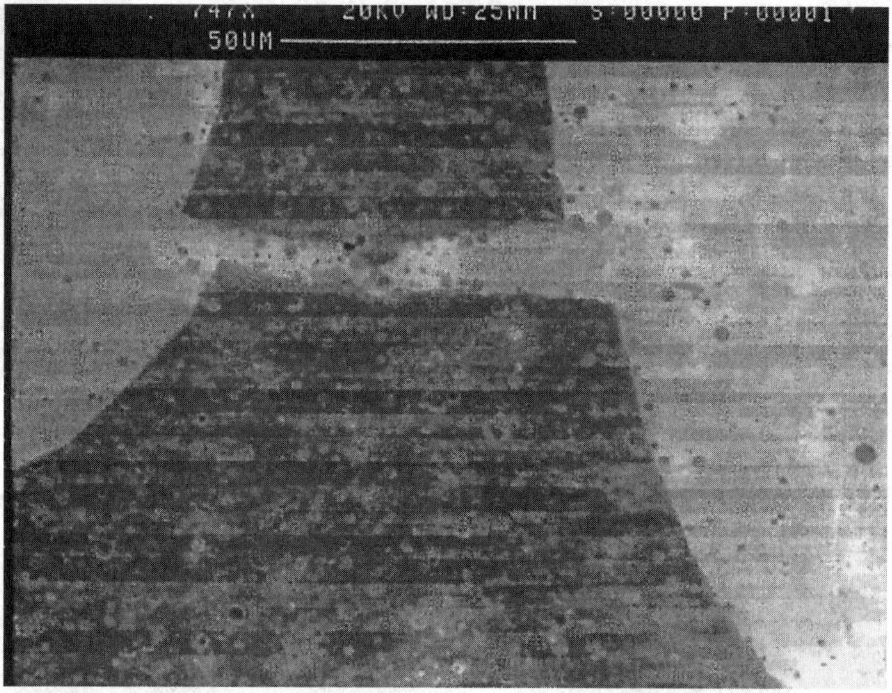

FIGURE 2.53 Flat section (die polished away) showing crack and solder extrusion in underfill between adjacent solder balls after 3200 thermal shock cycles.

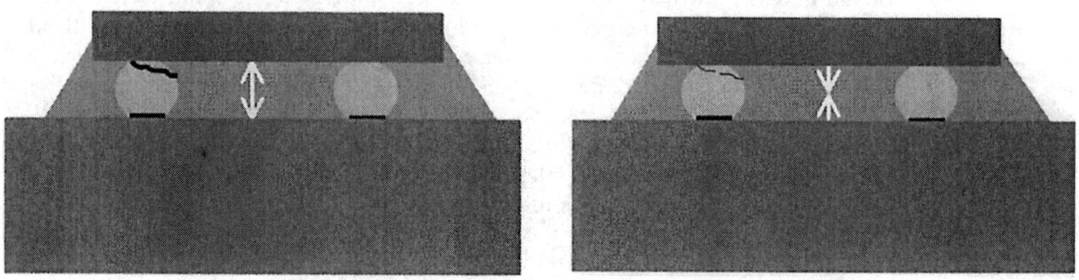

FIGURE 2.54 Illustration of the need for in situ monitoring.

also formulated to be chemically inert after the reflow cycle. A common test is 85 percent relative humidity at 85°C and 5 V for 1000 h. Failure is detected by electrical failure, shorts, and opens. Humidity and temperature also can result in degradation of underfill adhesion. Delamination can occur.

Autoclave and HAST. An *autoclave* is a pressure vessel used to accelerate humidity-temperature testing. A humidity level of 85 percent relative humidity can be achieved at 121°C under 2 atm of pressure. Other humidity, temperature, and pressure combinations also can be used. Adhesion loss and delamination are the common failure modes. Autoclave testing sometimes is used to rapidly screen underfill materials for moisture sensitivity. A highly accelerated stress test (HAST) is the addition of bias to the autoclave test. HAST accelerates humidity-temperature-bias aging tests.

High-Temperature Storage and Electromigration. High-temperature storage examines potential decomposition of the underfill and intermetallic growth in the solder joint. Figure 2.56 is a cross section of an FC solder joint after 3000 h of storage at 150°C. The intermetallic formation at the solder-to-PWB pad and the solder-to-underbump metallurgy (UBM) is evident. As mentioned previously, intermetallic formation can lead to dewetting, delamination, and embrittlement. Factors affecting reliability are UBM, UBM thickness, bump alloy, assembly processes, and service temperature. The UBM, bump alloy, and service temperature determine the intermetallic growth rate. The assembly process (number of reflow cycles, reflow profile, etc.) has an impact on the original intermetallic formation during assembly. Finally, the UBM thickness determines the amount of metal available for consumption. Since copper forms intermetallics with Sn/Pb eutectic solder at a faster rate than nickel, copper UBMs must be thicker than nickel UBMs.

If current flows through the solder bumps at elevated temperature, electromigration may occur. *Electromigration* is the movement of metal by momentum transfer from electrons. Increasing the current density (number of electrons per unit area) and temperature increases electromigration. In Sn/Pb eutectic solder, the Sn and Pb grains are separated by electromigration due to their different masses, as evident from Fig. 2.57. The concentration of Pb moves from the bottom of the solder ball to the top when the direction of current flow is reversed. From the plot, the maximum current a solder ball can carry decreases with increasing temperature and decreasing UBM/via size (increased current density).

Alpha Particles. An *alpha particle* is the nucleus of a helium atom (He^{4+}) that is emitted through radioactive decay. Trace amounts of thorium (Th), uranium (U), polonium (Po), and lead isotopes (^{214}Pb and ^{210}Pb) are the primary sources of alpha particles in tin-lead solder. A high-energy alpha particle (8 MeV) could generate up to 2.5×10^6 electron-hole pairs and cause soft errors in a sensitive semiconductor device, particularly memory. The *soft error* is not permanent damage to the device, but the data stored in the device are corrupted. To reduce soft errors, low-alpha-emission lead or lead- and bismuth-free solders can be used.

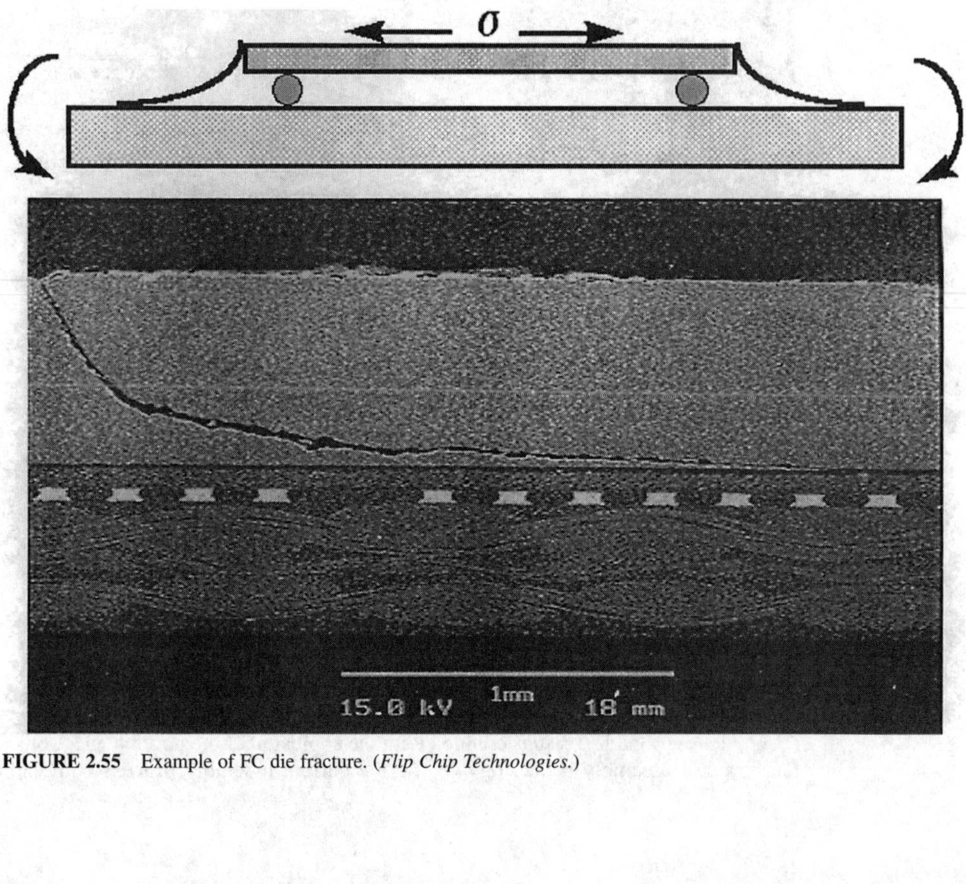

FIGURE 2.55 Example of FC die fracture. (*Flip Chip Technologies.*)

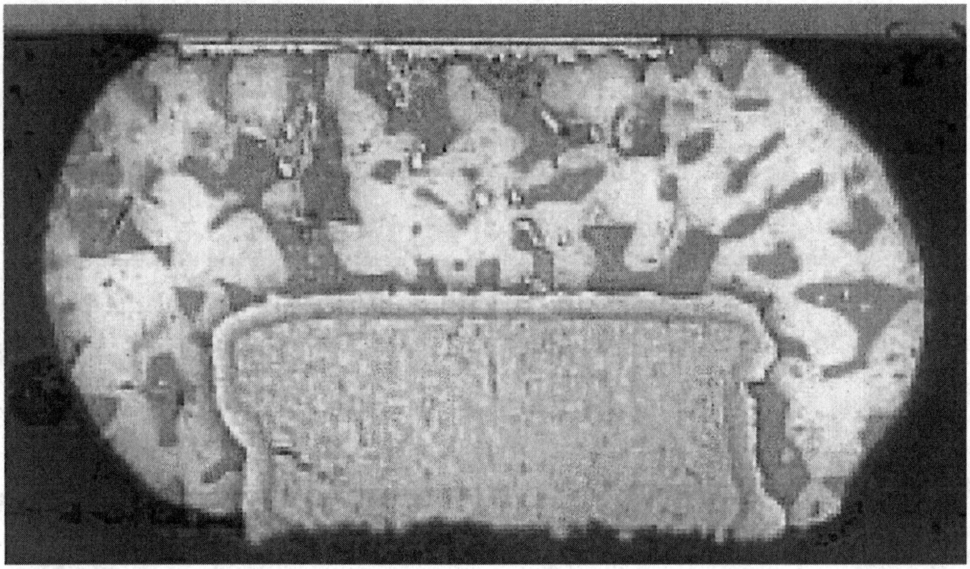

FIGURE 2.56 Cross section of solder ball aged at 150°C for 3000 h. (*Flip Chip Technologies.*)

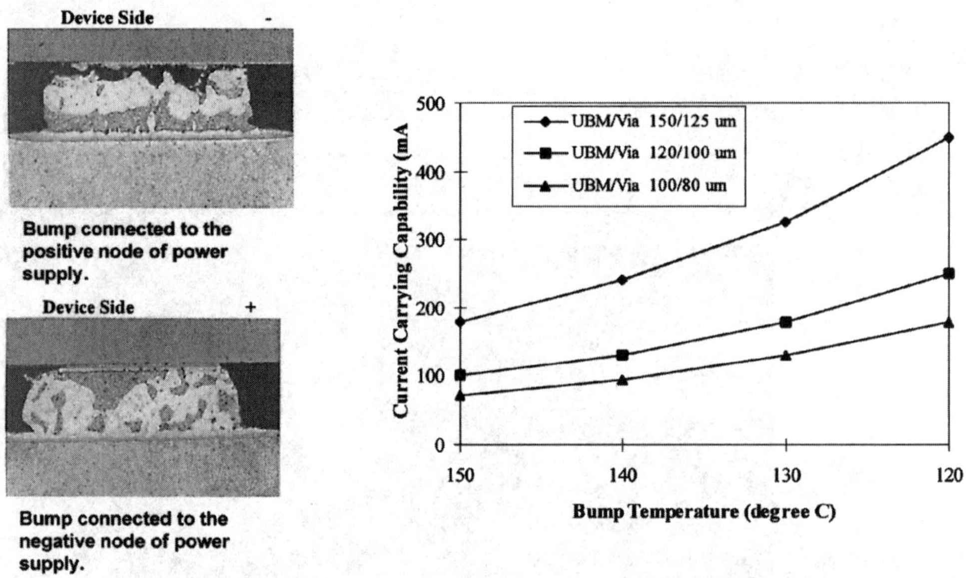

FIGURE 2.57 Electromigration test results and cross sections. (*Flip Chip Technologies.*)

2.7 SUMMARY

FC technology has been in use for nearly four decades. The continuing increases in semiconductor speed, die size, and I/O counts coupled with the explosion of the portable electronics market make FC an attractive assembly technology. While the current assembly process with capillary underfills is more complex than standard SMT assembly with packaged components, advances in materials and processes such as wafer-applied underfills may soon address this. Another limiting factor, the availability of bumped dies, is being addressed by the third-party bumping companies. Redistribution of perimeter I/O dies designed for wire bonding significantly adds to the bumping cost. In the future, more dies will be designed for FC assembly instead of wire bonding. This has already happened with microprocessor dies to meet the I/O requirements. Testing and burn-in of bumped dies still remain a hurdle to widespread use of FCs, but this too is being addressed by the semiconductor manufacturers.

The use of FC assembly will continue to grow as other approaches reach their limits and the issues mentioned herein are resolved.

2.8 REFERENCES

1. Totta, P. A., and Sopher, R. P., "SLT Device Metallurgy and Its Monolithic Extension," *IBM Journal of Research and Development,* May 1969, pp. 226–238.
2. Tummala, R. R., and Ramaaszewski, E. J. (eds.), *Microelectronics Packaging Handbook,* Van Nostrand Reinhold, New York, 1989, pp. 376–380.
3. Suryyanarayana, D., Hsiai, R., Gall, T. P., and McCreart, J. M., "Enhancement of Flip Chip Fatigue Life by Encapsulation," *IEEE Transactions on Components, Hybrids, and Manufacturing Technology,* 14 (1), March 1991, pp. 218–223.
4. Suryyanarayana, D., Wu, T. Y., and Varcoe, J. A., "Encapsulants Used in Flip Chip Packages," *IEEE Transactions on Components, Hybrids, and Manufacturing Technology,* 16 (8), December 1993, pp. 858–862.
5. Clementi, J., McCreary, J., Niu, T. M., et al., "Flip Chip Encapsulation on Ceramic Substrates," in *Proceedings of the 43rd Electronics Components Conference,* Orlando, FL, June 1–4, 1993, pp. 175–181.
6. Hu, K. X., Yeh, C. P., Doot, R., Skipor, A. F., and Wyatt, K. W., "Die Cracking in Flip-Chip-on-Board Assembly," in *Proceedings of the 45rd Electronics Components Conference,* Las Vegas, NV, May 21–25, 1995, pp. 293–304.
7. Novak, P., "Process Limit Testing on Fluxes Used for Flip Chip Soldering," in *Proceedings of the 1998 Surface Mount International,* San Jose, CA, August 23–27, 1998, pp. 275–280.
8. Qi, J., Johnson, R. W., Yaeger, E., et al., "Manufacturability Issues in Flip Chip on Laminate Assembly," *International Journal of Microcircuits and Electronic Packaging,* 22 (3), 1999, pp. 270–279.
9. Sellers, R., Gibson, M., and Gullion, D., "Cleaning for High Reliability Flip-Chip-on-Laminate Assembly," in *Proceedings of the 1998 Surface Mount International,* San Jose, CA, August 23–27, 1998, pp. 325–336.
10. Lewis, B., "Process Characterization and the Effect of Process Defects on Flip Chip Reliability," in *Proceedings of the Electronics Assembly Process (APEX) Conference,* Long Beach, CA, March 12–16, 2000, pp. AP2/3-1–AP2/3-6.
11. Schwiebert, M. K., and Leong, W. H., "Underfill Flow as Viscous Flow Between Parallel Plates Driven by Capillary Action," *IEEE Transactions on Components, Packaging and Manufacturing Technology, Part C,* 19 (2), April 1996, pp. 133–137.
12. Emerson, J. A., and Jones Adkins, C. L., "The Tools You Need for Development and Production When Using Underfill Materials," in *Proceedings of the 1998 Surface Mount International Conference,* San Jose, CA, August 23–27, 1998, pp. 270–274.
13. Crane, L., Torres-Filho, A., Ober, C., and Johnson, W., "Development of Reworkable Underfills, Materials, Reliability and Processing," *IEEE Transactions on Advanced Packaging,* 22 (2), 1999, pp. 111–115.
14. Kulkarni, P., Johnson, R. W., Yaeger, E., et al., "Manufacturing of Flip Chip-on-Laminate," in *Proceedings of the APEX Technical Program,* Long Beach, CA, March 12–16, 2000, pp. AP3/2/1–AP3/2/5.
15. Pennisi, R., "Adhesive and Encapsulant Material with Fluxing Properties," U.S. Patent No. 5,128,746, July 7, 1992.
16. Katze, D., "No-Flow Fluxing Underfill Process Characterization," in *Proceedings of the APEX Technical Program,* San Diego, CA, January 16–18, 2001, pp. MT2-21–MT2-29.
17. Previti, M. A., "No-Flow Underfill: A Reliability and Failure Mode Analysis," in *Proceedings of the APEX Technical Program,* San Diego, CA, January 16–18, 2001, pp. MT2-11–MT2-15.

18. Pascarella, N. W., and Baldwin, D. F., "Compression Flow Modeling of Underfill Encapsulants for Low Cost Flip Chip Assembly," *IEEE Transactions on Components, Packaging, and Manufacturing Technology, Part C,* 21 (4), 1998, pp. 325–335.
19. Leider, P. J., and Bird, R. D., "Squeezing Flow Between Parallel Disks. I. Theoretical Analysis," *Industrial Engineering Chemistry Fundamentals*, 13, 1974, pp. 336–341.
20. Leider, P. J., and Bird, R. D., "Squeezing Flow Between Parallel Disks: II. Experimental Results," *Ind. Eng. Chem. Fundam.,* 13, 1974, pp. 342–346.
21. Chason, M., Danvir, J., Yala, N., et al., "Development of Wafer Scale Applied Reworkable Fluxing Underfill for Direct Chip Attach," in *Proceedings of the APEX Technical Program,* San Diego, CA, January 16–18, 2001, pp. AT6-11–AT6-15.

CHAPTER 3
BGA AND CSP REWORK: WHAT IS INVOLVED?

Paul Wood and Howard Rupprecht
Metcal, Inc.

3.0 INTRODUCTION

As array package components become increasingly accepted in mainstream production, one of the most commonly asked questions is, "How can I rework these new components?" Many existing installed QFP rework machines may not have the required features for successful ball grid array (BGA) or chip scale package (CSP) rework. While rework of these components is not difficult, there are certain pitfalls that can be avoided if the engineer understands the method of interconnection, the different rework options, and the application of the product being reworked.

3.1 REWORK PROCESS CONSIDERATIONS

3.1.1 The Need for Process Control

If we compare a peripheral lead package such as a QFP with an array package such as a BGA or CSP, the fundamental difference is that one has leads on the perimeter and the other has an array of solder balls on its underside—functionally the packages are the same. If we are reworking a QFP and a defect occurs, it has always been possible to "touch up" that defect with a conventional soldering iron, repairing the fault. Engineers traditionally have had this safety net that often has meant that less process control was applied to the rework operation.

With any array package, when it is assembled, there is no possible way to "touch up" a defective solder joint—making complete package removal mandatory. Since none of us want to be perpetually removing and replacing devices, a greater level of process control is required to ensure that we achieve the perfect result on the first attempt.

Packaging trends such as increased input-output (I/O) count and reduced pitch have heightened the need for this level of process control, with components on the market with 1600 solder balls and pitches as low as 0.3 mm. On a package such as this, low DPM levels are absolutely essential for successful rework as well as production.

Another thing to consider is that rework normally is carried out at the end of the production process, possibly at final test stage. At this point any printed wiring board (PWB) assembly has reached its highest built-in value, making it a significant financial risk if a board is scrapped through uncontrolled rework.

3.1.2 What Is the Interconnection Method?

The primary consideration for any engineer who is reworking an array package is to know how that package is interconnected to the PWB or other substrate. There are many different package types from various manufacturers on the market, and understanding these is the key.

We have all heard the acronyms PBGA, CBGA, CCGA, CSP, and TBGA, but what do they mean in terms of interconnection? Components generally fall into one of two categories: those with a collapsible (melting) solder ball and those with a nonmelting ball or column.

Collapsible Solder Balls. The most common type of interconnection currently in use is the collapsible solder ball (Figs. 3.1 and 3.2). This ball generally is made from eutectic tin-lead solder (63/37 typically) and is designed to melt during the assembly reflow process. The actual component will "float" on the array of molten solder balls during reflow but will solidify to support the component. This is currently the accepted method on most plastic BGAs, most CSPs, and some tape BGAs, but there are exceptions to the rule.

Nonmelt Solder Balls and Columns. These normally are made from 90Sn/10Pb solder and generally are used in applications where the component is physically heavy, is in a high-reliability environment, or has extraordinary heat-dissipation requirements. In production reflow, these solder balls remain solid (they melt at approximately 300°C) and serve to support the component at the standoff height intended (Fig. 3.3). The solder paste printed on the PWB will melt and form the necessary intermetallic connection.

FIGURE 3.1 Eutectic solder ball and paste before reflow.

FIGURE 3.2 Eutectic solder ball and paste after reflow.

FIGURE 3.3 Nonmelt solder ball interconnection after reflow.

FIGURE 3.4 Reflowed solder joint with flux only.

This type of connection generally is found on ceramic BGAs, ceramic column grid arrays, and some tape BGAs and CSPs. Again, there are exceptions to the rule, and the component specification sheet should always be checked if one is unsure.

3.1.3 Do I Use Solder Paste or Flux?

The primary choice an engineer has when reworking is whether or not he or she uses solder paste or flux for component reconnection. In all cases, the main manufacturing process would be to use solder paste, since array package assembly is essentially the same process as surface mount technology (SMT).

If the solder ball/column of the component is nonmelt solder, then solder paste must be applied to make the solder joint (Fig. 3.4). The engineer can either print or dispense solder paste onto the pad area prior to component replacement. This is directly copying the production process.

If the component has a collapsible eutectic solder ball, the engineer has a choice of using either paste or flux (Fig. 3.5), depending on his or her own considerations. Things to consider:

- *Standoff height.* The height that the component stands off the board will depend mainly on the volume of solder. Normal standoff height of a 30-mil solder ball attached with paste would be about 20 mils after reflow. If only flux was used, this may reduce to 16 mils or less. If you have any concerns about meeting JEDEC standards, cleaning under the package, or heat dissipation, then this may not be acceptable.
- *Reliability.* Various independent tests have shown that there will be a difference in long-term reliability depending on standoff height. As a rule, it is generally considered that the higher the solder joint, the more compliant or flexible it will be.

FIGURE 3.5 Reflowed solder joint with solder paste.

FIGURE 3.6 Popcorn effect, delamination of component.

In practical terms, if you are working with a product that is destined for domestic use for a limited time, then flux probably will prove the easiest to apply and will offer more than adequate long-term reliability. Conversely, if you manufacture long-service, harsh-environment products, such as products for use in the aerospace industry, then flux probably would not be an acceptable method.

3.2 BASIC REWORK STEPS

3.2.1 Prebaking Components and PWBs

This step is often overlooked due to its time-consuming nature, but it can be extremely important in successful rework. Most plastic packages and PWBs are hydroscopic and will absorb moisture from the atmosphere during their life cycle. If the assembly is heated rapidly, as it would be during a rework operation, there may be a possibility of the moisture expanding to delaminate the component internally. This effect is known as *popcorning* and will lead to component failure.

If a board or component is suspect, it should be baked in an oven at low heat (≈100°C) for a number of hours to allow the parts to dry slowly prior to any rework operation. A number of large computer manufacturers rigorously enforce this process in their repair centers.

Prevention is always better than cure, and the key with components is to ensure that they are stored correctly. Most components are supplied in sealed packages, and once the packages are opened, the components should be used quickly or resealed for long-term storage.

Popcorning is characterized by a blister on the underside of the component as a result of the internal expansion (Fig. 3.6). Under x-ray inspection, you would normally see solder shorts toward the center of the device.

3.2.2 Component Removal

The primary heating method used by most rework systems is forced air/gas convection, although there are still some infrared systems available. Rework has followed reflow oven technology, where convection is preferred due to better temperature uniformity across dissimilar surface textures and colors. Convection is also considered to be easier to control for profiling and offering compatible ramp rates and dwell times.

To remove a component, use of a reflow nozzle that surrounds the component is preferred (Fig. 3.7). Generally, adjacent components around the nozzle should not reflow, although this will be impractical on very densely packed boards such as mobile phones and pagers. Using a low-velocity airflow system can minimize the problem of adjacent reflow (Fig. 3.8).

The removal process should be designed carefully so that the PWB does not become thermally stressed or warped. Generally, a removal profile can be shorter than a replacement profile because we are not concerned about solder paste characteristics, and in some cases, we will be disposing of the removed component. Care should be taken to allow even preheating on thermally demanding PWBs. This ensures good temperature uniformity, especially on the inner layers of a multilayer PWB (Fig. 3.9).

Most rework systems use a vacuum pickup to physically lift the component at the end of the removal cycle. It is important to ensure minimal vacuum pressure because this may cause solder collapse and extra cleaning operations.

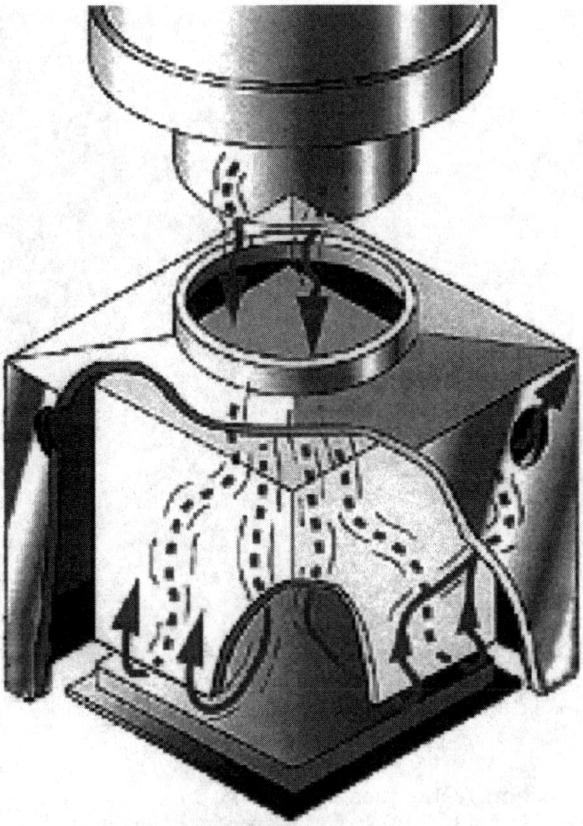

FIGURE 3.7 Principle of microoven reflow head.

3.2.3 Pad Cleaning and Preparation Methods

Once a component has been removed, generally, a certain amount of cleanup will be required to prepare the area for reapplication of solder paste or flux. Ironically, this is the process that presents the greatest risk to the PWB, normally as a result of pad, track, or solder resist damage (Fig. 3.10).

There are four recognized methods of solder removal:

1. *Desolder gun.* This method is time consuming, and direct contact with the conduction tip may cause solder resist damage.
2. *Solder wick with blade tip.* This method is used commonly and can be very successful with a skilled operator using a controlled soldering iron (Fig. 3.11). Pad finish generally is flat, making the surface ideal for stencil printing.
3. *Solder reapplication with hoof tip.* This method allows the operator to reapply solder to the pads using a concave soldering tip (Fig. 3.12). This method is suitable for flux-only applications and on small pads; a high degree of uniformity can be achieved.
4. *Solder "scavengers."* These systems effectively hot-air-level the pads while sucking off any excess solder. Results are very consistent, although the cost of equipment may be an issue. This type of system is ideal for removal of nommelt solder balls.

Once excess solder removal is complete, a cleaning operation will be required to remove flux residues prior to paste printing or new flux application.

FIGURE 3.8 Microoven on densely packed PWB.

3.2.4 Material Deposition Techniques

Once the decision to apply either solder paste or flux has been made, the engineer has to decide the most appropriate method of material deposition. There currently are four recognized methods for doing this.

1. *Brush application.* This method is the quickest option but is only suitable for applying fluxes. In most cases, a gel flux should be used because this will provide some adhesion to the component during reflow and will not evaporate too quickly. Brush application has the advantage of being quick and requires no component-specific tooling. Unfortunately, this process will leave large amounts of residue, probably necessitating a subsequent cleaning operation. The amount of flux applied is also imprecise and operator-dependent—too much flux may cause the component to float, and too little will result in unsoldered joints.
2. *Dispensing.* This process can be done either manually or automatically and can be used for both solder paste and flux. Manual dispensing is considered impractical due to its time-consuming nature and operator dependency. Automatic dispensing with an *xy* table will yield a consistently good result, but equipment costs may make this economically unfeasible.
3. *Printing.* This method can be done either manually or using vision assistance and generally is used for solder paste deposition. The process requires component/board-specific stencils that match the land pattern and any surrounding components. Advances have been made in single-component printing, making it relatively easy to print solder paste even on fine-pitch CSPs. A system that uses vision-mounted stencils will be much easier to align under high magnification. Coplanarity should be able to be adjusted to ensure absolute flatness of the stencil to the PWB. The actual print cycle is done using a metal squeegee, manually drawn across the stencil surface (Figs. 3.13 and 3.14). With minimal operator training, a high degree of consistency can be achieved.

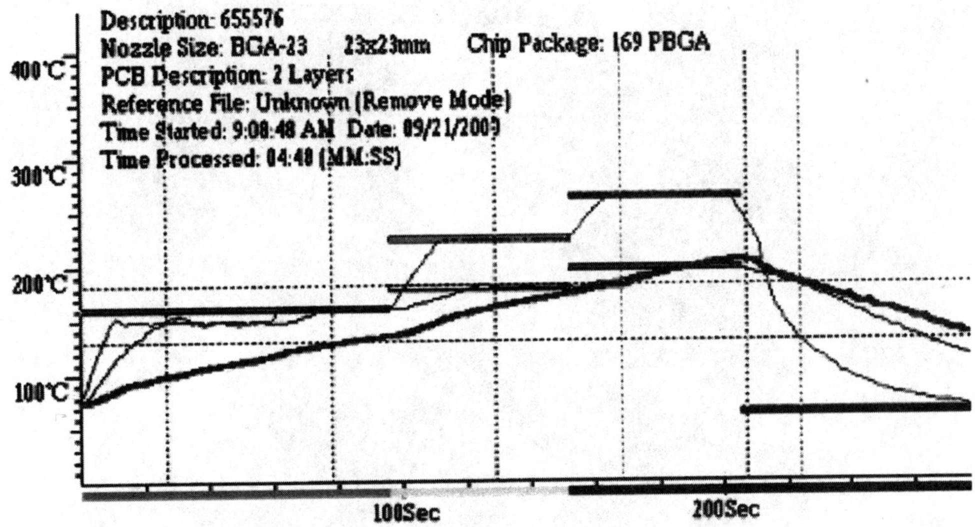

Time	Air	T.C.	PreH
00:27	150°C	100°C	152°C
01:18	161°C	129°C	162°C
02:09	225°C	162°C	181°C
02:48	264°C	181°C	186°C
03:27	241°C	205°C	195°C
03:44	130°C	184°C	180°C

FIGURE 3.9 Graph of typical reflow profile.

FIGURE 3.10 Solder resist damage.

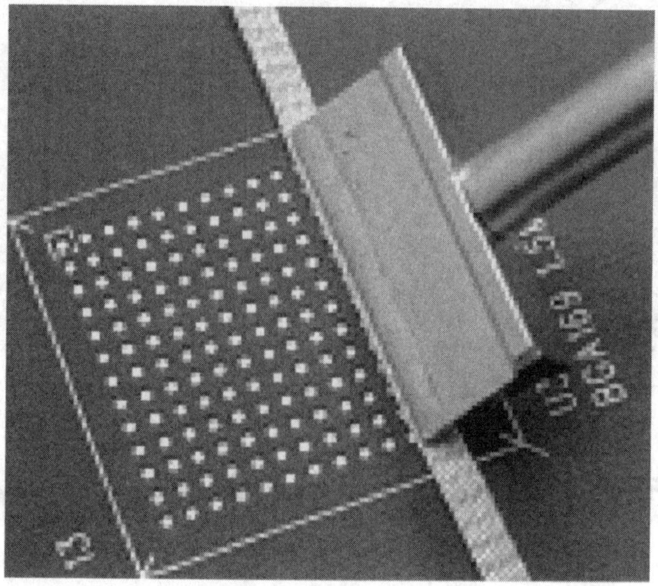

FIGURE 3.11 Removal of excess solder using blade tip.

FIGURE 3.12 Using a concave soldering tip to clean pads.

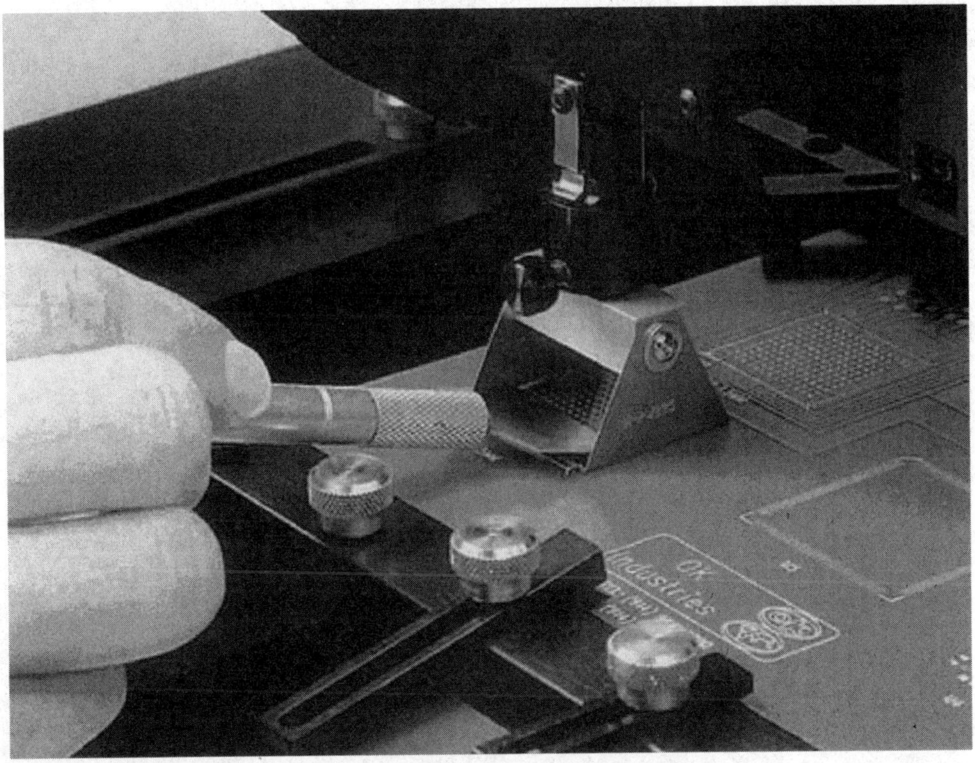

FIGURE 3.13 Manual solder paste printing.

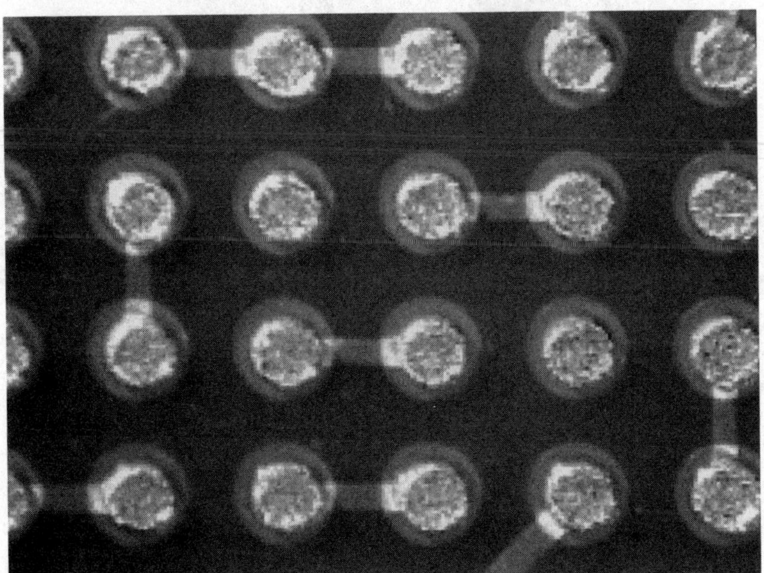

FIGURE 3.14 Photograph of successful manual print.

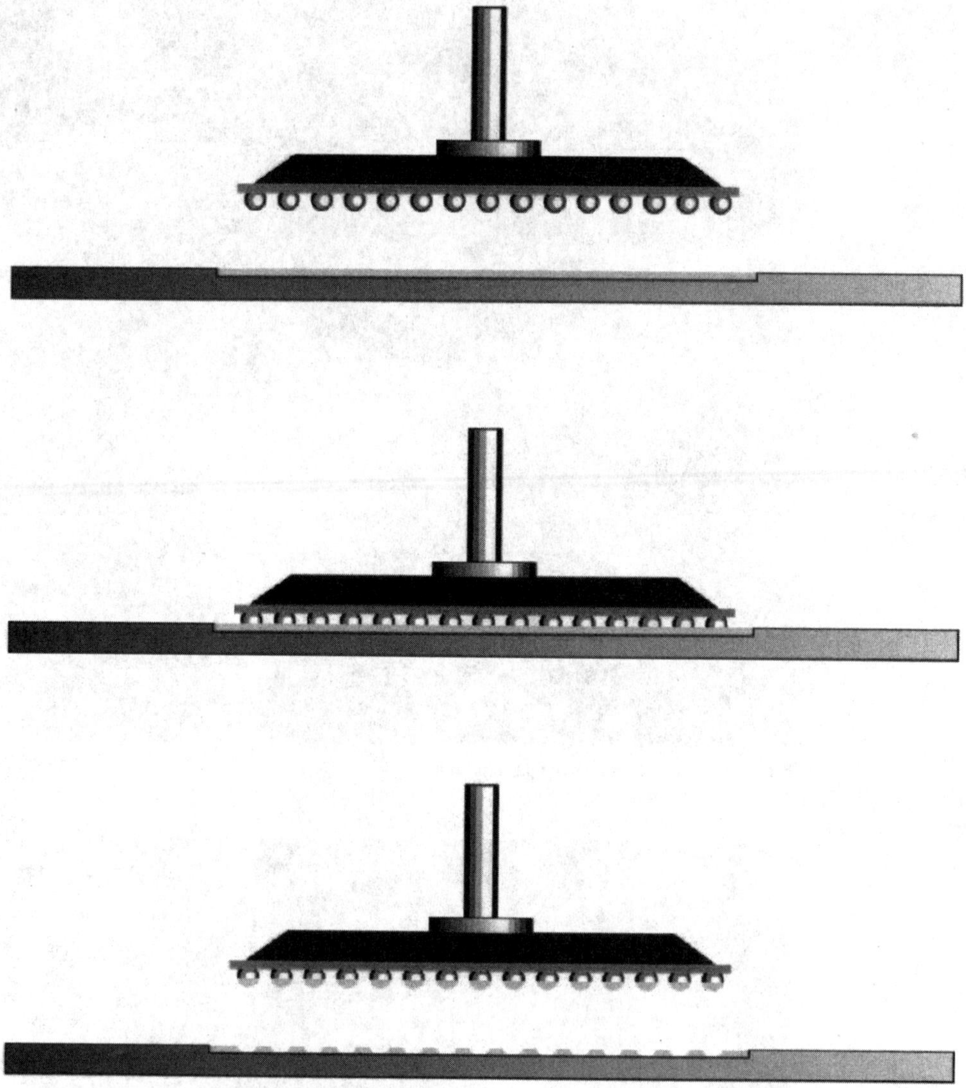

FIGURE 3.15 The steps of the dip transfer process.

4. *Dip transfer techniques.* This process involves dipping the component to be placed into a known depth of gel flux (results with solder paste are not practical) (Fig. 3.15). The technique is relatively new, being pioneered for use in this application by Salford University and OK International/Metcal, Inc. The advantages are that a very consistent flux deposit can be achieved very quickly. The process is becoming widely accepted as an efficient means of applying flux.

3.2.5 Component Placement

Array packages initially had relatively large ball pitches (typically 60 mils); this, combined with the strong self-centralization characteristics of the BGA, made hand placement a workable option for simple components. As component ball pitches continue to get smaller, vision-assisted placement is becoming essential in reliable rework.

The most common placement method in rework is to use a split-field prism system (Fig. 3.16). This allows the user to look simultaneously at the underside of the component and the topside of the PWB. The two images can be superimposed and accurately aligned on a monitor before placement. Care should be taken to ensure that the component is placed at the correct pressure.

3.2.6 Component Reflow Profiling

During component replacement, we try to emulate the original production reflow profile to match the required profile characteristics of the solder paste or flux being used. This normally would involve designing a four-stage profile as follows:

1. *Preheat stage.* The purpose of this stage is to get heat into the assembly and bring it up to flux activation temperature. A typical target would be to reach 120°C within 30 to 90 s, depending on the type of board.
2. *Soak stage.* The purpose of the soak stage is to allow the fluxes to activate at an elevated temperature, removing oxides and contaminants prior to soldering. Time is important because an overly extended soak time may "dry" out the solder paste, causing reoxidation. A typical soak would range from 120 to 170°C and would last between 30 and 90 s.
3. *Reflow stage.* In the reflow stage, the requirement is to bring the component to a temperature of between 200 and 210°C to allow soldering (63Sn/37Pb melts at 183°C). This time generally would be between 30 and 60 s.
4. *Cooling stage.* Having a cooling stage is important because overly extended reflow times will cause changes in the solder joint structure that could affect reliability. There is also a practical

FIGURE 3.16 A split-field imaging system.

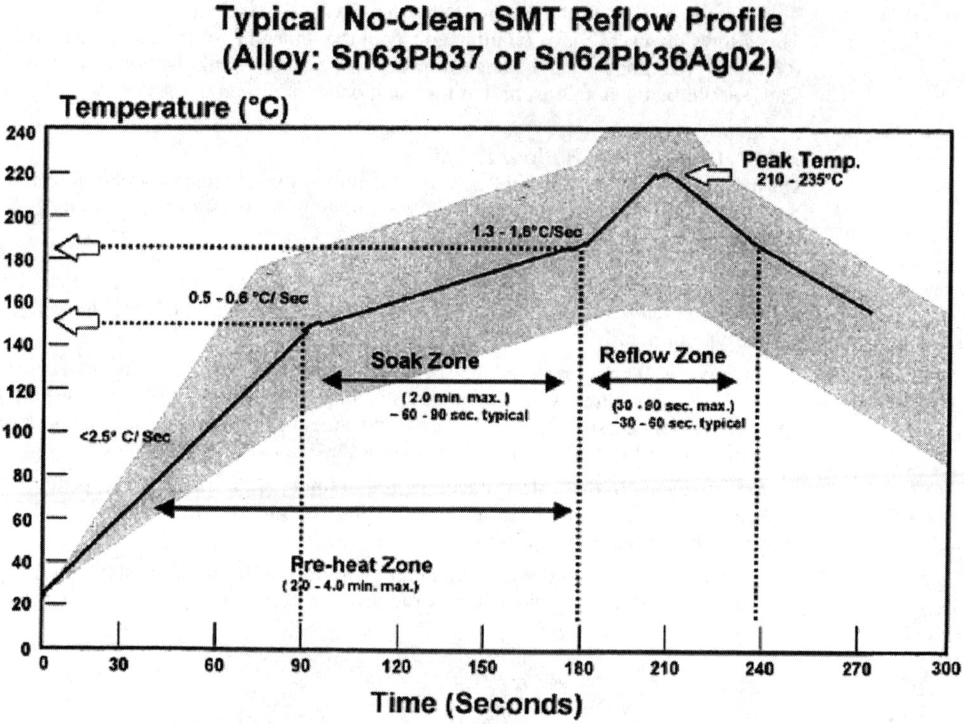

FIGURE 3.17 Typical solder paste process window. (*Kester solders.*)

side to ensuring that the board is back below reflow temperature before operator handling. The rework system should direct cold air through a separate air chamber because continuation through the heaters, even after they have been switched off, will promote extended reflow (Fig. 3.17).

3.3 IDENTIFICATION OF PROCESS YIELD PROBLEMS

If we consider profiling a PWB for reflow in production, we generally are concerned with achieving an even ΔT across all the joints being soldered. This is accomplished by evenly heating both the top and bottom sides of the whole board, generally through a multizoned convection reflow oven. The capacity of the oven would be appropriate to the size or thermal mass of the PWB and the production throughput requirements.

When we consider reworking an array package such as a BGA or CSP, it is recognized that process control is of paramount importance to achieve a successful result. As discussed earlier in this chapter, the normal way to approach the rework is to try to emulate the production reflow profile for the individual component being removed or replaced. The nature of the operation dictates that we only wish to reflow the component being reworked, and therefore, we have to heat a specific area of the circuit board. Selective heating of an area of board can inadvertently cause process failure even when all the parameters appear to be correct.

3.3.1 Balancing Profile Settings

On any reflow soldering system that heats both sides of the PWB assembly, the required solder-joint temperature is a function of how much heat is applied to both the top and bottom sides. It is quite

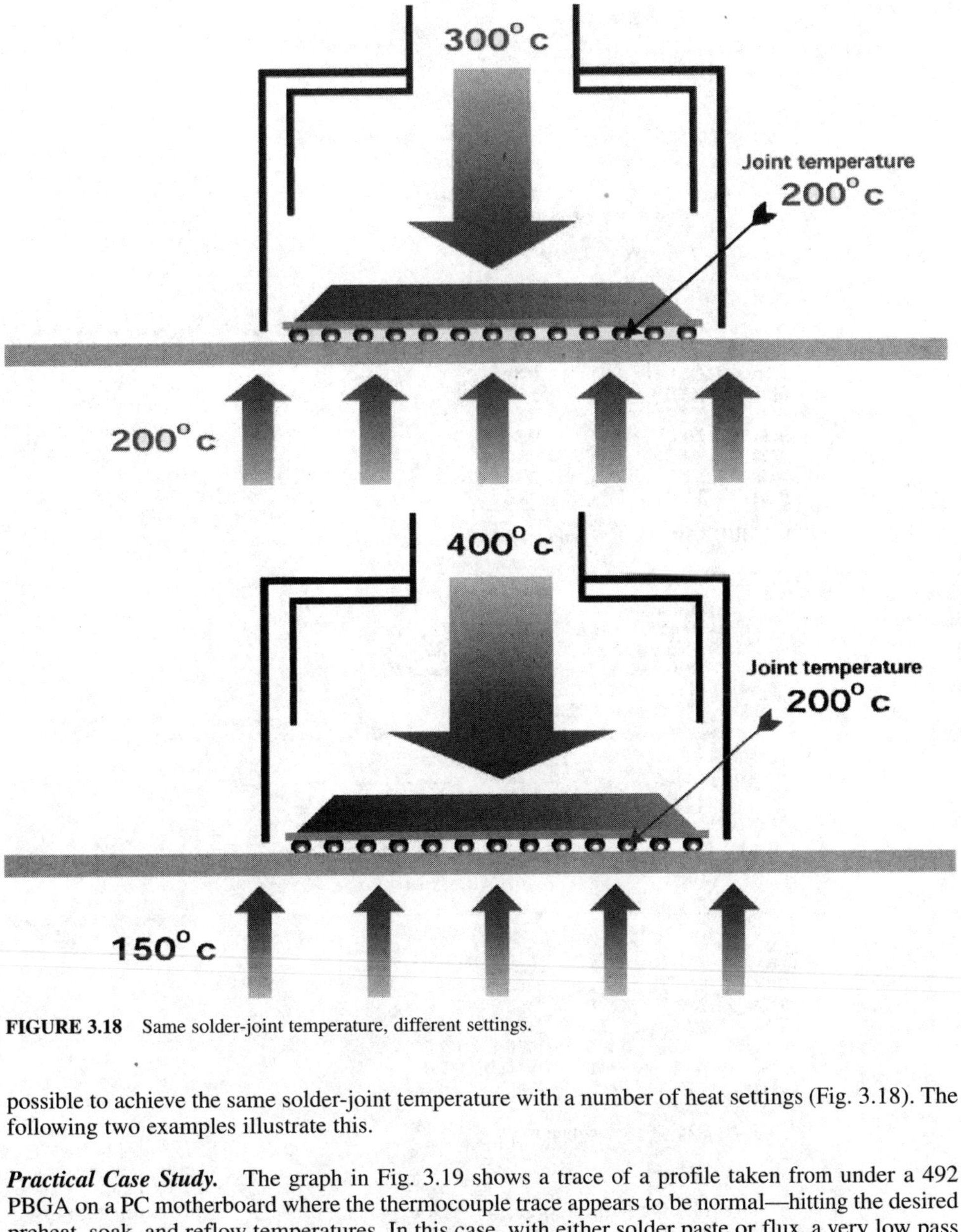

FIGURE 3.18 Same solder-joint temperature, different settings.

possible to achieve the same solder-joint temperature with a number of heat settings (Fig. 3.18). The following two examples illustrate this.

Practical Case Study. The graph in Fig. 3.19 shows a trace of a profile taken from under a 492 PBGA on a PC motherboard where the thermocouple trace appears to be normal—hitting the desired preheat, soak, and reflow temperatures. In this case, with either solder paste or flux, a very low pass rate was achieved.

The graph in Fig. 3.20 is the final profile that was used and has the original trace overlaid. As you can see, the two traces are very similar, but the heating has been done in a totally different way. In this case, we have reduced the top air temperature significantly while increasing the under-board heat to the whole board.

The root of the yield problem was that the PBGA warped during reflow. An excessive source temperature on the topside of the PWB caused this.

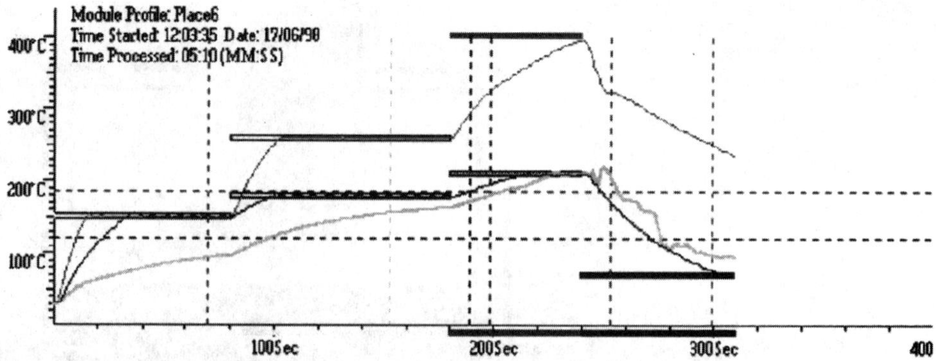

Time	Air	T.C.	PreH
01:10	150°C	093°C	151°C
02:32	259°C	155°C	180°C
03:09	295°C	171°C	185°C
03:18	324°C	179°C	195°C
04:13	322°C	206°C	170°C
04:59	249°C	098°C	075°C

FIGURE 3.19 Unsuccessful profile.

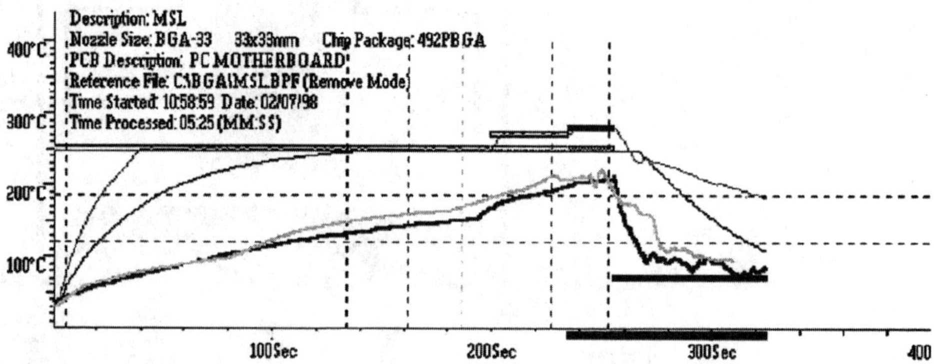

Time	Air	PreH	New	Old
00:06	078°C	057°C	042°C	040°C
02:14	249°C	245°C	131°C	147°C
02:42	249°C	246°C	142°C	157°C
03:06	249°C	246°C	149°C	168°C
03:47	269°C	246°C	190°C	210°C
04:13	278°C	246°C	206°C	206°C

FIGURE 3.20 Successful profile.

The following subsection offers an explanation of the kind of faults that can occur when everything appears to be set correctly (i.e., perfect screen print, no board damage, and accurate placement) but the yields are low.

3.3.2 Problems Caused by Warping

A major cause of defects is warpage of either the PC board or the component during reflow. Even minor deformation can lead to defects such as unsoldered joints and solder shorts. Component lifting by even 0.005 in across the device is enough to cause an open circuit. Another factor is that the larger the device, the more prone the device will be to these problems. Even if the board/component survives the process with no apparent defect, then the joint will be constantly under strain as the board returns to its normal shape—causing long-term reliability problems.

Figure 3.21 shows unsoldered joints caused by PWB warpage, the defect being caused by component warpage. The joints in the center of the component have lifted above the PWB and were not able to make the solder joint. This is more typical when only flux is used because there is no paste to make up the void area. Another characteristic is that the opens generally will be located toward the center of the device.

Figure 3.22 shows solder shorts. In this case, the deformation has caused the solder paste and the solder balls to combine into solder shorts. This is more typical with solder paste and normally is characterized by shorts on the device periphery, especially in the corners when the device becomes "dog-eared" during reflow.

The solution to the problem of warping generally is related to the profile settings, particularly insufficient under-board preheat temperature combined with too high a top temperature. Some boards, however, are badly balanced thermally, and this also can lead to problems.

On large PWBs, under-board support also can be a factor, with the PWB sagging under its own weight. Note that downward pressure should not be applied by eiher the reflow head or vacuum because this will exacerbate any problems.

3.3.3 Simultaneous Reflow Issues

The reason that manufacturers strive to attain simultaneous reflow is that having a large ΔT across a component can lead to rework and production defects. Some factors can be influenced directly by the profile settings, whereas others are determined by the physical makeup of the PWB assembly. The first factor that will influence how "simultaneous" the reflow is will be profile temperatures at the point of reflow—more specifically, how quickly we pass from solid to liquid state of the solder (Fig. 3.23).

Figure 3.24 shows a graph section of a component's solder joints going from solid to liquid at a melting temperature of 183°C. We are assuming that the ΔT achieved across the component is 10°C.

The rate of temperature rise (ramp rate) will determine the duration of reflow from start to finish. This is directly proportional to the input temperatures set on the profile. In this case, the ramp rate is 0.5°C/s.

The solution would involve changing the profile input temperatures to achieve a higher ramp rate. This, in turn, would produce a more simultaneous reflow, as illustrated in Fig. 3.25.

There are other factors that will influence ΔT on any PWB assembly, and these are often out of the control of the rework machine manufacturer. Position of the component on the PWB is a major

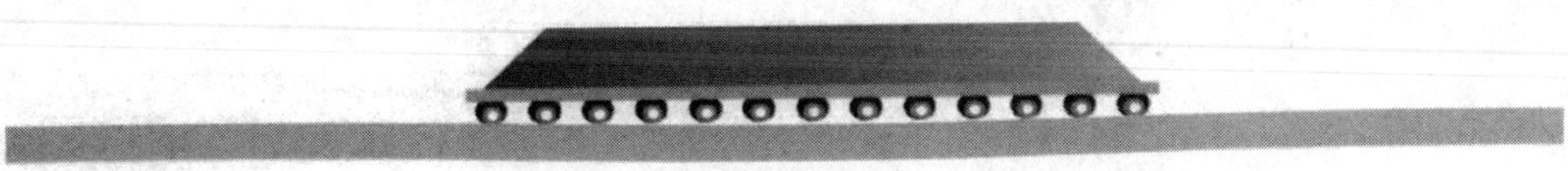

FIGURE 3.21 Unsoldered joints caused by PWB warpage.

FIGURE 3.22 Solder shorts caused by PWB warpage.

FIGURE 3.23 Uneven reflow caused by poor temperature uniformity.

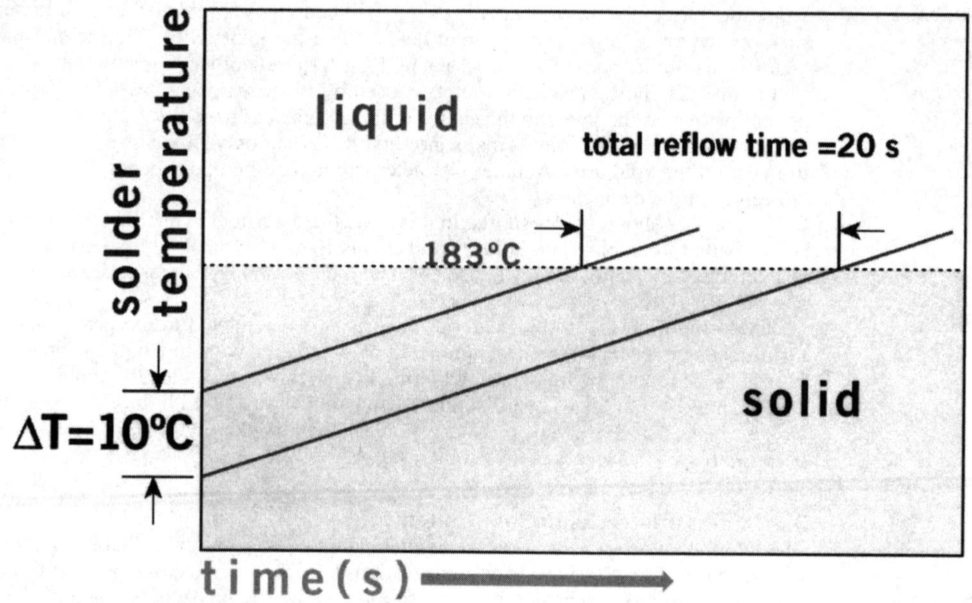

FIGURE 3.24 Slow transition from solid to liquid.

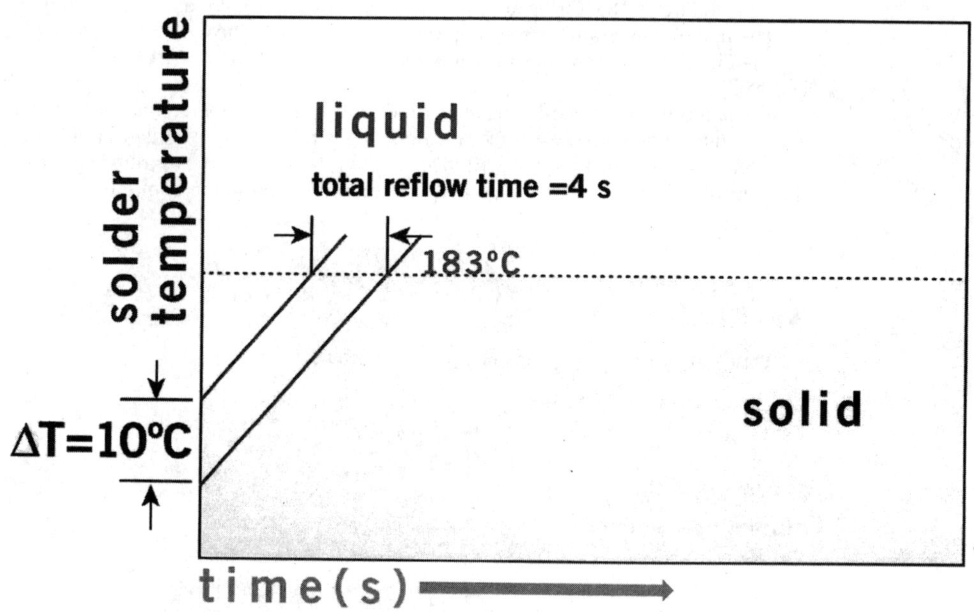

FIGURE 3.25 Faster transition from solid to liquid, more simultaneous.

influence; if a component is placed at the edge of a board, the temperature of the outer two sides generally will be greater than the temperature of the inner sides. This is due to greater heat dissipation toward the center of the PWB. Another factor is the area to which some solder balls are connected. It is quite common on a BGA for a center array of solder balls to be connected to a ground plane specifically to dissipate heat from the component die. These balls may require extra time to reflow.

3.4 COMPONENT REBALLING CONSIDERATIONS

One of the most commonly asked questions is, "Can I reuse components, and how do I reball the components?" While this is obviously relevant, there are very few companies who are currently reballing, and the following things should be considered before starting.

Component reusability questions need to be answered by the component supplier. The supplier will know what amount of thermal cycling its components will stand. If we consider a BGA assembled on a double-sided SMT PWB that was removed, reballed, and replaced, it would have had to undergo potentially six reflow cycles during this process, in addition to any ball reflow during the component manufacture:

1. Reflow assembly, topside
2. Reflow assembly, bottom side
3. Removal of component
4. Removal of excess solder from component
5. Reflow of new solder balls
6. Replacement of component

Another consideration would be the amount of handling and potential electrostatic discharge (ESD) damage that could occur during the reballing process.

Considering the preceding, in many cases, the reballing process is not viable commercially due to the extended labor times. We do, however, come across cases in which, because component value is very high or because of component unavailability, it is necessary to reball.

There are two basic methods most commonly used for reballing array packages: preforms and reballing fixtures. These methods can be used in conjunction with the features of the chosen rework machine.

3.4.1 Reballing Using Solder Preforms

One method for reballing involves using solder preforms (normally an array of solder balls to match the component, either held in a paper sheet or joined together). These are available from some of the solder manufacturers; there is also a product available called SolderQuik (Fig. 3.26). These preforms, in the case of eutectic balls, need to be soldered to the device in a controlled environment (reflow oven or head) using flux. When using high-temperature nonmelt solder balls, the preform needs to be attached to the device using screen-printed or dispensed solder paste. The typical process steps for attaching an array of solder spheres using the SolderQuik BGA preform are listed below:

1. *Removal of residual solder.* The pads on the part need to be prepared for the replacement solder balls using the methods described earlier in this chapter.
2. *Flux application.* Prior to placement, the correct flux should be applied to the pad areas by dispensing, printing, or more commonly, brush application.
3. *Preform placement.* Placement of the new preform is generally a manual operation where alignment is controlled by either edge registration or a fixture to align the component pad with the preform.
4. *Reflow.* After the part is fluxed and the preform placed, the next stage is to reflow the part. Again, reflow should be multizone controlled to match the profile of the flux or solder paste being used. When the reflow cycle is complete, the new array of solder balls should have been attached successfully.

FIGURE 3.26 SolderQuik and other preforms. (*Winslow Automation and the Indium Corporation.*)

5. *Carrier removal and component cleaning.* At this point, the carrier material needs to be removed, which can be accomplished by letting the carrier soak in some deionized water for a few seconds and then peeling away. The part then needs to be cleaned of flux residue, which can be done by almost any standard aqueous method. If you remove the paper with a batch cleaner, this step probably will not be necessary. The paper also can be removed easily when hot if a no-clean flux gel is used. This prevents the need to soak the paper off and is of great benefit because no prebaking of the component will be needed.

3.4.2 Reballing Using Fixtures

This method normally involves component-specific stencils and tooling to allow paste or flux to be printed and loose solder balls to be located in the correct position before being reflowed—again in a controlled environment. Solder balls of different sizes and compositions are available from most of the solder paste manufacturers.

Figures 3.27 through 3.29 show the process being used with printed solder paste and nonmelt, high-temperature solder balls on a CBGA component.

1. *Stencil printing of solder paste.* Figure 3.27 shows solder paste being printed onto the component. The part is held in a fixture, and using a chemically etched stencil, the solder paste is applied manually.
2. *Application of solder balls.* In the second stage, Fig. 3.28 shows that the paste-printing stencil has been removed and has been replaced with a larger stencil that will dictate the position of each solder ball. Loose solder balls are then poured onto this stencil and fall into their intended positions above each printed pad. When all the apertures have been filled, any excess solder balls simply can be brushed off or poured away.
3. *Reflow.* After all the balls have been placed, the component is ready for reflow. The method shown in Fig. 3.29 is to completely remove the stencil and use forced convection from above to reflow the solder paste attaching the nonmelt solder balls.

Some methods advocate leaving the ball-location stencil in position during the reflow operation because a ball's natural reaction is to self-centralize; this will make little or no difference. The stencil also may act as a heat sink, which will increase the required reflow temperature.

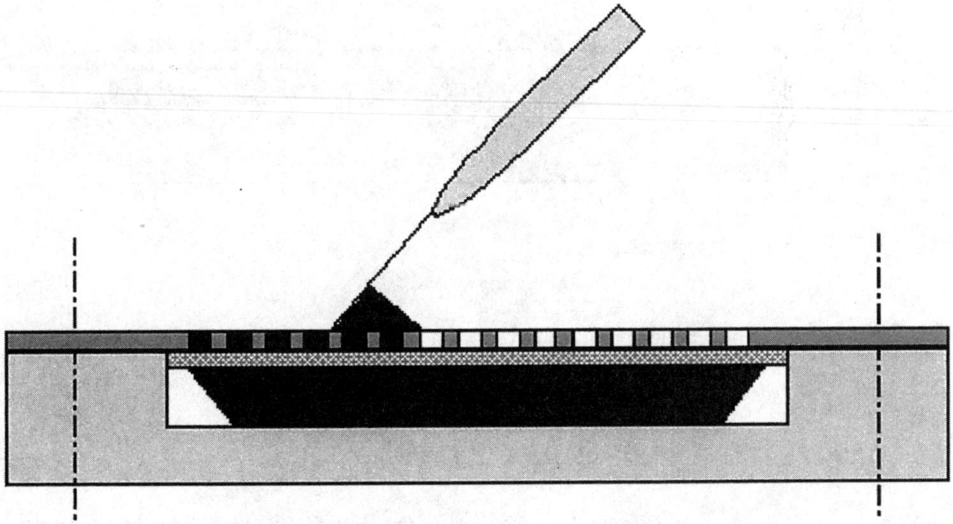

FIGURE 3.27 Printing solder paste onto the component.

FIGURE 3.28 Application of loose solder balls.

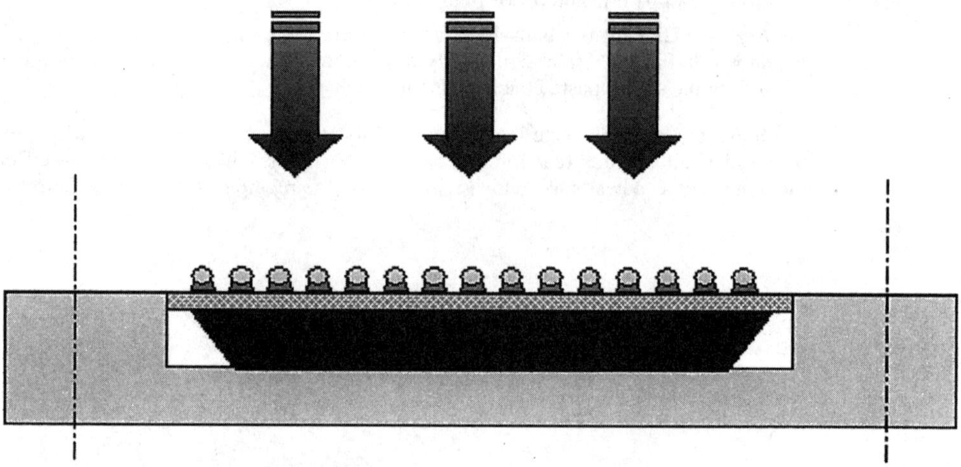

FIGURE 3.29 Solder ball reflow.

The reballed part is removed from the fixture and is ready to be attached to the PWB as a normal component.

3.5 ANALYSIS USING X-RAY INSPECTION

On an array package, because most of the solder joints are hidden from view, the only comprehensive way to evaluate results is to use an x-ray inspection system. The following subsections illustrate some of the common visible faults and their respective causes.

3.5.1 Solder-Joint Size Variability

When looking down on a BGA assembly, the uniformity of the solder joints can be a good indicator of process problems (Fig. 3.30). In a perfect result, we would expect all solder joints to be approximately the same size and shape. Variability could indicate the following defects:

- Component or PWB warpage
- Popcorning
- Insufficient solder paste deposits
- Excessive solder paste deposits

3.5.2 Solder-Joint Shape

Solder-joint shape can be an indicator of process or design problems but also can act as a tool to identify wetting. On an assembly with round pads that are "solder mask–defined," we would expect all joints to be round in shape. If the pads were teardrop-shaped, it could be indicative of the following defects:

- PWB solder resist misalignment (Fig. 3.31)
- Damage to solder resist causing solder migration along tracks

On an assembly with pads that are not solder mask–defined, we would expect to see a less uniform joint shape, where solder has wetted adjoining tracks. This would not be considered a defect but would indicate different PWB characteristics. This effect can be used to the engineer's advantage if pads are deliberately designed to be teardrop-shaped. In this case, a teardrop-shaped joint would indicate successful wetting of the solder joint.

3.5.3 Voids

The effect of voids in a solder joint depends very much on the size and quantity present. The main factor that contributes to generating voids is an incompatibility between the reflow profile and the materials used (Fig. 3.32).

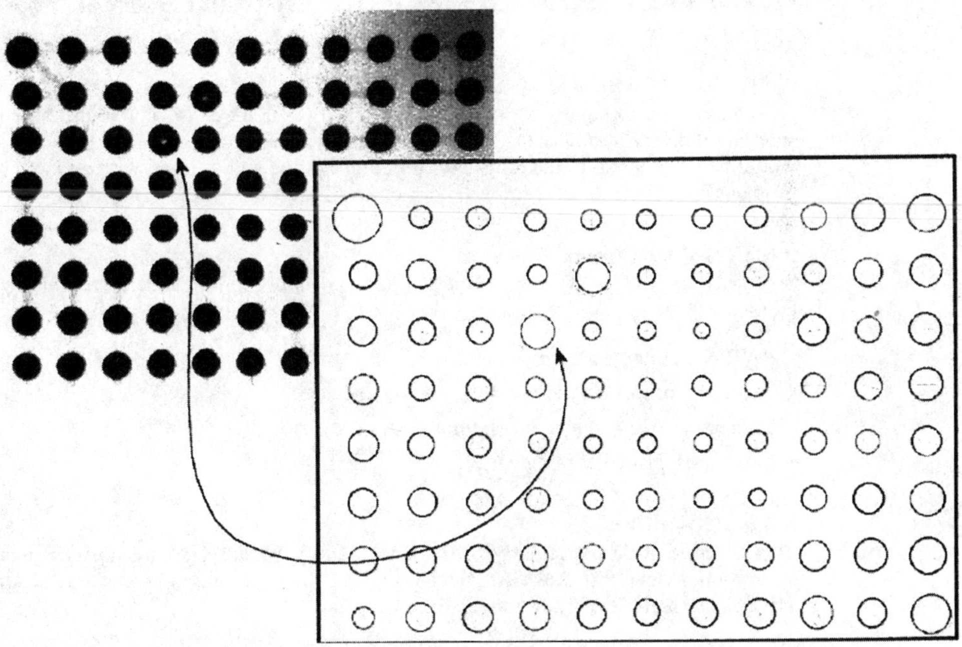

FIGURE 3.30 Variable solder-joint sizes. (*Glenbrook Technologies.*)

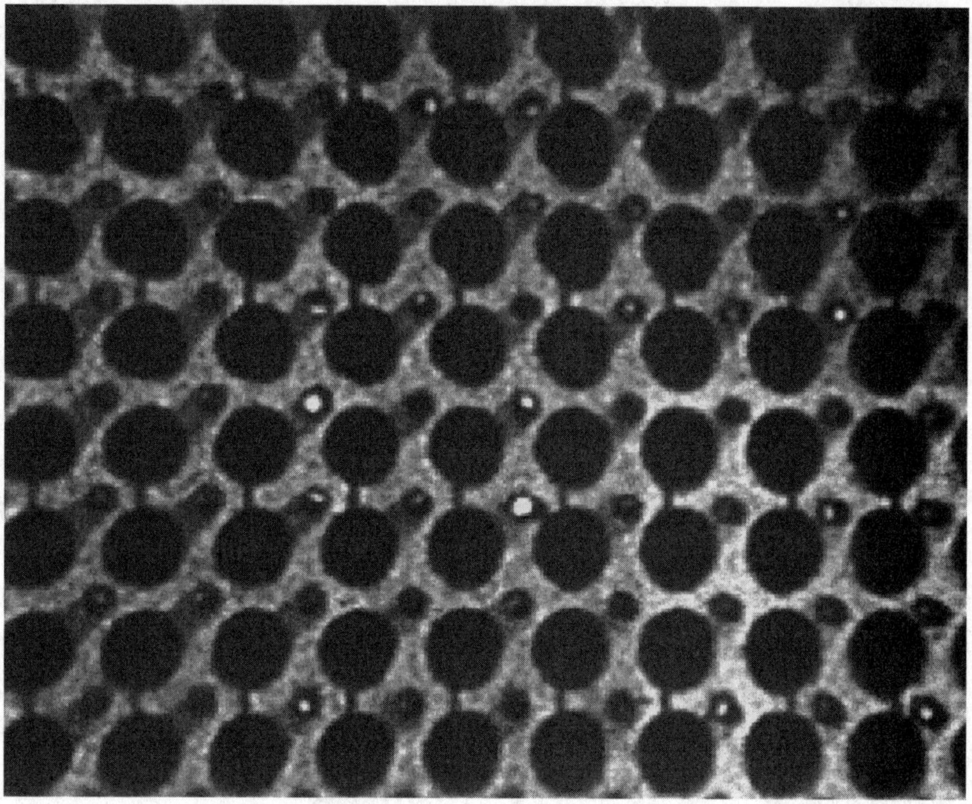

FIGURE 3.31 Misshaped joints due to misalignment. (*Glenbrook Technologies.*)

If a profile is too quick, you may see a number of small, consistent voids, although these are considered relatively harmless to long-term reliability. If large and inconsistent voids are observed, this would be detrimental to the long-term reliability of the solder joint, with an increased risk of cracks propagating throughout the joint. Incorrect profile settings would cause this.

3.5.4 Solder Shorts

Solder shorts (Fig. 3.33) represent a complete assembly failure and can be caused by the following factors:

- PWB or component warpage
- Excessive or inconsistent solder paste deposits
- Damage to solder resist on adjoining tracks
- Incorrectly placed components

3.5.5 Open Circuits

This defect can be particularly difficult to identify because it may not be apparent from a two-dimensional view (Fig. 3.34). As stated previously, inconsistent joint size may be an indicator of this. The primary causes of opens are as follows:

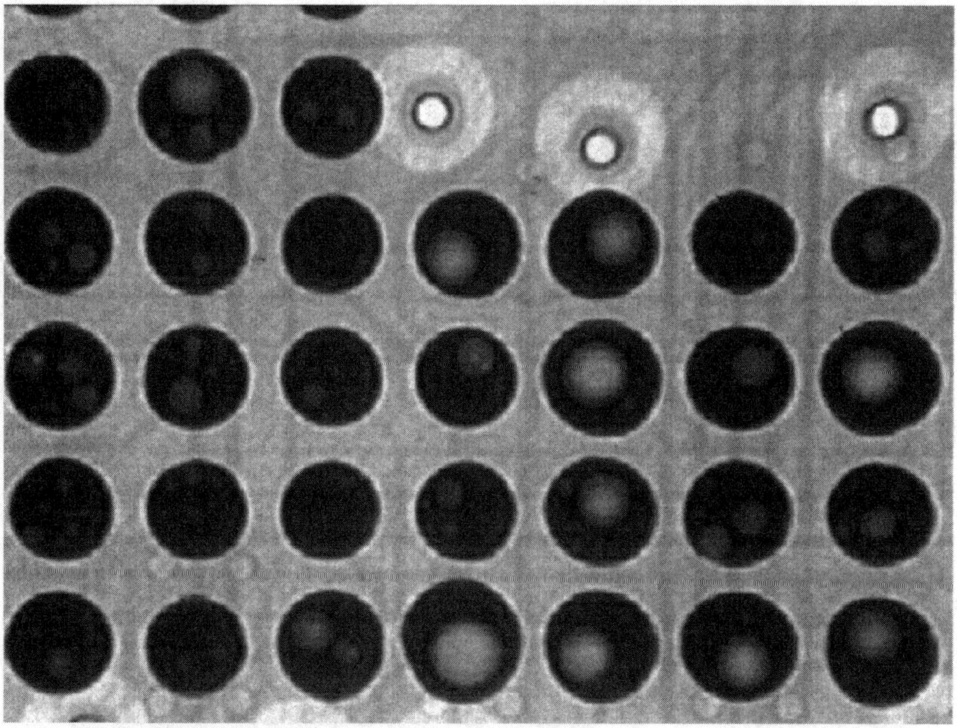

FIGURE 3.32 Voids in solder balls. (*Glenbrook Technologies.*)

- PWB or component warpage
- Insufficient solder paste or flux deposit
- Missing solder balls on the component

3.6 PWB DESIGN GUIDELINES FOR REWORK

As manufacturers of rework machines, we find it unfortunate that in most cases design to facilitate easy rework is a rare occurrence. If the following rules are observed, then rework problems often can be avoided:

1. *Clearance for reflow nozzles.* If some clearance is left around the perimeter of the device, it will be much easier to locate the nozzle and shroud the component completely. This will result in better ΔT characteristics and minimal adjacent reflow. A suggested keep-out zone would be 80 mils. Where a product demands a tightly packed board, this may not be possible. In this case, the engineer will have to accept adjacent reflow as a consequence.
2. *Clearances for stencils.* If you intend to print solder paste in rework, it is imperative that adequate clearance is allowed for the stencil. A suggested keep-out zone of 100 mils normally would be acceptable. If this is not provided, it may mean that the operator has to remove surrounding components to facilitate printing.

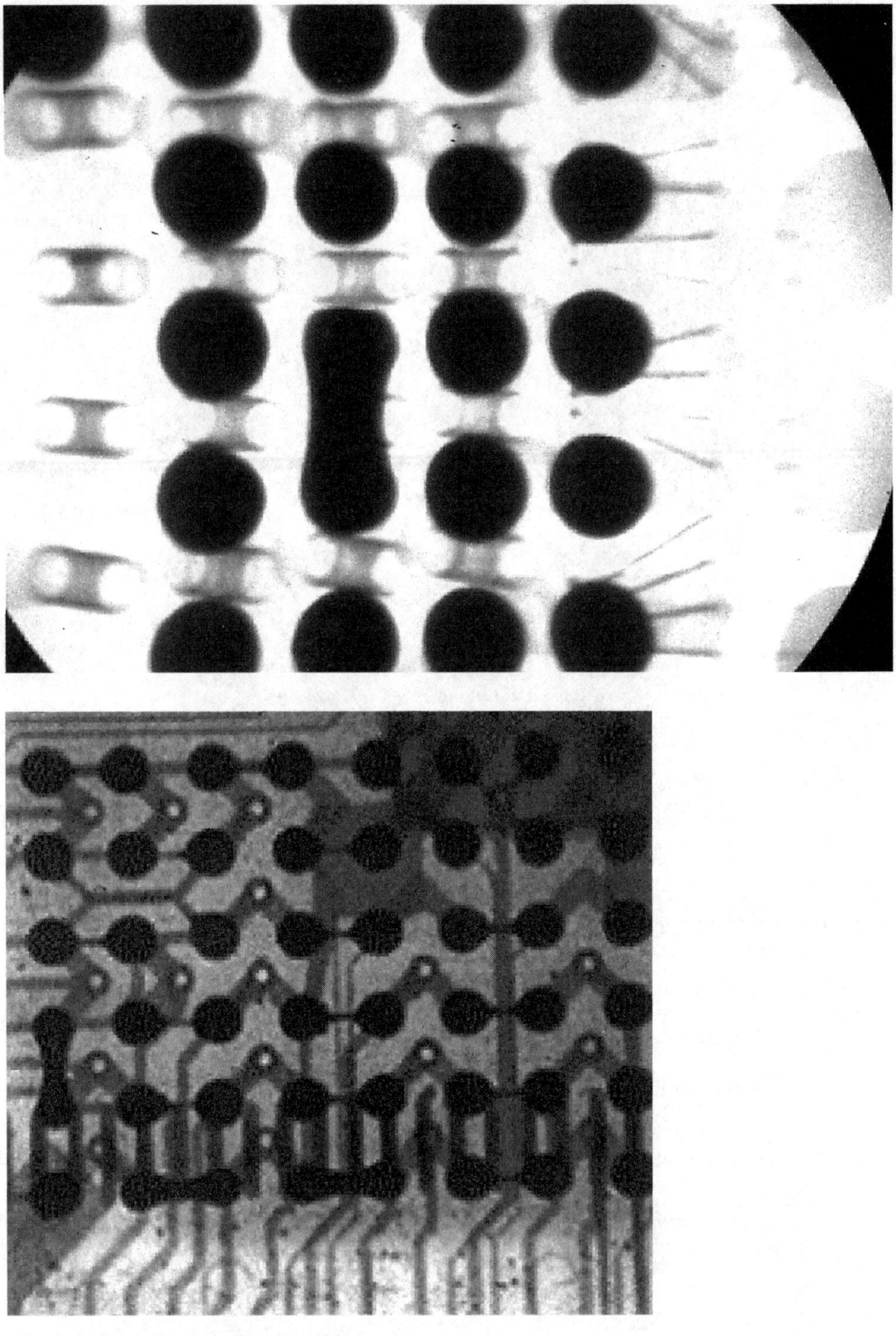

FIGURE 3.33 Solder shorts. (*Glenbrook Technologies.*)

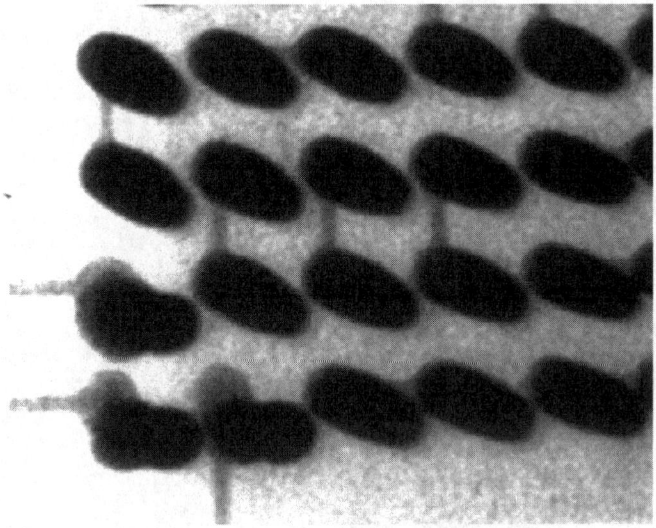

FIGURE 3.34 Formation of an open circuit (3D view).

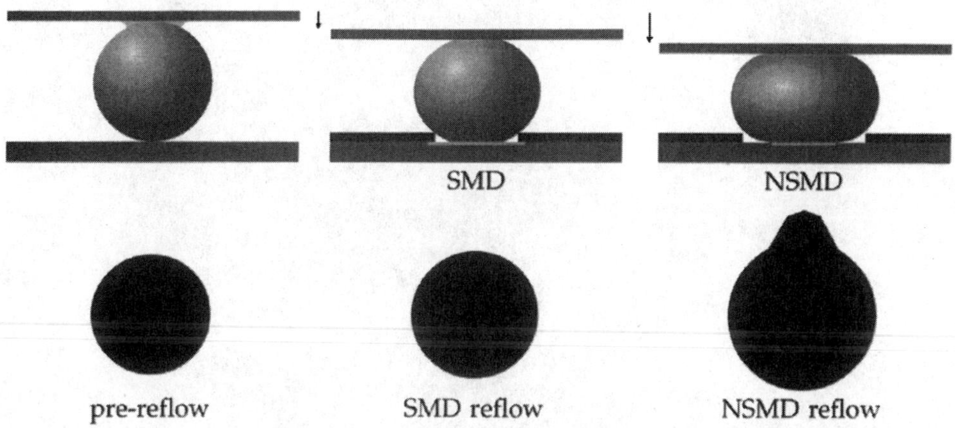

FIGURE 3.35 Pad solder mask definition can facilitate inspection. (*Glenbrook Technologies.*)

3. *Pad design.* As stated previously, pad design can affect wetting and inspection characteristics. It also can be of benefit when removing nonmelt solder ball components. If the pad on the board is smaller than the pad on the component, then the solder balls will tend to adhere to the component during removal, making the cleaning operation easier (Fig. 3.35).
4. *Via holes.* Open via holes in pads will cause migration of solder and low standoff heights after reflow.
5. *Positions of components.* The position of an array component may influence the ΔT characteristics during reflow. It may be a problem to position a large array at the edge of a PWB because simultaneous reflow will be impractical.

CHAPTER 4
BGA ASSEMBLY RELIABILITY

Raza Ghaffarian
Jet Propulsion Laboratory

4.0 INTRODUCTION

The dominant trend within electronics today is the drive toward increasingly smaller feature sizes on integrated circuit (IC) chips. Ten years ago, feature sizes of 1 μm were the norm; by the mid-1990s, 0.35 μm was the norm, and research and development ICs are at 0.25 μm and under. This has led to a drastic increase in the number of input-outputs (I/Os) on the chip, particularly for application-specific ICs (ASICs). Indeed, 300 to 500 I/Os per chip are no longer uncommon, and now ASICs having up to and higher than 1000 I/Os are under development. Figure 4.1 shows different surface-mount components from plated through-hole (PHT) to chip scale packages (CSPs) in the trend toward miniaturization.

This great increase in I/Os pushes on the packaging engineer to arrive at a suitable way of interconnecting the IC to a printed wiring board (PWB) or to a ceramic substrate. Mounting on an organic substrate, typically referred to as a PWB, is much less expensive than mounting on a ceramic substrate. Concomitant with the increase in I/Os, or pins, is the drive to achieve higher component densities, thereby increasing board functionality. The current trend in electronic packaging is

- Increasing pin count
- Increasing component density (more functionality per unit area)
- Lighter weight

Several different approaches are being employed to meet the challenge of mounting high-pin-count ICs to substrates and packages on PWBs. These approaches can be considered a form of direct chip attach (DCA), flip chip (FC), and use of package ICs rather the use of bare IC dies. Each of these approaches has its own merits and drawbacks.

- Attach the IC die to the bare PWB and wire bond from the die bonding pads directly to bonding pads of the PWB. This is commonly referred to as *chip-on-board* (COB) or *chip-and-wire DCA*. ICs can be mounted first on polyimide tape, and then the ICs can be gang bonded to the PWB using the hot-bar soldering technique. This is commonly referred to as *tape automated bonding* (TAB) or *tape carrier package* (TCP) *DCA*.
- Permanently attach small solder bumps to the bottom of the IC die, flip it over, and then mount it onto suitable mounting pads on the PWB. This is commonly referred to as *direct flip chip* (FC).
- Mount the IC internally, wire bond or FC, on a flexible/rigid organic or ceramic substrate and package the chip into a suitable package material. The package then has small solder bumps applied to the bottom, is flipped over, and mounted onto suitable mounting pads on the PWB. This is commonly referred to as *ball grid array* (BGA) technology when the ball pitches (distance

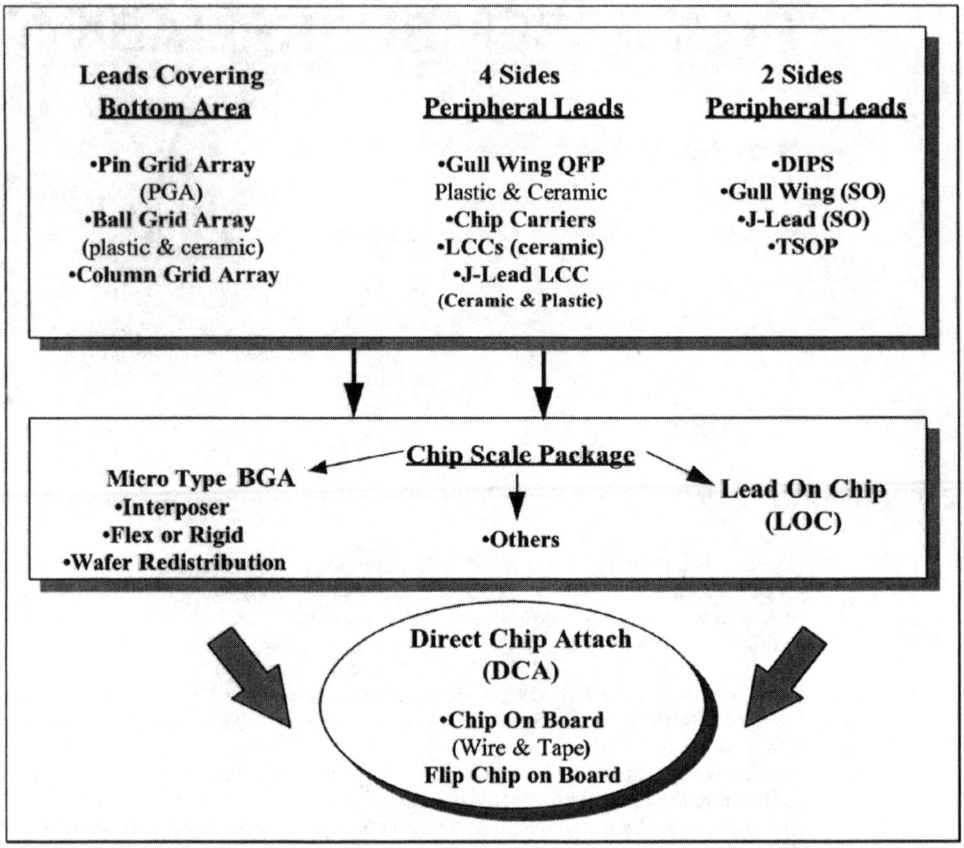

FIGURE 4.1 Electronics package miniaturization trends.

between two adjacent ball centers) equal 1 mm and higher, normally 1.27 mm. If the package dimensions are very nearly those of the IC itself, this technology is being called *chip scale packaging* (CSP) with pitches equal to or less than 0.8 mm. The principal advantage of BGA and CSP is their combination of a package to protect the IC and their close similarity to FC. Figure 4.2 shows a comparison of a COB component with a CSP.

Each approach has, of course, its own unique set of advantages and drawbacks. Many technical advances in materials now allow the use of these technologies in more demanding environments requiring higher reliability.

4.1 BALL GRID ARRAY (BGA)

BGA combines the best features of FC but ensures that the die is fully enclosed by a suitable packaging materiel to protect it from hostile environments. BGAs are the package analogue of FC technology (controlled collapse chip connection, or C4) developed by IBM for denser interconnection at the IC packaging level. The typical pitch for these is 1.27 mm (0.050 in).

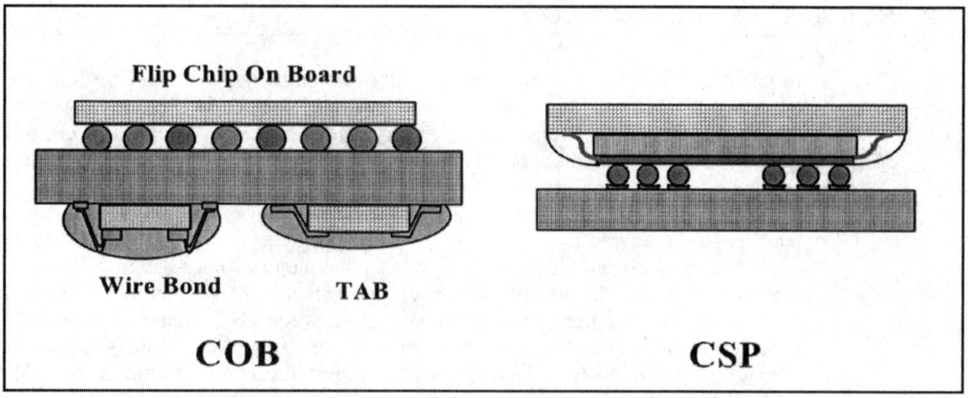

FIGURE 4.2 Comparison of a COB and a CSP.

Because the IC is mounted within a suitable packaging material, the chip can be burned in as any conventional package. In addition, the package provides direct protection of the IC, and it can be made to provide hermeticity if necessary. Because the package is an area array, it uses less PWB real estate than peripheral array-mounted packages such as the quad flatpak (QFP). Generally, an underfill is not used. On the other hand, conventional QFPs, with their gullwing leads, consume considerable board real estate.

BGAs offer several distinct advantages over FP (fine pitch) and UFP (ultrafine pitch) surface-mount components (SMCs) having gullwing leads, including

- BGAs are capable of high pin counts, generally more than 200.
- BGAs have larger lead pitches, which significantly reduces the manufacturing complexities for high-I/O parts.
- Higher packaging densities are achievable because the lead envelope for the gullwing leads is not applicable in the case of BGAs; thus it is possible to mount more packages per board.
- BGAs have faster circuitry speed than gullwing SMCs because the terminations are much shorter.
- BGAs have better heat dissipation than gullwing SMCs.
- Conventional surface-mount technology (SMT) production techniques such as stencil printing and component mounting can be employed.

BGAs are also robust in processing. This stems from their higher pitch (1.27 mm, 0.050 in, typical), better lead rigidity, and self-alignment characteristics during reflow processing. This latter feature, self-alignment during reflow, is very beneficial and opens the process window considerably.

BGAs, however, are not compatible with multiple solder processing methods, and individual solder joints cannot be inspected and reworked using conventional methods. In high-reliability applications, inspection of solder joints visually has been a standard practice and is a key factor for providing confidence in solder-joint reliability. Advanced inspection techniques including x-rays need to be further developed to provide such confidence for BGAs and grid CSPs.

To summarize, the three main drawbacks of both FCs and BGAs are similar:

- Lack of direct visual inspectability
- Lack of individual solder-joint reworkability
- Interconnect routing between the chip and the PWB necessitating a multilayer PWB

4.2 BGA AND QFP COMPARISON

Figure 4.3 shows a size comparison of a 256-pin gullwing QFP and a 256-pin BGA. In this figure, the top of the 256-pin QFP and the bottom of the 256-pin BGA are shown. A reduction by 2.5 times can be achieved when comparing BGA parts with QFP parts. When considering PWB area requirements, the area required on the PWB for solder lands is included. For a QFP, this area would be calculated by measuring from lead tip to lead tip of opposing sides.

The 256-pin BGA shown is a plastic package with a peripheral solder-ball array. When BGA packages first entered the market, they were configured in a full solder-ball array layout. Thermal cycling tests of various package sizes showed that the full solder-ball array layout was susceptible to high stresses under the area of the die periphery resulting in premature failure of the solder joints in this area. Also, the fatigue life of the solder spheres for a BGA part is directly related to the die size used. These high stresses are caused by the constraining effect of the semiconductor die (typically silicon) on the laminate part package. The highest stresses are found in the solder spheres located under the corners of the silicon device.

A silicon die has an approximate coefficient of thermal expansion (CTE) of 2 to 3 ppm/°C, whereas the plastic part package has a CTE of approximately 17 to 20 ppm/°C. The silicon die strains the part laminate material, which has a close CTE match with the board material. To aid in reducing the stresses induced under the semiconductor die, peripheral solder-ball array BGA parts were developed. In the peripheral solder-ball array parts, the solder spheres are placed around the periphery of the package, away from the die edges. This reportedly results in better reliability performance during thermal cycling.

Figure 4.4 shows the difference in pitch for the 256-pin QFP and the 256-pin BGA. In the figure, the BGA has been placed directly on top of the QFP. The QFP requires a pitch of 0.020 in to achieve a lead count of 256, whereas the BGA part requires 0.050 in to achieve this same lead count. The greater pitch of the BGA relative to the equivalent QFP package allows for a far more robust

FIGURE 4.3 Comparison of a 256-I/O plastic BGA and 256-I/O ceramic QFP.

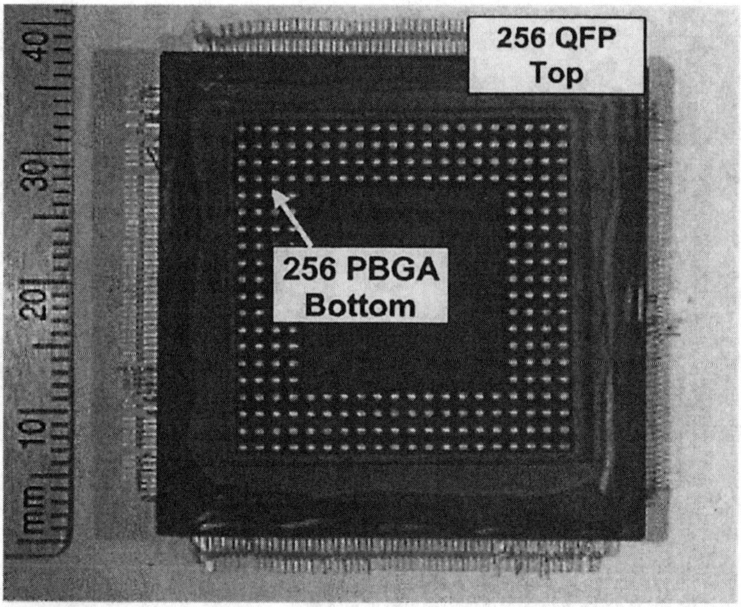

FIGURE 4.4 Pitch difference for BGAs (1.27 mm) and QFPs (0.508 mm), same I/O count.

manufacturing cycle. Solder shorting and assembly part placement issues are substantially reduced with BGA packages.

Significant differences also exist between fine-pitch (0.020 in) QFPs and BGA parts in manufacturing processing. QFPs require lead forming, lead degolding, lead tinning, coplanarity inspection, coplanarity adjustment, specialized packaging and handling procedures, and specialized PWB placement processes. BGAs require package flatness inspection and solder sphere planarity inspections (both part types require PWB flatness verification). The handling and transport of QFPs are extremely critical because the leads are damaged very easily, with skew and nonplanarity resulting. BGA leads are very robust when handled with reasonable care.

The downside to BGAs for high-reliability applications is the inability to inspect all solder joints. Since conventional visual inspection techniques are not adequate for BGA solder joints, process and material controls combined with simple verification techniques following assembly are required to be performed to ensure the quality of solder-joint interconnections.

4.3 RELIABILITY OF BGA ASSEMBLY BY JPL-LED BGA CONSORTIUM

Reliability of BGAs was assessed as part of the Jet Propulsion Laboratory (JPL) BGA consortium effort. Nearly 200 test vehicles, each with four packages, were assembled and tested using an experimental design (Figs. 4.5 and 4.6). The most critical variables incorporated in the experiment were package types (ceramic and plastic), board materials (FR-4 and polyimide), surface finishes (OSP, HASL, and Ni/Au), solder volumes (low, standard, and high), and environmental conditions.[1–3]

The two test vehicle assembly types were plastic (PBGAs) and ceramic (CBGAs) packages. Both FR-4 and polyimide PWBs with six layers 1.57 mm thick (0.062 in) were used.

Plastic packages covered the range from overmolded pad array carrier (OMPAC) to superBGAs (SBGAs). These were (Table 4.1):

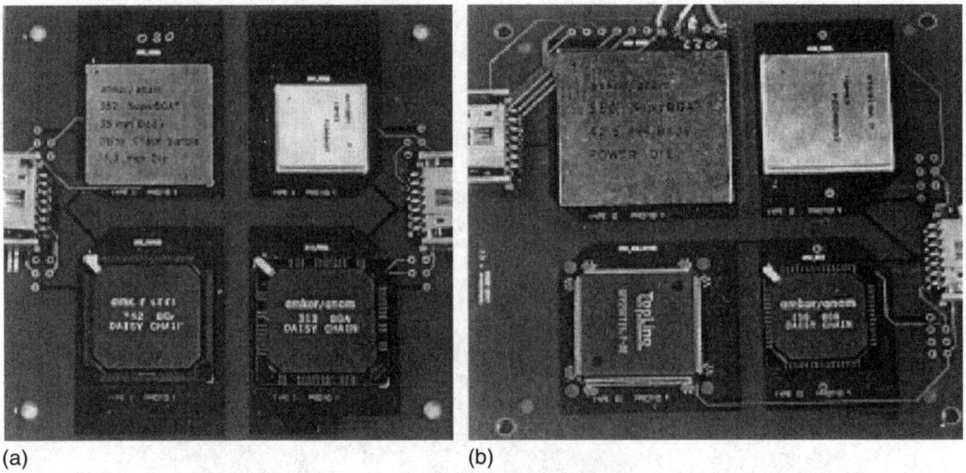

FIGURE 4.5 Test vehicles: (*a*) type 1 (300 I/Os) and (*b*) type 2 (600 I/Os).

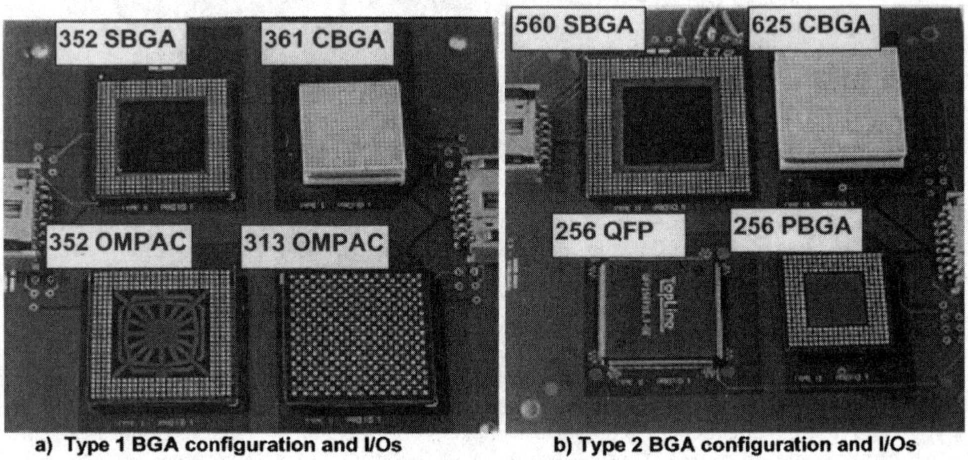

FIGURE 4.6 Test vehicles with unassembled parts topping assembled ones: (*a*) type 1 BGA configuration and I/Os; (*b*) type 2 BGA configuration and I/Os.

TABLE 4.1 Packaging Characteristics

Part type	I/O	Materials configurations	Size (mm^2)	Wiring	Pitch	Selected die size
PBGA 300	352	Plastic/peripheral/superBGA	35	Daisy	1.27	13.3
PBGA 300	352	Plastic/peripheral/OMPAC	35	Daisy	1.27	13.3
CBGA 300	361	Ceramic/full array	25	Daisy	1.27	
PBGA 300	313	Plastic/full array/OMPAC	35	Daisy	1.27	13.3
PBGA 600	560	Plastic/peripheral/superBGA	42	Daisy	1.27	15.25
CBGA 600	625	Ceramic/full array	32.5	Daisy	1.27	
PBGA 300	256	Plastic/peripheral	27	Daisy	1.27	10.8
GW	256	Plastic/gull wing	30.6	Daisy	0.4	

- Two peripheral SBGAs, 352 and 560 I/Os
- Peripheral OMPAC, 352 I/Os, PBGA, 352 and 256 I/Os
- Depopulated full-array PBGA, 313 I/Os
- 256-I/O QFPs, 0.4 mm pitch

In SBGAs, the IC die is attached directly to an oversized copper plate, providing better heat dissipation efficiency than standard PBGAs. The solder balls for plastic packages were eutectic (63Sn/37Pb).

Ceramic packages with 625 and 361 I/Os also were included in my evaluation. Ceramic packages had high-melting solder balls with a composition of 90Pb/10Sn and 0.035-in diameters. These balls were attached to the ceramic substrate with eutectic solder (63Sn/37Pb). During BGA assembly, package-side eutectic solder and the PWB-side eutectic paste are reflowed to provide the electromechanical interconnects.

Plastic packages had dummy dies and daisy chains, with the daisy chains on the PWB designed to complete the package chain loop to be able to monitor critical solder-joint regions. Most packages had four daisy-chain patterns, the 560 I/O package had five, and the QFP had one.

Full assembly was implemented after process optimization from the trial test. The following procedures were followed:

- PWBs were baked at 125°C for 4 h prior to screen printing.
- Two types of solder pastes were used, a rosin mild activation (RMA) and a water-soluble one.
- Pastes were screen printed, and the heights were measured by laser profilometer. Three levels of paste were included in the evaluation: standard, high, and low. Stencils were stepped to 50 percent to accommodate assembling ceramic, plastic, and fine-pitch QFP packages in the type 2 test vehicle.
- A 10-zone convection oven was used for reflowing.
- The first assembled test vehicle (TV) using an RMA reflow process was inspected visually and x-rayed to check solder-joint quality.
- All assemblies were x-rayed.

Two test vehicles were assembled:

- Type 1, ceramic and plastic BGA packages with nearly 300 I/Os.
- Type 2, ceramic and plastic BGA packages with nearly 600 I/Os. Also used were a 256-I/O leaded ceramic and a 256-I/O plastic BGA package for evaluating and directly comparing manufacturing robustness and reliability.
- Assemblies with water-soluble flux were cleaned using an Electrovert H500. Those with RMAs were cleaned used isopropyl alcohol (IPA) and a 5% saponifier.
- All fine-pitch QFPs had to be reworked for bridges.

4.4 BGA THERMAL CYCLING

For cost-effectiveness and engineering efficiency considerations, it was desirable to have the shortest possible test period. However, adequate high-temperature dwells are necessary for solder creep and relaxation to occur. On the other hand, excessively high-temperature dwells may cause excessive aging, resulting in brittle fracture. The ramping rates of heating and cooling usually are limited by the capacity of the environmental test chamber.

Four different thermal cycle profiles were used (Figs. 4.7 through 4.10):

- *Cycle A.* The cycle A condition ranged from −30 to 100°C and had an increase/decrease heating rate of 2 to 5°C/min and dwell of about 20 min at the high temperature to ensure near-complete creeping. The duration of each cycle was 82 min.

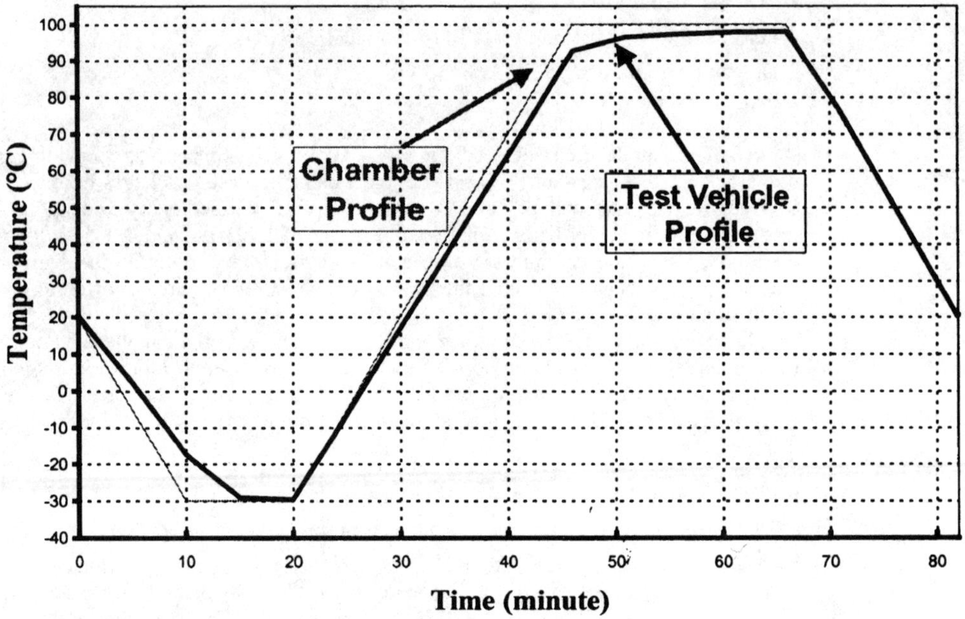

FIGURE 4.7 JPL thermal cycle profile (−30–100°C)—cycle A.

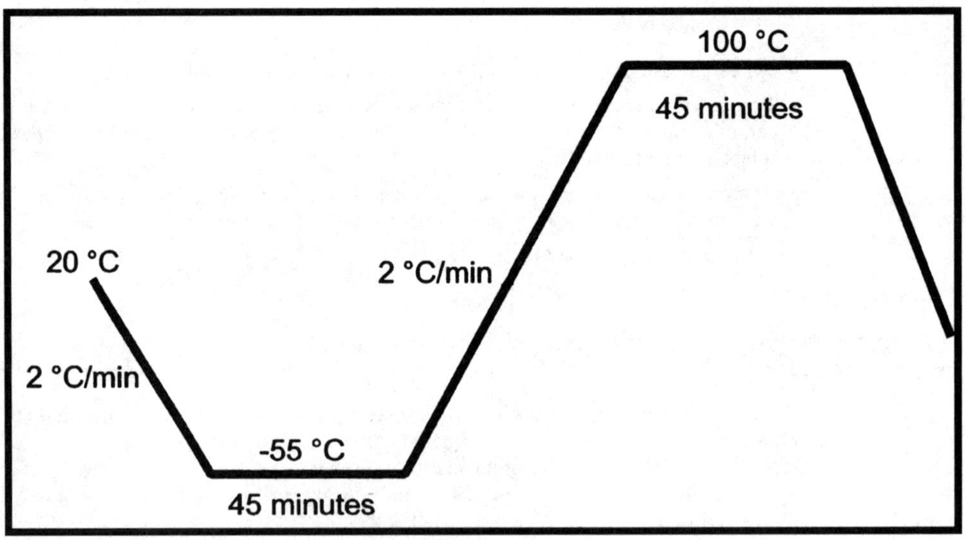

FIGURE 4.8 JPL thermal cycle profile (−55–100°C)—cycle B.

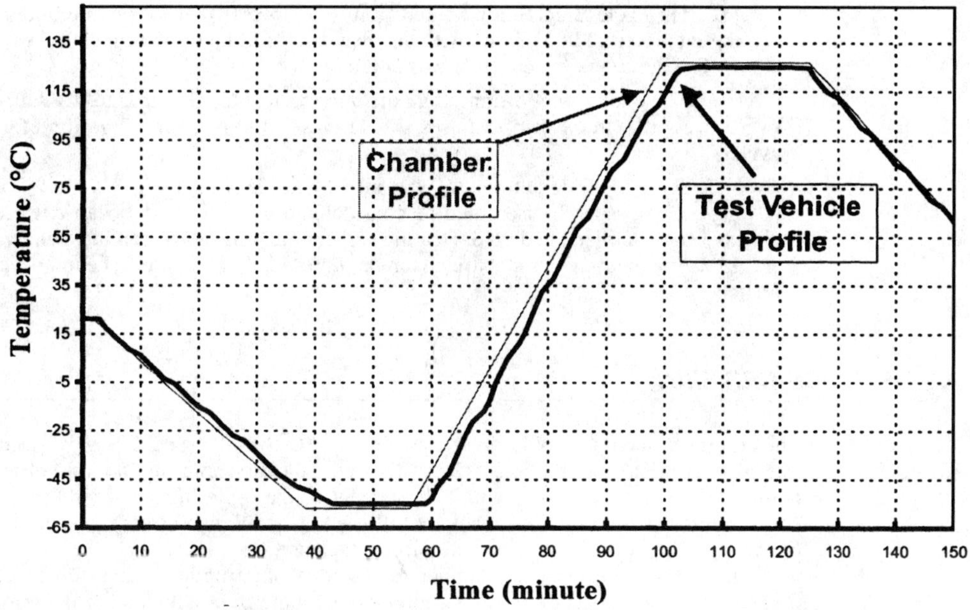

FIGURE 4.9 JPL thermal cycle profile (−55 to 125°C)—cycle C.

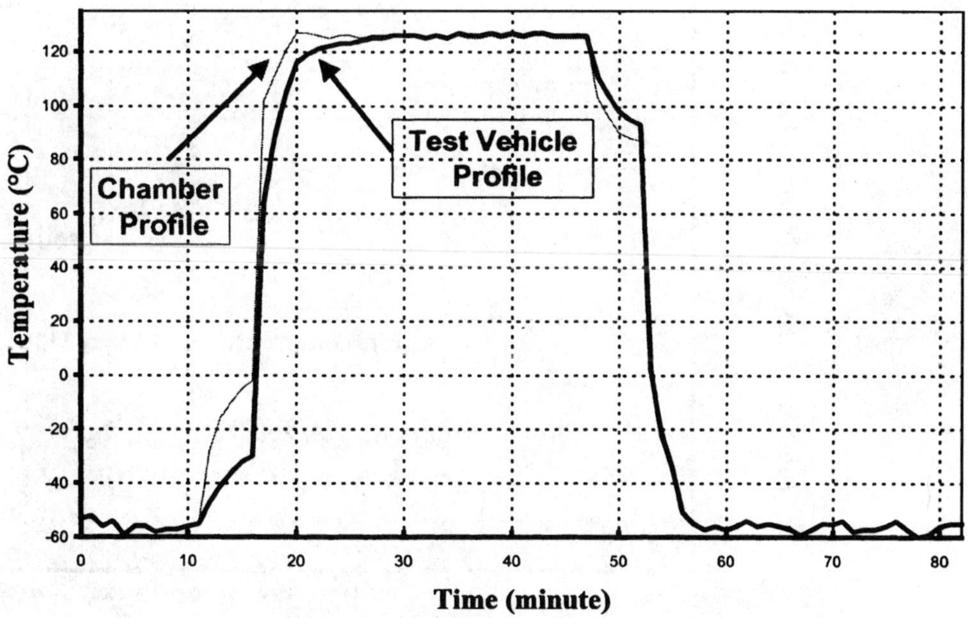

FIGURE 4.10 Boeing thermal cycle profile (−55 to 125°C)—cycle D.

- *Cycle B.* The cycle B condition ranged from −55 to 100°C with long duration. The heating and cooling rates were 2 to 5°C/min with an oven dwell setting of 45 min at the two extreme temperatures. The duration of each cycle was 246 min.
- *Cycle C.* The cycle C condition ranged from −55 to 125°C with 2 to 5°C/min heating/cooling rate. Dwell at extreme temperatures was at least 10 min, with a duration of 159 min for each cycle.
- *Cycle D.* The cycle D condition ranged from −55 to 125°C, the same as cycle C but with very high heating/cooling rates. It also could be considered a thermal shock because it used a three-region chamber: hot, ambient, and cold. Heating and cooling rates were nonlinear and varied between 10 and 15°C/min with dwells at extreme temperatures of about 20 min. The total cycle lasted approximately 68 min.

4.5 FAILURE CRITERIA

The criteria for an open solder joint specified in IPC-SM-785, Sec. 7.8, were used as guidelines to interpret electrical interruptions. Generally, once the first interruption was observed, there were many additional interruptions within 10 percent of the cycle life. In several instances, a few nonconsecutive early interruptions were not followed by additional interruptions until much later stages of cycling. This was found more with plastic packages.

Figure 4.11 plots the total number of electrical interruptions (daisy-chain opens) versus the number of cycles for the cycle D condition. For ceramic BGAs, the rate of interruptions (slope) was

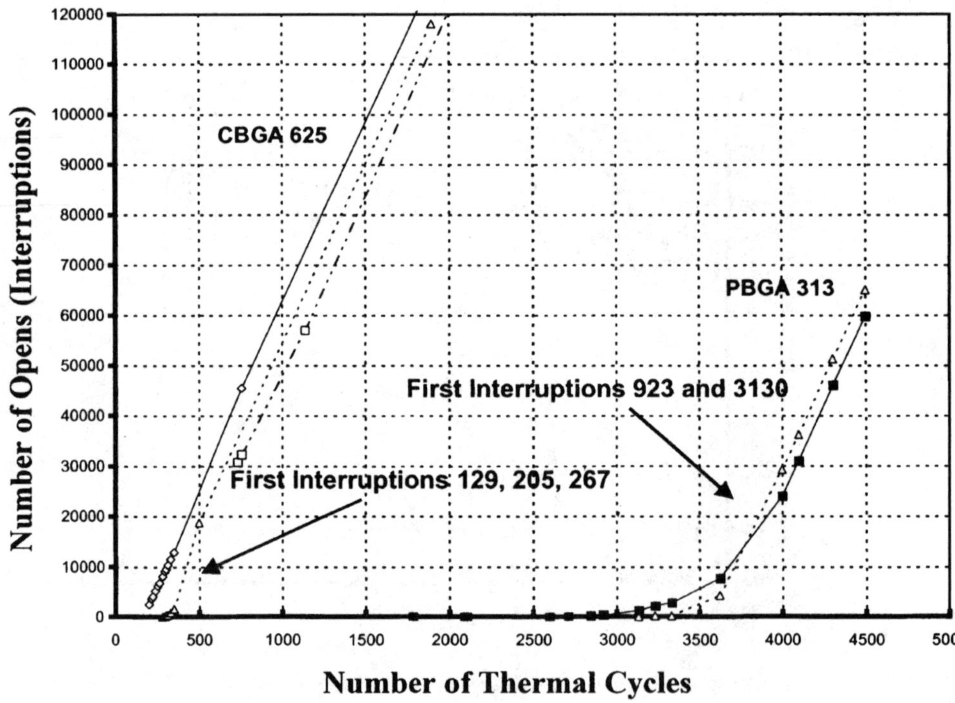

FIGURE 4.11 Number of electrical interruptions detected with thermal cycles (−55 to 125°C) for ceramic and plastic BGAs.

approximately constant. This was not the case for the plastic BGAs. For a PBGA with 313 I/Os, the first interruption was at 923 cycles, with no additional interruptions until about 3000 cycles. Further erratic interruptions were observed between 3000 and 3700 cycles, followed by linear interruptions having approximately the same slope as the CBGAs. Another PBGA showed the first interruption at 3130 cycles and become open at about 3300 cycles.

4.6 CBGA FAILURE MECHANISMS

4.6.1 Damage Monitoring

For conventional SMT with peripheral interconnections, the pass/fail criteria for high-reliability applications rely on visual inspection of solder joints at 10× to 50× magnifications. For the BGAs on the test vehicles, only edge balls, those not blocked by other components, were inspected visually. A series of single assemblies that were cut from the test vehicles also were used for both visual and scanning electron microscope (SEM) inspection to better define visual criteria for acceptance of solder joints as well as to monitor damage progress under different cycling environments.

4.6.2 Failure Mechanisms for Ceramic Assemblies

Figure 4.12 shows representative SEM and cross-sectional micrographs for a CBGA 625-I/O package after 348 cycles of cycle A conditions. Cross-sectional photos are for the perimeter balls, two corners, and a center ball. SEM photos for package and board sides of a corner (maximum distance to neutral point, or DNP) and the edge center ball are also included. Note the large and small pin holes in the eutectic solder of the corner pin near the board site (lower right SEM photo). Generally, pin holes formed prior to crack initiation.

Figure 4.13 compares the corner solder joint of Fig. 4.6 and a typical failure photo for the B cycling conditions for a CBGA 361-I/O package after 350 cycles. Note that the failures for the cycle A

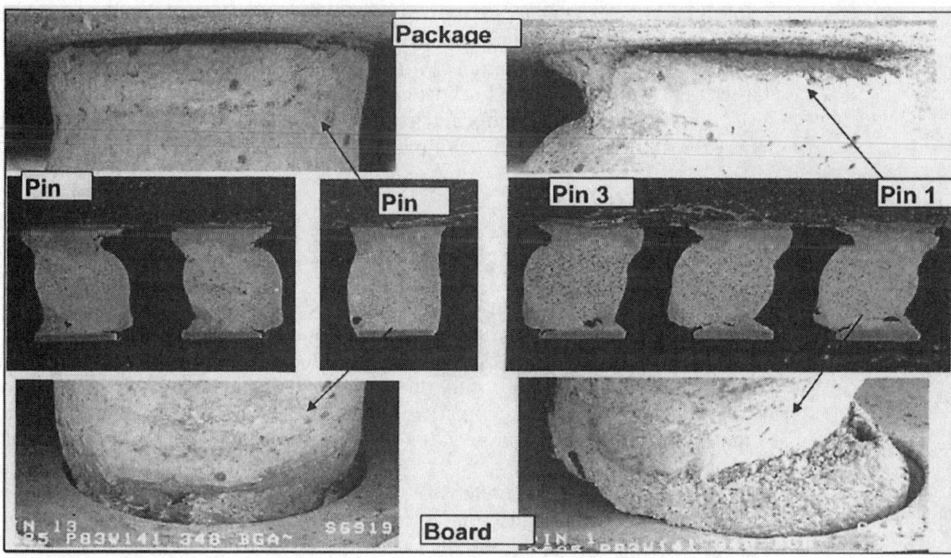

FIGURE 4.12 SEM and cross-sectional SEM photographs of CBGA 625 after 348 cycles (−30 to 100°C). Balls (pins) 1 and 25 are the corner balls with the maximum DNP, and ball 13 is the center of the edge balls.

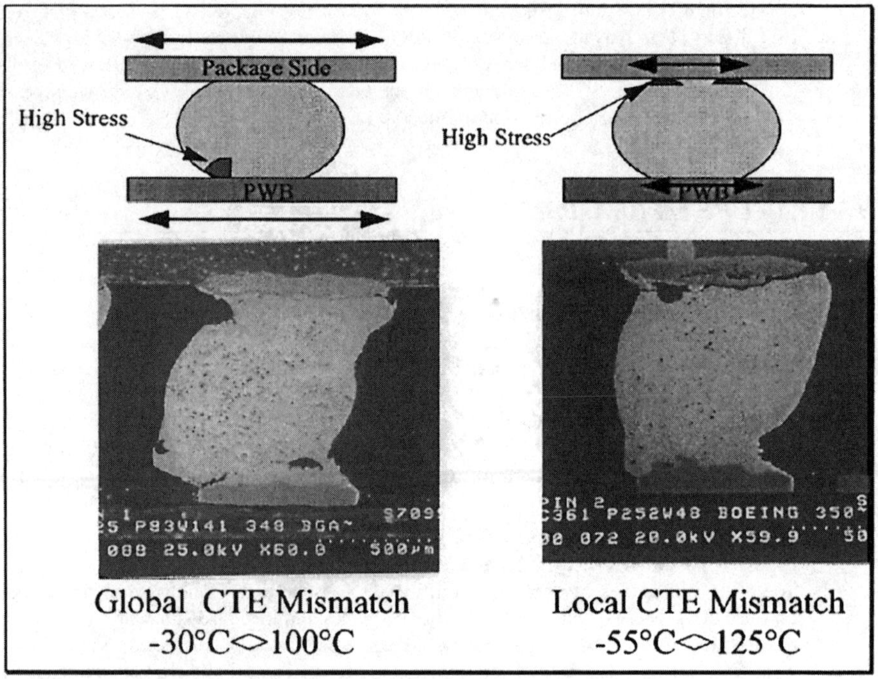

FIGURE 4.13 Cross sections of failure sites for CBGA 625 after 350 cycles under cycle A and D conditions.

conditions were from the PWB and for the cycle B conditions were from the package sites. These were the typical failure mechanisms for the two conditions and are similar to those reported in literature.[4]

Failure mechanism differences could be explained by global or local stress conditions. Modeling indicates that the high-stress regions shifted from the board to the packages themselves when stress conditions changed from global to local. For A cycling with slow heating/cooling ramping, which allowed the system to reach uniform temperature, damages could indicate a global stress condition. For D cycling with rapid heating/cooling, damage could indicate a local stress condition.

4.7 PBGA FAILURE MECHANISMS

Among the plastic packages, the PBGA 313-I/O package with depopulated full-array balls was first to fail under both D and A thermal cycling conditions. Figure 4.14 shows various cross sections of this package's balls from corner to center at 4682 A cycles. These photographs are of the balls under the die, where most damage occurs due to local CTE mismatch. Photographs with and without voids also were included for comparison. Voids appear to have concentrated at the package interface under the die. Cracking propagation occurred at package or board interfaces for sections with or without voids.

The sections with voids were opens, indicated by seepage of mounting materials into the voids. Except for the interface-connecting cracks, there appeared to be no crack propagation among the voids. The solder joints for this package also were inspected visually to identify indicators of defects. None were apparent. Even when these were inspected at 100× by SEM, no gross damage was observed, as evidenced by the top left SEM photograph of Fig. 4.14.

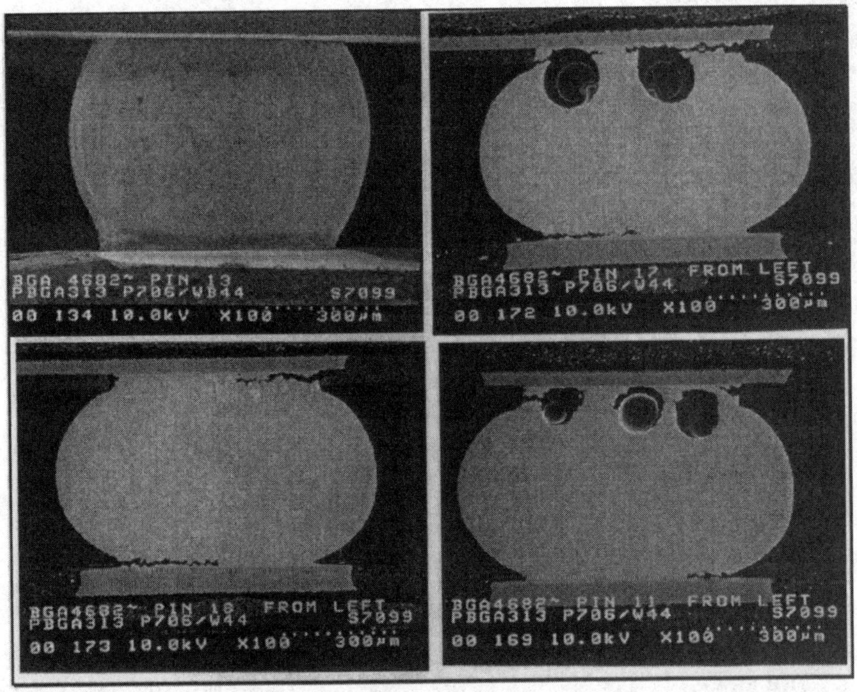

FIGURE 4.14 SEM and cross sections of PBGA 313-I/O package after 4682 cycles under cycle A conditions (−30 to 100°C).

Package attachment become very weak after 4500 D cycles, and the package could be detached easily from the board. Packages were detached, and images of balls that remained on package were mapped. The ball map distributions for several detached PBGA 313-I/O packages are shown in Fig. 4.15. All joint failures were not from the board interface, and there appeared to be a trend in their distribution.

For this depopulated full-array package, most balls were separated from the package interface, as evidenced by the low number of dark spots that represents the balls on the package. This is even more noticeable for the balls under the die. More joint failures at the package under the die portion could have occurred from the extreme local CTE mismatch of these joints. Similarly, mapping for other plastic packages was performed, and trends are being studied. For ceramic packages, mapping was not possible because most of the packages were already separated from the board at 4500 cycles. Failures were mostly at the package interface, since detached packages generally had no balls on them. In addition, several test vehicle boards had a large number of balls remaining at the ceramic package locations.

Figure 4.15 shows cycles to first failure for four daisy chains of PBGA 313. The chains were for balls under the die, the edge die row, outer/inner rows, and outer rows. As expected, the majority of the cycles to first failure were from the inner row, those under the die. Those which lasted longest were from the outer rows. There were overlaps, however, in distribution, and distinctions between sites of failure cannot be clearly defined. Similar plots also were generated for other packages.

For SBGA 352, package daisy chains were divided equally into four regions, each covering a corner of the package. For this package, cycles to first failure were random from any of the four daisy chains. For PBGA 256 and SBGA 560, both with peripheral-array configuration, it appeared that cycles to first failure were not from the inner to outer rows.

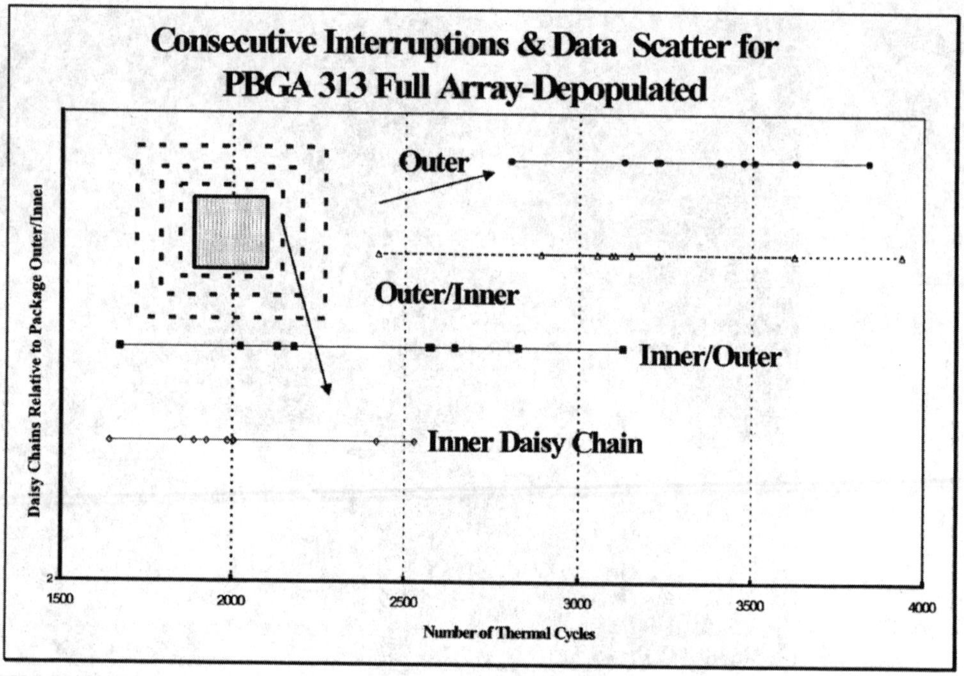

FIGURE 4.15 Cycles to first failure for inner to outer daisy chains of PBGA 313.

4.8 PBGA THERMAL CYCLING RESULTS

Figure 4.16 shows cycles to first failure for all plastic packages: PBGA 313 and 256, OMPAC 352, and SBGA 352 and 560. The results are for assemblies on FR-4 and polyimide PWBs. Preliminary results for two assemblies showed generally higher cycles to first failure for polyimide under cycle B conditions. The most recent PBGA 313 assemblies that failed under cycle A conditions are also included in the plots for comparison.

These plots were generated by ranking cycles to first failures from low to high and then approximating the failure-distribution percentiles using a median plotting position, $F_i = (i - 0.3)/(n + 0.4)$,[5] where i is the ranking number and N is the total of test vehicles. No attempts were made to generate Weibull parameters because of sample size and wide distribution.

Several trends were identified. PBGA 313, with a depopulated full array with solder balls under the die, was the first plastic package to show signs of failure at both cycling conditions. The scatter for both was narrow, indicating a large m Weibull parameter.

The next package assembly failure was SBGA 560. This package and the PBGA 256 with failure data at the extreme cycles showed very large scatter, indicating a low m value. Scatter in PBGAs, in general, could be attributed to slow failure progress. Scatter in failure data for SBGA 560 could be due to the immaturity of the package at the time of use. These packages were taken from those assembled in the early stage of production. The reasons for a large scatter for the PBGA 256 cycles to first failure data are not known at this time. Both OMPAC and SBGA 352 failed at later cycles than SBGA 560, and SBGA 560 was between a low tail for PBGA 313 and a high tail for PBGA 256. The 352-I/O package showed low scatter in cycles to first failure data.

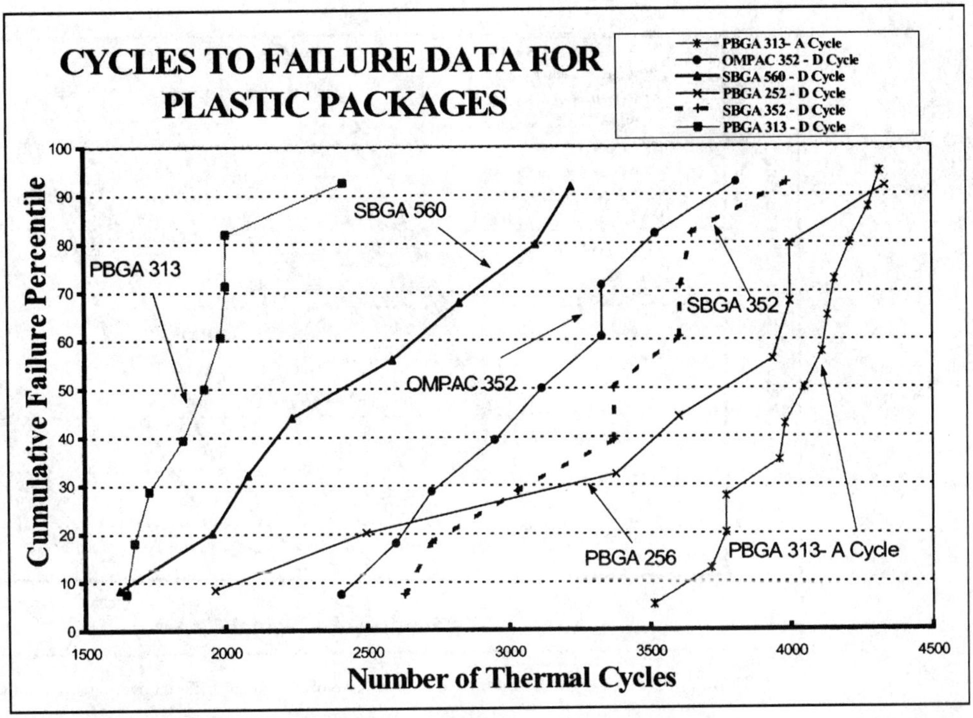

FIGURE 4.16 Cumulative cycling to first failure for PBGA assemblies (cycle B, −55 to 125°C) and PBGA 313 (cycle A, −30 to 100°C).

4.9 CBGA THERMAL CYCLING RESULTS

Figure 4.17 shows cycles to first failure for CBGA 625 under four different thermal cycling conditions. Again, to generate plots, the cycles to first failure were ranked from low to high, and failure distribution percentiles were approximated using a median plotting position, $F_i = (i - 0.3)/(n + 0.4)$.

Weibull parameters for cycles to first failure also were generated and plotted in continuous graphs. Weibull parameters for the four conditions are listed in Table 4.2. In addition to cycling conditions, it includes manufacturing and assembly variables that possibly affected the m values.

The two-parameter Weibull distribution was used to characterize failure distribution. The Weibull cumulative failure distribution was used to fit cycles to failure data. The equation is

$$F(N) = 1 - \exp[-(N/No)^m]$$

where $F(N)$ = cumulative failure distribution function

N = number of thermal cycles

No = scale parameter that commonly is referred to as characteristic life and is the number of thermal cycles with 63.2 percent failure occurrence

m = shape parameter—a large m is approximately inversely proportional to the coefficient of variation (CV) by $1.2/CV$; that is, as m increases, spread in cycles to failure decreases

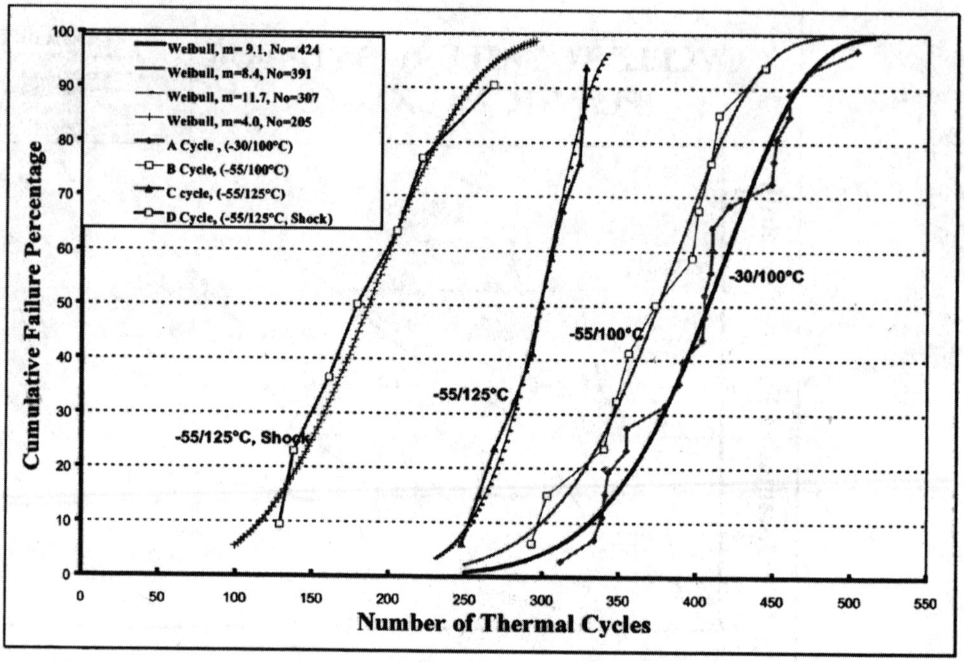

FIGURE 4.17 Cycles to first failure for CBGA 625 assemblies under different cycling conditions.

TABLE 4.2 Weibull Parameters for CBGA 625 Under Different Thermal Cycling Conditions

Cycling condition	Weibull *No* (N50%)	Weibull *m*	Comments
A (−30–100°C)	424 (407)	9.1	23 test vehicles (TVs), includes FR-4, polyimide PWB, and three solder volumes
B (−55–100°C)	391 (373)	8.4	11 TVs, FR-4 and polyimide
C (−55–125°C)	307 (298)	11.7	11 TVs, different assembler
D (−55–125°C, near thermal shock)	205 (189)	4.0	7 TVs, 3 on FR-4, 3 on polyimide, 1 with high solder volume

This equation, in double logarithm format, results in a straight line. The slope of the line will define the Weibull shape parameter.

The widest spread, i.e., $m = 4$, was observed for those assemblies cycled under condition D. The other three conditions had m values ranging from 8.4 to 11.7, with the narrowest spread, $m = 11.7$, for condition C. Near-thermal shock with two hot and cold extreme temperatures and the use of a limited number of test samples are possible causes of low values for m under condition D. Only seven test vehicles were tested: three on polyimide, three on FR-4, and one with solder volume above the norm. Test vehicles tested under condition C were manufactured at another facility under different manufacturing conditions.

4.9.1 Fatigue Equation

The Coffin-Manson relationship is a simple relationship in which cycles to first failure are inversely proportional to creep strain. Its modified version includes the effects of frequency as well as the maximum temperature in the form of

$$(N_1/N_2)\ \alpha\ (\Delta\gamma_2/\Delta\gamma_1)^{\beta}\ (f_1/f_2)^{\kappa}\exp\ [1414\ (1/T_1\ -\ 1/T_2)]$$

- N_1 and N_2 are cycles to first failure under two plastic strain conditions. β is the fatigue exponential and generally is assumed to be equal to 1.9.
- $\Delta\gamma$ is proportional to $(DNP/h)\ \Delta\alpha\Delta T$, where *DNP* is the distance from neutral point, *h* is equal to the solder-joint height, $\Delta\alpha$ is the difference in the CTE of package and PWB, and ΔT is the cycling temperature range.
- f_1 and f_2 are frequencies. κ is the frequency exponential varying from 0 to 1, with value 0 for no frequency effect and 1 for the maximum effect, which depends on materials and testing conditions. A value equal to ⅓ is commonly used to extrapolate the laboratory accelerated thermal cycles to first failure data with short duration (high frequency) to on/off field operating cycle with long duration (low frequency), i.e., a shorter field cycles-to-first-failure projection.
- T_1 and T_2 are absolute maximum temperatures under the two cycling conditions.

4.9.2 Projections

The preceding relationship was used to correlate the test results for CBGA 625 generated under the preceding four thermal cycling conditions. Cycles to 50 percent failure data are shown in Fig. 4.18. Projections to lower temperature ranges from two data points and a single data point using the preceding relationship also were plotted. Projections were compared with a test result for the same package thermal cycled in the range of 0 to 100°C (Martin et al., *SMTA International,* 1997). The following key points were considered:

- *Effect of frequency.* The effect of frequency was considered to be minor because frequency ranges in our study were very similar and their dwells at maximum temperatures were at least 10 min.

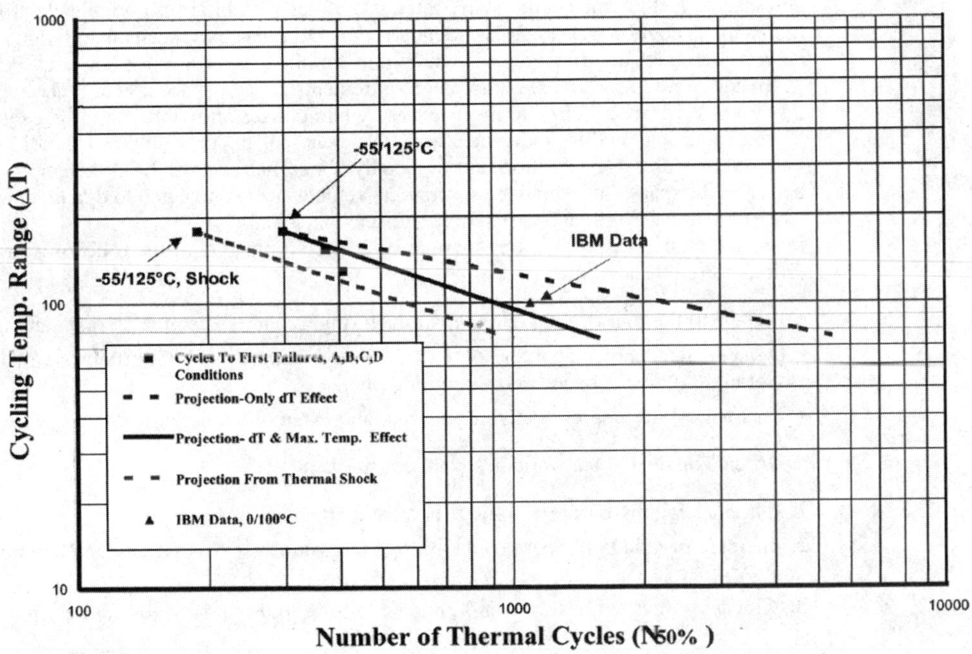

FIGURE 4.18 Effect of temperature range on 50 percentile failure cycles for CBGA 625 and projection using a modified Coffin-Manson relationship.

Creep was assumed to be almost complete for the two maximum cycling temperatures of 100 and 125°C. These temperatures are well above the creep temperature of the eutectic Pb/Sn solder.

- *Projection from cycles C and B (−55 to 125°C and −55 to 100°C) to lower cycling ranges.* The projection lines use the Coffin-Manson relationship. It includes plots from projection with and without considering the effect of maximum temperature. Projections with correction for maximum temperature better approximate the test results under conditions C and A as well as an IBM data point.
- *Projection from near-thermal shock data.* Cycles to first failure under thermal shock conditions were much lower than those under low-heating/cooling-rate cycling; therefore, projection to other temperature ranges provides conservative failure cycles. Characteristic changes in solder behavior under rapid heating and cooling as well as failure mechanism changes for thermal shock conditions may not realistically represent most field applications.
- *Projection from −55 to 100°C and −30 to 100°C to lower temperatures.* This projection is not shown in the figure, but it is conservative because the slope between the two data points is much deeper than the slope connecting data sets for the −55 to 125°C and −55 to 100°C cycles. This means that damage induced by fatigue due to a decrease in cold temperature is not as severe as creep induced by an increase in hot temperature.

4.10 SHOCK AND THERMAL CYCLING SYNERGISM EFFECTS ON RELIABILITY OF CBGAs

For electronics in commercial applications, commonly, thermal cycling tests are performed to simulate on/off conditions. However, most electronic systems are exposed to other environments, including mechanical fatigue and random vibration. For example, vibration occurs during transportation, and mechanical fatigue occurs by repeated use of key punching for portable electronics. Occasional high shocks could occur due to accidental drops. Drop test and mechanical fatigue are beginning to be considered for qualification of electronic assemblies, especially for newer CSP assemblies.

In addition to much harsher thermal requirements for aerospace and military applications, generally assemblies are required to meet severe dynamic loads and vibration fatigue cycling. Therefore, there is a strong need to understand assembly behavior under such stress conditions. Very limited data are available for vibration and especially shock behavior of BGA assemblies. To understand behavior for space environments, several test vehicles were subjected to dynamic testing representative of a launch, 3 min at 3 axis vibration/shock.

Synergistic stresses including thermal cycles representative of the following environmental conditions were considered:

- Prelaunch thermal cycling due to manufacturing, repair, screening, storage, and transportation
- Dynamic testing representative of the launch environment, including sinusoidal vibration, transient vibration, pyroshock, and acoustic vibration
- Thermal cycling representative of internal and external temperature changes of a spacecraft

Effects of several test variables were considered:

1. Thermal cycling before vibration/shock and after
2. Increase in rigidity of board by addition of thin rigid strips in one case and bonding to a rigid plate in another case
3. Change in thermal history condition by exposure to thermal cycles before vibration/shock

This section presents experimental results as well as analyses for two CBGA assemblies (361 and 625 I/Os) subjected to either or both thermal cycling and vibration.

4.10.1 Literature Review of Dynamic Behavior of BGAs

Cycles to first failure after vibration of CBGAs and column ceramic grid arrays (CCGAs) were presented by M. Cole of IBM.[6] Balls and columns were 90Pb/10Sn and were 0.89 mm (0.035 in) in diameter for CBGAs and 0.5 mm (0.02 in) in diameter for CCGAs and 2.2 mm (0.050 or 0.087 in) in height. Packages were assembled on FR-4 boards with single test specimens mounted on 110 × 90-mm boards, clamped by screws at 101 × 75-mm locations. Mil-STD 810E was used to generate impact and vibration data for the test vehicles.

No failures of CBGAs and CCGAs were observed after vibration with a heat sink of 73 g. CCGAs with heat-sink weights of 100 and 150 g failed, whereas CBGAs did not. Cracks were induced in CBGAs in the eutectic solder either in package or board sites when subjected in 20 to 2000 Hz with 7.73 g. It was reported that for CBGAs, crack initiations were similar to those of accelerated thermal cycling (ATC) but with no deformation typically present in ATC. Also, thermal mismatch induced both shear and tensile stress, but vibration induced primarily tensile stress and did not cause deformation.

It was observed that CBGA assemblies with heat sinks of lower weight than 150 g that were subjected to shock and vibration did not show any degradation in thermal fatigue life. No statistical differences between those with initial shock and vibration and those without any a priori test were observed.

In another study,[7] CBGAs with 256, 625, and 1089 I/Os were subjected to thermal cycling and vibration evaluation. The 1089-I/O package was fabricated with balls rather commonly supplied in columns because they were used on metal matrix-restraining core rather than polymeric board such as FR-4. For restrained board with a CTE (6 ppm/°C) close to the ceramic package (7.2 ppm/°C), local mismatch between solder and package/board was considered to play a primary role rather than global mismatch. Assemblies on restrained board were subjected to thermal cycling in the range of −55 to 125°C (a 5°C/min ramp and a 30-min dwell at each extreme). No failures were reported to 1500 cycles.

Vibration tests also were performed on three CBGAs at three levels:

- Level I was the military avionics equipment workmanship vibration level of environmental stress screening (ESS).
- Level II was an enveloped subassembly based on several avionics programs.
- Level III was the same as level II with power spectral density (PSD) increased by 3 dB.

Test results are listed in Table 4.3. All assemblies passed the level I, whereas only CBGAs with 254 I/Os passed the level II vibration test. Finite-element prediction for level I agreed with the experimental results but for other levels had mixed agreement.

4.10.2 Vibration

Three test vehicles with three levels of rigidity representing three loading conditions were stacked with spacers and subjected to random vibration. The bottom TV was bonded to an aluminum plate, the middle had center and edge stiffener strips, and the top had no stiffeners. The test vehicles were clamped from the two sides of the PWB, the sides with no connectors, as shown in Fig. 4.19. The block of three stacked test vehicles was mounted on a very stiff, thick aluminum plate on the shaker table with their natural frequencies well above the test vehicle assembly range.

TABLE 4.3 Vibration and Thermal Cycling for CBGAs

Vibration level	Duration (min)	Input level (g)	Response (g)	CBGA 256 I/O	CBGA 625 I/O	CBGA 1089 I/O
I	10	6.06	21	Pass (4/4)	Pass (2/2)	Pass (3/3)
II	60	12.8	37	Pass (4/4)	Pass (0/2)	Pass (0/3)
III	120	18.07	53	Pass (1/4)	N/A	N/A

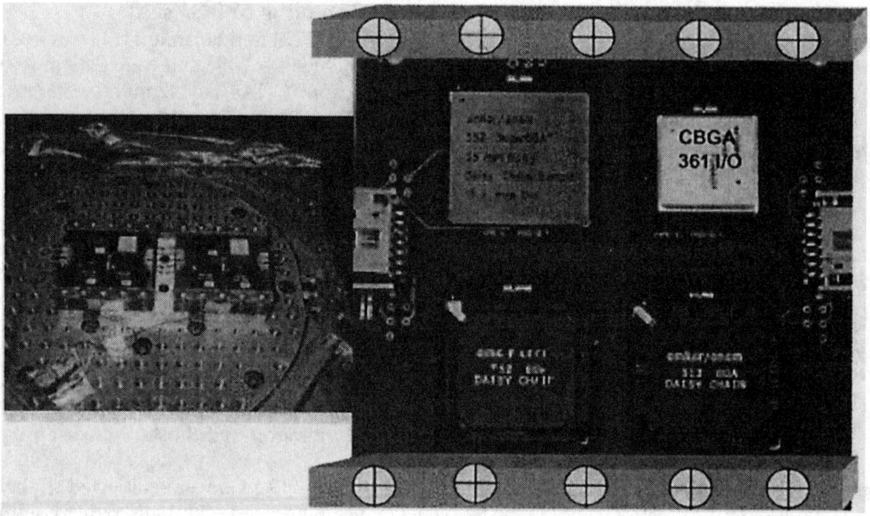

FIGURE 4.19 Random vibration test setup, (*Left*) Two three-stack of test vehicle on vibration table. (*Right*) Enlarged test vehicle with the CBGA 361 at the top right corner.

Initially, a very low vibration spectrum was sweeped to determine natural frequencies for the test vehicle. The board was subjected to a vibration spectrum in the range of 200 to 2000 Hz.

4.10.3 Natural Frequency Measurement/Projection Test Results

Plots for natural frequencies were generated. A simplified finite-element analysis was performed to predict natural frequencies and stress/strain conditions at the solder joints.[8] Ten percent model damping was used for the random response calculation. Projections for the first, second, and third natural frequencies are compared with measured values, as listed in Table 4.4.

4.10.4 Damage Induced by Thermal Cycling Alone

Both board and package interface cracking was observed with increasing number of thermal cycles. Figure 4.13 shows typical failures for the two cycling conditions, A and D. Failure under cycle A conditions generally was from the PWB and for cycle D conditions from the package sites. Failure mechanism differences could be explained either by global or local stress conditions.

Modeling indicates that the high-stress regions shifted from the boards to the packages themselves when stress conditions changed from global to local. For cycle A conditions, with slow heating/cooling ramping, which allowed the system to reach a uniform temperature, damage could indicate a global stress condition. For cycle D conditions, with rapid heating/cooling, damage could indicate a local stress condition.

TABLE 4.4 Measured and Finite-Element Projection of Natural Frequencies and Stress/Strain for CBGA 361

TV condition	Frequency mode 1 (modes 2,3)	Amplitude (g^2/Hz)	Maximum stress (psi)	Maximum strain (μstrain)
Measured values for no stiffener	356 (688,1700)	190	N/A	N/A
FEA for TV with no stiffener	381 (594,795)	170	945	315
FEA for TV with stiffeners	593 (820,1625)	43	558	185
FEA for TV bonded to Al plate	3155 (3520,4700)	N/A	25	8

4.10.5 Damage Induced by a Priori Vibration

Damage induced by vibration was different in some cases, as shown in Figs. 4.20 and 4.21. Appearance of tensile deformation from the center of high balls as shown was significantly different from that observed for thermal cycling conditions. Similar to thermal cycle conditions, damage was more dominant for the balls with higher distance to neutral point (DNP), especially for the corner balls. However, additional microcracks in eutectic solder joints, different from the norm for thermal cycling, were induced by tensile and shear load during random vibration.

4.10.6 Synergism of Vibration and Thermal Cycling

Cycles to first failures for three assemblies with levels of rigidity and a priori vibration conditions are shown in Table 4.5. Test data for cycles to first failure after vibration for the most severe conditions,

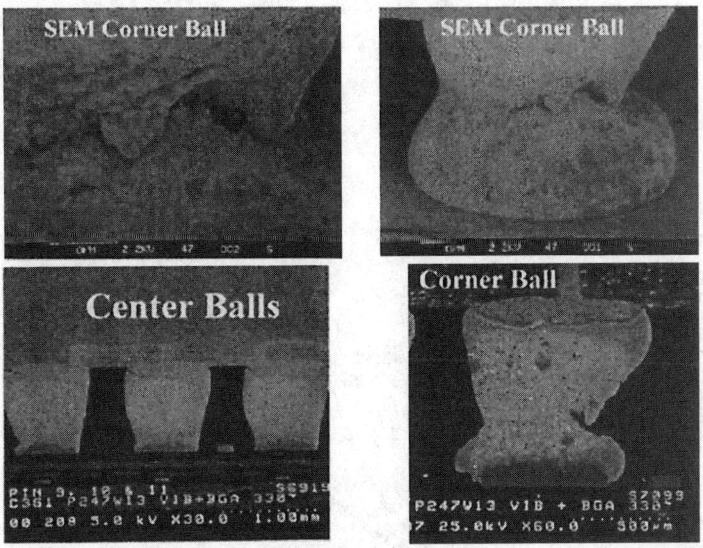

FIGURE 4.20 Thermally cycled samples after vibration. Note tensile failure for the corner ball and minimum damage for the center balls.

FIGURE 4.21 Thermal cycled samples after vibration. Note mechanical crack propagation at two locations and minimum damage at a center ball.

TABLE 4.5 Cycles to First Failure after Random Vibration

TV condition	ID, vibration (3 axes), and thermal cycle type	Thermal cycles to failure
No stiffener under vibration	4, vib. + cycle B	4, failed between 292 and 326 cycles
	50, vib. + cycle B	50, failed at 400 cycles
	34, vib. + cycle A	34, no failure to 434
With stiffeners	33, vib. + cycle A	33, failed between 292 and 326 cycles
	13, vib. + cycle A	13, failed at 330 cycles
Bonded to Al plate	31, vib. + cycle A	31, No apparent change to virgin test vehicles

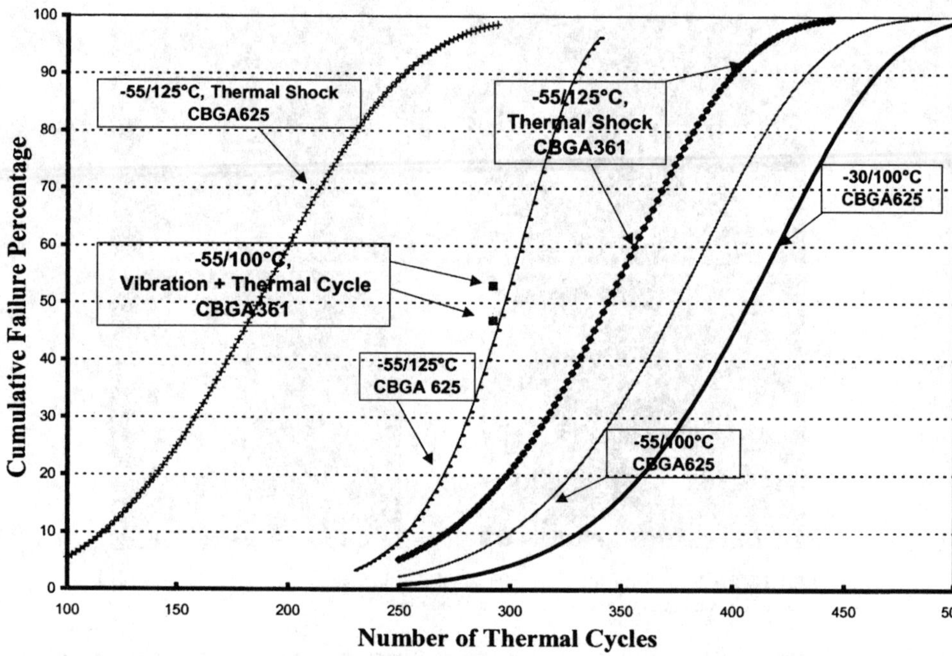

FIGURE 4.22 Cycles to first failure data for two CBGAs under four thermal conditions with two assemblies with a priori vibration conditions.

i.e., the test vehicles with no rigidity enhancement, were marked in Fig. 4.22 for comparison only. It is difficult to draw a statistically meaningful conclusion because of insufficient sample size.

4.11 BGA CONCLUSIONS FOR QUALITY AND RELIABILITY ASSURANCE PROGRAMS

- Plastic assemblies with I/O counts from 254 to 560 on FR-4/polyimide PWBs showed an order of magnitude larger number of cycles to first failure than ceramic assemblies with I/O counts from 361 to 625. They were in thousands of cycles for plastics, whereas in hundreds for ceramics, in cycling temperature ranges of −55 to 125°C, −55 to 100°C, and −30 to 100°C. Not reported here, vibration and mechanical shock exposures prior to thermal cycling were shown to have varying effects on the subsequent cycling to failure.

- Near-thermal shock conditions induced the most damage on CBGA assemblies compared with thermal cycling. Up to 50 percent reductions in cycles to first failure were observed when heating/cooling rates significantly increased for the thermal cycling in the range of −55 to 125°C.
- Failures for thermal cycled CBGAs were either from board or package sites in the eutectic solder joints. Those subjected to vibration showed tensile deformation in high-melt balls and tensile and shear failures in eutectic solder.
- An increase in the maximum cycling temperature from 100 to 125°C had a more damaging effect than a decrease in the minimum temperature by the same amount (−30 to −55°C), keeping the ΔT constant. Fatigue, rather than creep, is considered to be the main damage mechanism below one-half the absolute melting temperatures.
- Cycles to first failure can be projected reasonably using a simple modified Coffin-Manson relationship and accelerated thermal cycle test results.
- Assemblies with three-level rigidity passed a launch random vibration/shock condition. However, cycles to first failures after vibration were affected by the rigidity of the board, a significant reduction (>50 percent) for less rigid and minimum reduction for highly rigid assemblies after vibration.

4.12 REFERENCES

1. Ghaffarian, R., "Ball Grid Array Packaging Guidelines," distributed by Interconnect Technology Research Institute (ITRI), August 1998; available at *http://www.ITRI.org.*
2. Ghaffarian, R., and Kim, N., "Ball Grid Reliability Assessment for Aerospace Applications," in *Proceedings of the 30th International Symposium on Microelectronics,* Philadelphia, PA, October 1997, pp. 396–400.
3. Ghaffarian, R., "CBGA/PBGA Package Planarity and Assembly Reliability," in *The 1997 International Flip Chip, Ball Grid Array, TAB and Advanced Packaging Symposium Proceedings,* Semiconductor Technology Center, Neffs, PA, February 17–19, 1997.
4. Lau, J. (ed.), *Ball Grid Array Technology,* McGraw-Hill, New York, 1995.
5. Kapur, K. C., and Lamberson, L. R., *Reliability in Engineering Design,* John Wiley & Sons, New York, 1977.
6. Cole, M., Kastberg, E., and Matin, G., "Shock and Vibration Limits for CBGA and CCGA," in *Surface Mount International Proceedings,* Surface Mount Technology Assoc., Edina, MN, 1996, pp. 89–94.
7. Marsico, J. W., AIL Systems, private communications, 1999.
8. Pitarresi, J., University of New York at Binghamton, private communications.

ACKNOWLEDGMENTS

Part of the research described in this chapter was carried out by the Jet Propulsion Laboratory, California Institute of Technology, under a contract with the National Aeronautics and Space Administration.

Special thanks to Drs. D. Mulville, NASA HQ Code AE, M. Greenfield, NASA HQ Code Q, H. K. Detweiler, Director of Safety and Mission Assurance Directorate (SMAD), T. Gindorf, Manager of Assurance Technology Office, P. Barela, Manager of Quality Assurance Office, and P. Zulueta, Group Supervisor of Applications Engineering, for their vision, invaluable support, and continuous encouragement. I wish to acknowledge in-kind contributions and cooperative efforts of JPL-led BGA consortium team members with my deepest appreciation to others who contributed to meet the objectives of the programs.

CHAPTER 5
DIE ATTACH AND REWORK

Ken Gilleo
Cookson Electronics

5.0 INTRODUCTION

The semiconductor die or chip typically is bonded to a substrate for several critical reasons. First, an unattached die is subject to much greater mechanical abuse during handling of the package and assembly during manufacturing. The final end product can expect considerable abuse during its lifetime, mostly as mechanical shock. Portable products must be especially robust because users have come to expect a system that can be mistreated, including dropping. Attaching the die to a lead frame, chip carrier, or circuit board makes the system very much more robust and immune to most mechanical breakage of both the die and the interconnection structure, which can be very fragile. The attached die is also better able to dissipate heat to the substrate, particularly when a metal lead frame or thermally conductive metal or ceramic package is used. The die-attach material also can be designed with electrical conductivity in order to ground the back of the die.

There are three specific categories of die-attach materials: metallurgic, inorganic (ceramic), and polymeric. The earlier materials were either pure metals or alloys that provided an excellent thermal path. Metallic die-attach systems are still used but represent only a small niche today. Ceramic adhesives, most notably silver-glass, also have been used for some time. However, polymer adhesives are the dominant materials for today and are used widely for area array packaging. More than three decades ago, polymer-based adhesives began to displace metals as the preferred attachment materials. The polymer adhesives provided a simple, reliable, and automatable means of die attachment.

Polymers can be divided into two major classes, thermosets and thermoplastics. Both are used for die attach, and each has a unique set of useful attributes. Thermosets, the de facto standard, are easier to use but form permanent bonds that preclude repair. Thermoplastics are easily reworkable but may require modified die bonders or special equipment. Before moving into the various die-attach processes, a brief discussion of polymers will be provided.

5.1 POLYMER BASICS

Polymers are molecules made of long chemical chains with high molecular weights. A single polymer chain can be made up of hundreds or thousands of atoms, making polymers some of the largest molecules found in nature or synthesized. However, since the individual polymer chains in the thermoset class are interlinked, these polymers can contain millions of atoms. The long, chainlike structures are the key to the unusual strength, elasticity, and general resilience so common to many polymers. Many polymers, both natural and synthetic, are stronger than metals on a weight basis. High strength, low density, strong adhesion, and excellent electrical properties make them the best choice for die attach in the vast majority of applications.

5.1.1 Thermoplastic versus Thermoset Polymers

Polymers are easily classified into two major categories: thermoplastics and thermosets. Thermoplastics are capable of softening or even melting when heated but will return to a solid state when cooled. The process is repeatable and does not alter the polymer characteristics to any extent. However, heating can "anneal" films to reduce stress, but this is a physical rather than a chemical change. Heating simply imparts enough thermal energy (molecular vibration) to allow the long polymer chains to move freely past one another and take on new shapes. Cooling reduces molecular motion to a level where chains no longer move freely.

Thermosets do not display this reversible phase-change behavior. During the polymerization, or chain-building, step, thermoset polymers form links, or chemical bonds, between adjacent chains. The result is a three-dimensional network that is much more rigid than the linear thermoplastic structure. The interlinked chains are not free to move when heat is applied, and the thermoset, as the name implies, is "set" into a permanent shape after polymerization. The level of cross-linking can be varied. Materials with high cross-link densities are hard, rigid, and somewhat brittle substances. Heating to high temperatures can soften thermosets with low cross-link densities, but they do not melt, and their original shape is retained.

Both types of polymers are used in electronics, but the thermosets, like phenolic and epoxy resins, have a long history as wiring board and packaging materials. Thermoplastics, with the exception of polyester and polyimide film, popular flexible circuitry dielectrics, have only recently gained popularity in electronics. Advanced thermoplastics now can provide the high melting points and impressive dimensional stability that earlier polymers could not deliver. State-of-the-art thermoplastics now perform in the same temperature ranges as the thermosets. Low moisture absorption and low thermal coefficients of expansion are now possible, making thermoplastics the preferred type of polymer for such applications as molded circuits, encapsulation, and die attach. It is the *remeltable* quality that sets thermoplastics apart and provides special processing advantages. Table 5.1 compares thermoset and thermoplastic characteristics.

TABLE 5.1 Thermoplastic versus Thermoset Properties

Property	Thermoplastic	Thermoset
Initial physical state	Solid	Liquid or low-melting solid
Polymerization	Prepolymerized by supplier	Polymerized by user
Bonding temperatures	Low to high; near softening point	Can be at ambient
Cure time	0; already cured	Variable; minutes to hours
Bonding time	Can be < 1 s	> 1 minute
Ionic content	Near 0	Can be > 50 ppm
Storage	Room temperature	Usually freezer for one-part
Available forms	Solid film, solvent paste	Solventless paste, solid preform
Reworkability	Good to excellent	Generally poor
Property variability	Narrow range	Moderate, depends on cure
Types	One-part only	One- or two-part
Outgassing	Low	Low to high
Solvent resistance	Low to moderate	Typically high
High-temperature performance	Moderate	Moderate to high
Dimensional stability	Moderate	Can be high

5.2 DIE-ATTACH MATERIALS

5.2.1 Metallurgic; Solders

Gold has been used for several decades to bond silicon chips to metal lead frames. A common "solder" consists of 94 percent gold and 6 percent silicon. The common form is ribbon or preforms cut to the die size. Heat is applied by the die pickup collet that also vibrates in the *xy* plane to "scrub" the chip and metal together. This allows a hard bond to form. The bonding temperature is over 400°C, and the process takes 5 to 10 s. While the resulting bond has excellent thermal and electrical properties, the high processing temperature and relatively long processing time are considered to be deficiencies for metallurgic die attachment. However, performance is unmatched by other types of materials.

5.2.2 Glass Die-Attach Materials

Metal-glass materials emerged later as a specialty bonding system. Silver-glass blends have been used for some time and are still used today, although volume continues to diminish. The silver-glass adhesives have good thermal conductivity and high stability, thus providing outstanding reliability. However, the glass bonding agents also require high bonding temperatures, even though the temperature has been reduced over the years. The bonding temperatures are too high for use with organic chip carriers. Like metal, the glass adhesives are considered to be premier high-reliability systems and are used with some ceramic packages.

5.2.3 Polymer Die-Attach Adhesives

Metal Fillers. Silver-filled epoxy thermosets are the largest class of die-attach adhesives in use today. They provide efficient processing, low-temperature bonding, electrical conductivity, reasonably good thermal conductivity, and generally high reliability. Epoxies are the most commonly used binder or matrix material because of a good balance of properties. The most common form is a one-part paste that is thermally cured but stored in a freezer to prevent premature polymerization that can occur at ambient temperatures. Newer polymers are also used, including polyimides, cyanoacrylates, cyanate esters, and silicones. Some of the resins, such as the cyanate esters, can be coreacted with epoxies. However, the epoxies continue to dominate even though the alternates may have modestly better properties such as very low moisture absorption. However, the high demands required for the increasingly thin plastic ball grid arrays (BGAs) may open up a market for these more advanced die-attach systems. Lower moisture absorption, which would reduce explosive delamination (called *popcorning*) on soldering, may be more important if higher-temperature lead-free solders are implemented.

Silver-filled reworkable thermoplastics are also used and are supplied in paste, ribbon, and die-sized preforms. The ribbons and films, unlike the epoxy thermosets, can be stored at room temperature because they are already polymerized. The thermoplastic resins are often totally free of undesirable ionic contaminants that are nearly always present in epoxies due to their formation during synthesis. Gold is also used as a filler for very high reliability applications where silver's high chemical activity could be an issue.

Inorganic Fillers. Inorganic compounds such as aluminum nitride (AlN) are also used as fillers for die-attach adhesives where electrical insulation is required but thermal conductivity is important. Fillers also reduce the coefficient of thermal expansion (CTE), thus providing a closer thermomechanical match to the die. Silicon has a CTE of about 2.5 ppm/°C, whereas polymers can approach more than 100 ppm/°C. Epoxies fall within 70 to 90 ppm/°C range, and thermoplastics can be twice as high. The addition of fillers with small CTE values lowers the CTE value of the adhesive to a more reasonable number ranging from 10 to 20 ppm/°C.

Unfilled. Pure polymer die-attach materials are used in some cases, but these are mostly thermoplastics where temporary attach is the goal. The short duration required for testing may not require fillers. Absence of filler allows faster and lower-temperature bonding of thermoplastic films. Die removal of the tested and burned-in dies is often simpler without filler. However, filled die-attach adhesives are preferred for permanent bonds.

5.3 DIE-ATTACH APPLICATION METHODS

5.3.1 Needle Dispensing

Needle dispensing is by far the most popular process for applying polymeric die-attach adhesives. The dispenser can be a stand-alone automatic system or one that is integrated into the bonder. Force is applied to the adhesive paste by means of air pressure or a mechanical pump. Precision positive-displacement pumps offer the best control and repeatability. Modern equipment will deposit the desired amount of adhesive and also can be programmed to apply a specific pattern that can be useful for larger dies. Today, some area array packages contain multiple chips so that more sophisticated dispensing robots are required to place the correct amounts of paste in the desired locations. Figure 5.1 shows a needle-dispensing machine intended for high volume that can be used for die-attach adhesives (see Chap. 6 for more details).

5.3.2 Stenciling

Stenciling, also called *stencil printing,* can be used to mass-apply accurate amounts of adhesive paste for high-throughput applications such as area array packaging. A metal stencil is produced in a thickness equal to the desired deposition height. Openings are laser cut or chemically etched into the stencil metal that are equal to the die size and in the locations that match the array. Chip carrier substrate is often produced in sheets with hundreds of die bond sites. This is the ideal format for stenciling because all the bond sites can be coated accurately with die-attach adhesive in a single fast step. Area array substrate is also used in strip form, where a dozen or more die bond sites are presented. A

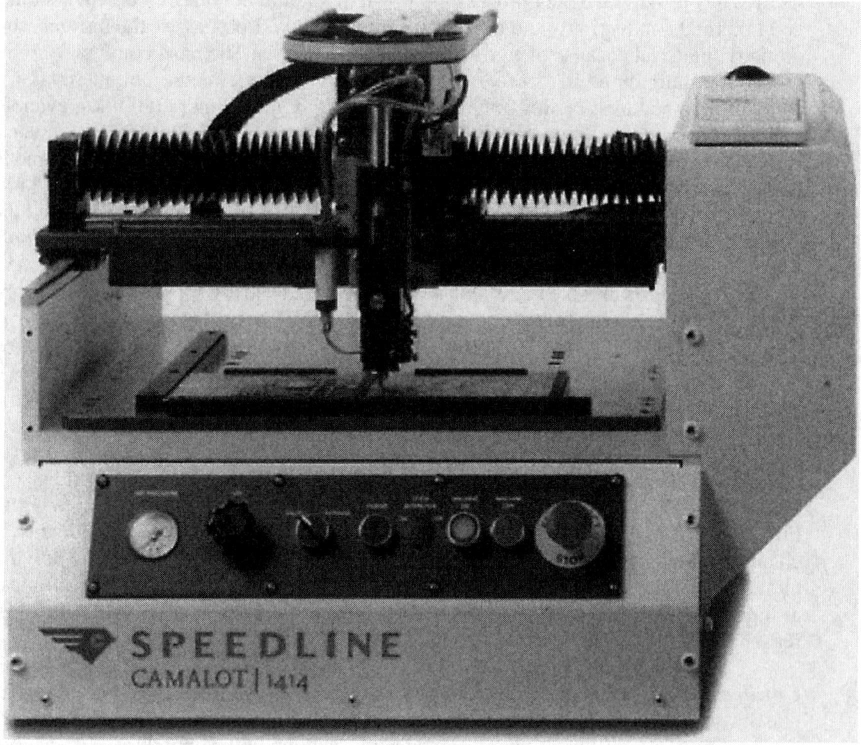

FIGURE 5.1 Needle-dispensing machine.

smaller stencil printer can be used that has automatic feed to accommodate the strip format (see Chap. 7 for more details).

5.3.3 Screen Printing

Screen printing is similar to the stenciling process and may use the same equipment. However, the printing tool has a continuous screen mesh across the holding frame. Openings are produced with a polymer emulsion applied over the entire screen and photoimaged into a pattern of openings. Printing occurs where emulsion has been removed to leave an opening in the screen mesh. Screen printing is more limiting than stenciling because the adhesive must be able to pass through the mesh and transfer cleanly to the substrate. Thermoplastic adhesives tend to be stringy and do not screen print very well, especially since dissolving the polymer in solvent makes the paste. Solvent evaporation of paste on the screen can cause blocking. Thermoset adhesives generally do not contain solvents but rather liquid reactants that do not tend to evaporate. This makes thermoplastics a better candidate.

Screen printing is more suitable for thinner deposits than stenciling and may be a better choice where less than 2 mils (50 μm) of deposit is needed. Stencils with a thickness of even less than 3 mils (75 μm) can be difficult to produce and handle. A 50-μm-thick stencil can be a hazard because such thinness effectively produces a knifelike edge. Screen printing is quite precise even down to 10 or 12 μm, which is not practical with stenciling.

5.3.4 Fluid Jetting

Fluid jetting is an off-contact dispensing process where a short-duration pressure wave or impulse is applied to fluid in a dispenser. A small needlelike orifice serves as the directing barrel. A piezoelectric "piston" can provide the right impulse to cause a droplet of adhesive to be ejected at the substrate. Pulses can be rapid, ranging up to many hundreds or thousands per second. The high rate in effect produces a seemingly steady stream of material that can be applied in the desired thickness and shape by moving either the head or the substrate. Presently, jetting is considered too slow for general die-attach adhesive use, but increasing rates may make it a better option.

5.3.5 Spin Coating on Wafer

Wafer coating can only be used with adhesives that can be hardened but remain bondable. While thermosets can be partially cured, or B-staged, to a solid that can be bonded later, the thermoplastics are the preferred class for this process. In fact, some thermoplastic die-attach adhesives are specifically designed for wafer use. Thermoplastic paste is deposited onto the back of the spinning wafer near the center. The spinning action produces flow out by means of centrifugal force to yield a thin, precision coating. The coating is then dried by solvent evaporation, typically by baking in an oven. The dried wafer now contains a solid die-attach material on the backside. The wafer can be singulated into individual dies by sawing. Many thermoplastic adhesives can be sawn without problems, provided that the saw blade does not overheat to the extent that melting occurs. The die is now ready to bond in an efficient process that only requires placing the chip with a heated pick tool, or thermode. Precoated dies can be bonded in less than 1 s, making this the fastest die bond method. The chip carrier may be preheated to increase the adhesion level and reduce bonding time. Stenciling and screen printing also can be used for wafer coating.

5.3.6 Film Application

Both thermosets (B-staged) and thermoplastics are available in films. The thermosets have a limited useful life unless kept in cold storage. The thermoplastic films, since they are already polymerized, have a nearly infinite life but should be used within 6 to 12 months because they can absorb contaminants and fillers can age. Room temperature storage in sealed plastic bags or in a dry box is preferred. The film typically is cut to match the die size before proceeding. The cut adhesive can be applied to the die or substrate using a process called *tack bonding.* Here, a lower temperature is used with just enough dwell time to cause the film to adhere. The cut film also can be placed on the substrate, the die placed on the film, and the stack can be bonded with heat and pressure. Films also can be applied at wafer level, but this is considered more difficult than liquid coating. The dies are singulated by sawing, as described previously.

Sizing, Cutting, Singulating. Film is available in several forms that can require cutting, blanking, shearing, and other mechanical processes. Rolls of film can be cut to reels of the desired width with automatic equipment. Some bonders can cut film to the right length to produce die-sized preforms, assuming that the film width is correct. Film punching equipment also can be used, but tooling is required for every size. Preforms are often placed in waffle packs for presentation to the bonder, and this can be done automatically by the film blanker. Vendors offer all forms that have been described.

Laser cutting has become increasingly popular because no tooling is required. Many die-attach films, especially the thermoplastics, cut very well. There can be problems with B-staged thermosets because the laser heating can polymerize the cut zone.

Film Attaching Process. The film is usually prebonded to the die or substrate before the die-attach bonding process, but this is not essential because the film also can be sandwiched between the substrate and die just prior to bonding. However, bonding the film accurately to the die is preferred because this will produce high placement precision and preclude the possibility of the loose film moving. The film tacking process requires just enough heat, pressure, and dwell to get some level of bonding. Since the thermoplastics are already polymerized, the process window can be quite wide.

Final Die-Attach Bonding. The final bond with film adhesives typically uses higher temperature and pressure to get complete die-to-substrate contact and to force any air pockets out to the die edge. The key parameters for dry-film bonding are temperature, pressure, and time. All three are interactive, and the user may trade one variable against another. For example, bond time can be reduced if the temperature or pressure is increased. Time can be reduced to a minimum, even down to milliseconds, if both pressure and temperature are increased. Conversely, increasing one or both of the other factors can reduce either pressure or temperature. This interactivity allows the process to be optimized over a wide range of products and applications. Some minimum pressure must be applied during the bonding process even though the value may be small; the weight of the die alone unfortunately is insufficient. Since common epoxy thermoset die-attach materials are fluids when the die is placed, no added pressure is required. Standard die-attach bonders therefore have been designed for minimal force application. Thermoplastic die-attach processing often requires equipment that can apply significant force. Although some existing die-attach equipment may apply adequate die bond pressure for thermoplastics, many are not suitable. Modified or totally new equipment may be needed. However, the need for equipment investment can be justified because the resulting benefits of reworkability, better precision, and faster bonding are realized.

Thermoplastic adhesives are suitable for all types of area array packages, especially multiple-chip types, where removing a die may be critical. The bonding surface mechanism is primarily a mechanical interlocking process and therefore is more dependent on morphology than on material composition. The triad of bonding parameters creates a wide processing window that enables thermoplastics to fit into various assembly manufacturing systems, particularly those with processing restrictions such as temperature sensitivity and time constraints. Thermoplastics thus can span the full range of packages: ceramic, metal, rigid organic, and flexible.

Adequate bonding is achieved by bringing the thermoplastic adhesive to the right viscosity range. Figure 5.2 relates viscosity to temperature and shows the generic bonding process window. Thermoplastic adhesives can bond over a broad process window with more than a 100°C span. Bonding is performed above the glass transition temperature T_g of the polymer but below the decomposition point. These two points determine the adhesive limits, but the substrate and die may set the upper bond limits if either is temperature-sensitive. Bonding is accomplished by applying pressure to the softened or liquefied polymer. It is important that adhesive be forced into the surface microstructure of the substrate and die for maximum bond strength. Bonding results from both atomic forces and mechanical linkage. The greater the penetration of the adhesive into the microstructure, the higher is the bond strength. Shortening the dwell time or reducing temperature and pressure can

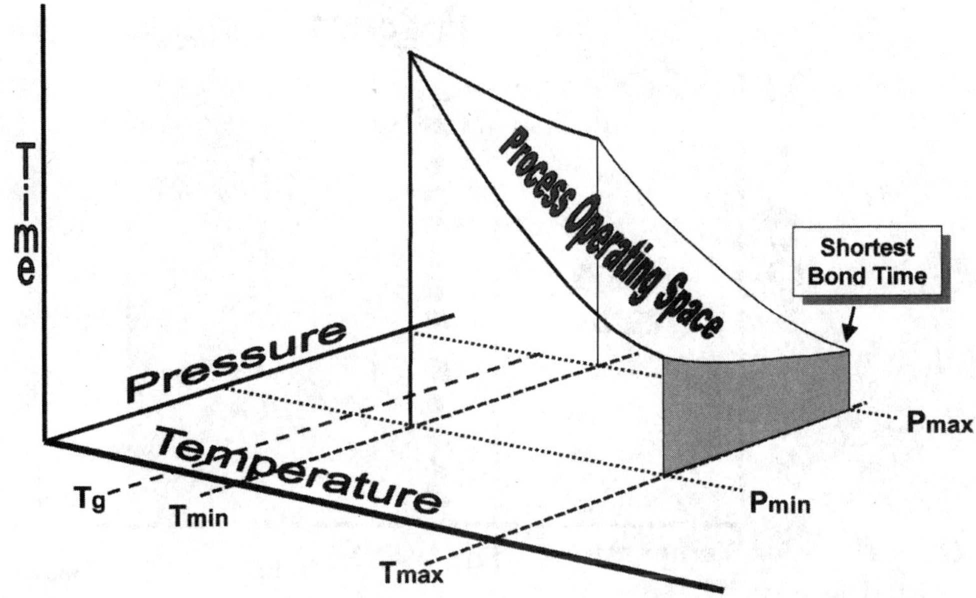

FIGURE 5.2 Thermoplastic bonding window.

reduce penetration and bond strength. The time factor, or dwell, allows heat to distribute at the interface and the fluid polymer to penetrate the surface microstructures. The process is fast, clean, and efficient, and the bond is stable and reliable.

Above T_g, the polymer changes from a harder, glassy state to a softer, rubbery form. Increasing the temperature continues to soften the plastic, which is seen as a drop in viscosity. Viscosity continues to drop as temperature is increased, but the effect will level off. Some thermoplastics become high-viscosity, deformable masses, whereas others change to thin fluids. However, it is not necessary to reduce the polymer to a fluid for bonding. A viscosity level must only be reached where the polymer can be forced into the substrate microstructure under pressure to wet these surfaces. As the plastic cools to ambient temperature, the viscosity increases, and the mechanical bond structure is locked in place. Under typical bonding conditions, very little squeeze-out occurs because the viscosity is still relatively high at suitable bonding temperatures. Filler also will tend to hold the adhesive in place and prevent undesirable squeeze-out. Thermoset adhesives, on the other hand, experience a much greater drop in viscosity on heating, and spreading can occur.

The triad of key bonding parameters—temperature, pressure, and time—is interactive, as stated earlier, but the parameters are also nonlinear, as can be seen from Fig. 5.2. Temperature has the most pronounced effect. The interaction between process parameters is shown in Fig. 5.3. The minimum operating temperature must be somewhere above the glass transition temperature in a region where viscosity begins to drop off (T_{min}). The upper limit (T_{max}) must lie some safe level below the polymer decomposition temperature. Pressure is determined by the minimum that will cause polymer flow, wetting, and subsequent bonding. The high-pressure limit is less well defined and may be limited by equipment, substrate, and components. At very high pressure, some squeeze-out can occur, however.

Thermoplastic paste adhesive bonding normally does not require pressure because the adherends are wet by the adhesive during component placement into the wet paste. When the thermoplastic adhesive is used in dry form, pressure is required to force the softened polymer into the surface microstructure. However, thermoplastic pastes are limited to small dies of about 3 to 4 mm (250 mils) because solvent must escape during the drying process. Larger dies can trap solvent or voids can form because the solvent is unable to vent through the longer path to the edge of the die. Figure 5.4 diagrams the various die bonding processes for thermosets and thermoplastics.

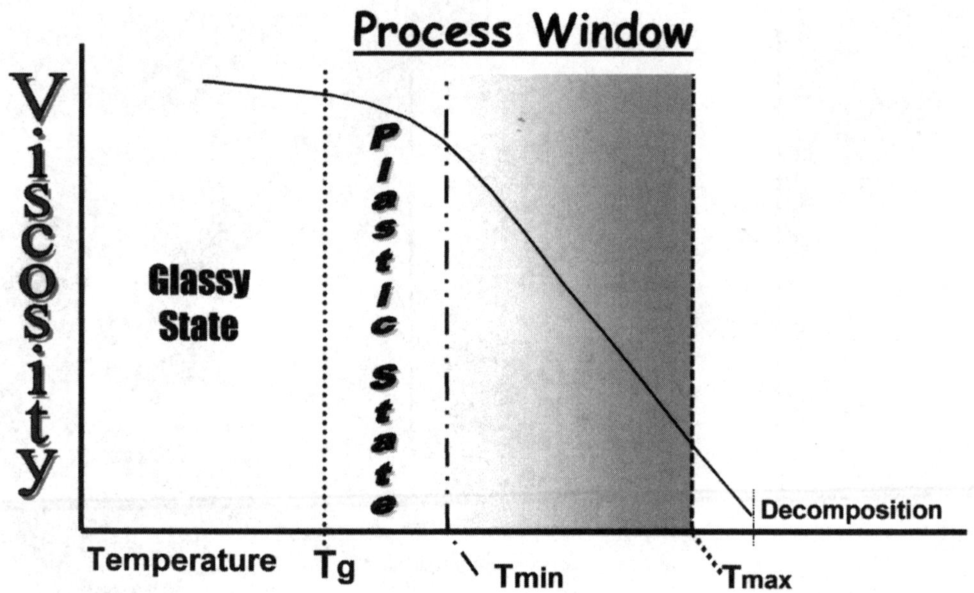

FIGURE 5.3 Graph of bonding parameters.

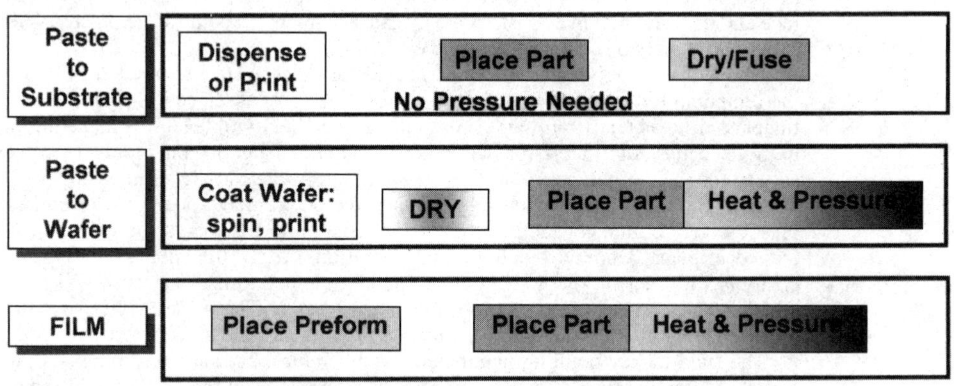

FIGURE 5.4 Block diagram of die-attach processes.

5.4 APPLICATIONS (USES)

5.4.1 BGAs

BGAs typically use dies that are adhesively bonded to the substrate, which can be organic, ceramic, or metal. Both thermoset and thermoplastic die-attach adhesives are used. Reworkable thermoplastic adhesives are especially useful for multiple-chip packages.

5.4.2 Lid Seal

While not a true die-attach process, the same adhesives are being used for lid sealing of hermetic-type packages instead of welding or soldering. However, the resulting package will not be a full hermetic because gases can move slowly through the polymer adhesive. Sealed packages using polymer adhesives are referred to as *near-hermetic* or *quasi-hermetic*. The use of die-attach-type adhesives simplifies the sealing process and reduces temperature. Many systems show high reliability, even with the slow gas leak resulting from the polymer's moderate barrier characteristics. Molecular scavengers, called *getters,* can be placed inside the package to absorb contaminants, including moisture, during the life of the package. The adhesives can be coated onto the lid and dried or provided as preforms that look like tiny gaskets. The precoated lids, as would be expected, offer a very manufacturing-friendly format, and the method has gained popularity.

5.4.3 Known Good Dies

Die testing before packaging remains a significant challenge as chips become more complex and speeds increase.[1] Testing can reveal bad dies but may not indicate the maximum performance speed. Chips often are rated by maximum clock speed and priced accordingly. A very high speed chip may be housed in a top-of-the-line package. It is therefore important to have a method that can ensure known good dies (KGDs) with a high level of confidence. Burned-in dies also may be required and generally command a premium. So-called infant mortality occurs in the early life of the semiconductor, and burn-in often can sort out this type of defective product.

Thermoplastic die-attach adhesives can be used to temporarily attach a die to a test platform for burn-in. The attached die is electrically connected by probes or by weak, temporary wire bonds. The chip is powered up and exposed to accelerated environmental aging conditions. This can be high-temperature storage, thermal cycling, or a combination. The die is given a final test and removed from the test platform. The die-attach adhesive is debonding by heating to its rework temperature as force is applied. Alternatively, a solvent-sensitive adhesive can be used that debonds with a specific solvent. More information is given in Sec. 5.7.

5.5 PROPERTIES

5.5.1 Bond Strength

Bond strength is a common metric for good adhesive performance. An attached die can be sheared or pushed off using a commercial die shear tester that slowly applies an increasing level of force to the side of the die until the bond is broken. Bond strength can range from an initial value of 2500 psi to slightly more than 4000 psi for thermoplastics. Thermoset materials fall into a similar range. However, heat aging often will boost thermoplastic strength, presumably by annealing the adhesive and/or allowing more microstructure penetration and bonding. Table 5.2 shows bond strengths for common thermoplastics.

TABLE 5.2 Die-Attach Adhesive Bond Strength*

Substrate	Conditioning	Initial bond strength, psi	Final bond strength, psi
Ceramic	150°C, 1000 h	2500	4800
Gold-plated ceramic	150°C, 1000 h	3100	4900
Aluminum	150°C, 1000 h	4100	4500
Ceramic	85%RH, 85°C, 1000 h	2500	2800
Gold-plated ceramic	85%RH, 85°C, 1000 h	3100	2600
Aluminum	85%RH, 85°C, 1000 h	4100	3600
Ceramic	−65 to +150°C, 500 cycles	2500	4000
Gold-plated ceramic	−65 to +150°C, 500 cycles	3100	3700
Aluminum	−65 to +150°C, 500 cycles	4100	4800

*Bonding conditions were 180°C for 10 s at 135 g on 80 × 80-mil Ni/Au-plated Kovar tabs.

5.5.2 Bonding Temperature/Curing

Thermoset die-attach adhesives cure at 150 to 175°C, and the bake time can be from 30 min to 2 h. "Snap" cure products can cure is as little as 5 min, however. Thermoplastics have bond times of only a few seconds or less because they area already polymerized. Table 5.3 shows bonding temperatures for a range of thermoplastics compared with T_g.

5.5.3 Outgassing

Outgassing is the release of vapors or gases from hardened die-attach adhesives that is mostly detrimental, especially if the agents can degrade the die or the electrical interconnect structure. Epoxies, because of their complex curing chemistry, can release a variety of materials. Unreacted hardener and resin also can be released, especially as the package is heated. It is nearly impossible for all the polymer reactants to be consumed completely during polymerization, so a trace amount may be released as volatiles. Anhydrides, common epoxy hardeners, can be a problem because they react with moisture to form organic acids that can corrode structures.

Thermoplastics, however, show very low outgassing. This is so because they have been polymerized in highly controlled reactors that generally preclude the presence of unreacted starting materials. The thermoplastics subsequently are manufactured into films or pellets, and this heating process further removes any volatiles that potentially could outgas.

5.5.4 Electrical Conductivity

Electrical conductivity is often desired to ground the die to a conductive platform. Silver is the most common filler used to achieve electrical conductivity, although gold powder may be used for very high reliability situations such as military packaging. Die-attach adhesives optimized for electrical conductivity have volume resistivity values in the range of 0.00001 to about 0.0005 Ω·cm.

5.5.5 Coefficient of Thermal Expansion

While it is theoretically desirable to match the adhesive's coefficient of thermal expansion (CTE) with that of the low-expansion die (~2.5 ppm/°C), this is not really necessary because of the elastic nature of many die-attach adhesives that can be expressed as a modulus. The actual CTE values for commercial products can range from a low of about 15 ppm/°C to a high of more than 200 ppm/°C. More common values are around 30 to 80 ppm/°C. Filler significantly reduces the CTE because the percentage of polymer is reduced and both metal and ceramic fillers have relatively low CTE values.

5.5.6 Glass Transition Temperature T_g

This value is often engineered to be above 125°C so that no phase changes occur during thermal cycling, which typically is run from −55 to 125°C. However, this is not an absolute rule, and good adhesives can have lower values. It was shown in Table 5.3 that T_g values could be well below ambient provided the modulus is also relatively low.

TABLE 5.3 Thermoplastic Bonding Temperatures

Fillers	T_g, °C	Bond temperature, °C
None	−40	100–150
Ag, AlN, none	25	150–200
Ag, AlN, none	45	160–220
Ag, AlN, none	85	160–250
None	145	200–230
Ag, Au, AlN, none	180	325–400
None	280	350–450
Thermoset (ref.)	100–150	150–175

5.5.7 Thermal Conductivity Properties

A good thermal pathway is often desired to allow die-generated heat to dissipate to the package. While solders can have values of 30 W/m·K or higher, polymer adhesives are an order of magnitude lower, ranging from 2 to 4 W/m·K. However, the large die area and the small distance between the die and substrate offset the concern over lower thermal conductivity. Thermal resistance is directly proportional to thickness, so thin bond lines make good pathways. Typical thermal fillers are silver and aluminum nitride (not electrically conductive). Beryllium compounds have conductivity values higher than many metals but are not common because of toxicity during manufacturing and other problems. Carbon nanostructures could find use as die-attach fillers in the future once better manufacturing methods are developed.

5.5.8 Thermal Stability

Both thermosets and thermoplastics can handle a minimum of 300°C before any decomposition begins. High-performance thermoplastic and thermoset systems may not decompose below 400°C. Special polyimides may exceed 500°C. However, modern area array packages typically are made from organic substrate that should not exceed about 260°C during processing. The die-attach adhesive will not be the weakest thermal link. The service temperature rating of die-attach adhesives is usually given as 150°C based on a standard test even though these materials may be able to handle higher values.

5.5.9 Modulus

Modulus, a measure of material stiffness, can be varied over a wide range for die-attach adhesives. Table 5.4 shows the range of values found in commercial thermoplastic products. Thermosets tend to be stiffer materials with values as high as 500,000 psi or even more. Adhesives with a modulus value of less than 100,000 psi are considered low-stress stain-relief materials.

5.5.10 Solubility/Chemical Resistance

Cured and hardened adhesives should not dissolve or even swell in common cleaning solvents. Today, solvent cleaning is much less common, especially for removing solder flux. Aqueous systems are now used where cleaning is deemed necessary, but no-clean solder pastes and fluxes are used commonly for environmental and productivity benefits. Thermosets are very resistant to even strong polar solvents. Thermoplastics typically will dissolve only in very polar solvents that are not used in packaging plants. Since most packages provide a barrier to the environment, die-attach adhesives would not be expected to encounter solvents once the area array package is completed. It may be noted that solvent-reworkable thermoplastic die-attach adhesives are designed to be sensitive to alcohols and are not necessarily sensitive to other solvents.

5.5.11 Debonding Characteristics

See Sec. 5.7.

TABLE 5.4 Thermoplastic Die-Attach Adhesive Modulus

Fillers	T_g, °C	Bond temperature, °C	Modulus, psi
None	−40	100–150	60,000
Ag, AlN, none	25	150–200	60,000
Ag, AlN, none	45	160–220	500,000
Ag, AlN, none	85	160–250	400,000
None	145	200–230	150,000
Ag, Au, AlN, none	180	325–400	360,000
None	280	350–450	370,000

5.5.12 Test Methods

MIL-STD-883D has been used commonly to test die-attach adhesives and is considered a more severe set of conditions than should be expected in commercial use. Most die-attach adhesive suppliers test and qualify to this specification.

5.6 RELIABILITY

The performance of thermoplastic adhesives is much more consistent and reproducible than that of thermosets for fundamental reasons. First, a thermoplastic has been fully polymerized by the manufacturer under highly controlled conditions of temperature, pressure, reaction rate, and time. Thermoset adhesives are polymerized by the end user, and final properties vary with conditions that cannot always be controlled. Thermoplastic adhesives can be bonded in a dry state that significantly reduces the possibility of voids so often experienced with epoxies and polyimide thermosets. Voids and adhesive stress are the major causes of failures related to die attach.[2] However, another increasingly important attribute of thermoplastic adhesives is their low modulus and corresponding elastic characteristics.[3] The result is very low stress on the die. The advent of larger and larger dies makes low stress imperative, especially when dies are mounted directly to organic circuit substrates. Table 5.5 gives the properties of a range of thermoplastics now in commercial use.[4]

5.6.1 Heat Aging

Heat aging is not a problem for most die-attach adhesives. The thermoplastics are stable, and the thermosets may cross-link slightly and increase the modulus, but this is acceptable. Thermoplastics, as was seen earlier, often will develop stronger bonds on heat aging due to annealing or increased microstructure penetration.

5.6.2 Thermocycling

Thermocycling should not have an adverse effect on die-attach bonds unless the material has a high modulus, the die is large, and the CTE is high.

5.6.3 Temperature and Humidity (T&H)

Humidity exposure is the real "litmus test" of polymer systems. Most die-attach adhesives will absorb some moisture, and this may lower the bond strength slightly. The effect is generally reversible. However, the T&H exposure also can boost strength for some systems, as seen in Table 5.2.

TABLE 5.5 Thermoplastic Die Attach Adhesive Properties

Fillers	T_g, °C	Bond temperature, °C	Rework, min·°C	Die shear, psi @25°C	Thermal conductivity, W/m · K	Modulus psi
None	−40	100–150	110	1700	0.20	60,000
Ag, AlN, none	25	150–200	160	2200	0.3–3.0	60,000
Ag, AlN, none	45	160–220	170	2600	0.3–3.0	500,000
Ag, AlN, none	85	160–250	170	3000	0.3–3.0	400,000
None	145	200–230	210	3800	0.22	150,000
Ag, Au, AlN, none	180	325–400	350	4000	0.3–3.0	360,000
None	280	350–450	400	4900	0.25	370,000

Note: Typical electrical conductivity is 1×10^{-4} Ω·cm for metal-filled and 1×10^{12} Ω·cm for nonmetal. Typical thermal conductivity is 3.0 W/m · K for silver and 1.0 W/m · K for AlN.

5.6.4 Failure Modes and Mechanisms

Bond failure of attached dies is uncommon and should not be expected in normal situations. Die cracking or bond delamination is an indication that the wrong adhesive has been used or that the process is out of control. Cracking suggests that a lower-modulus adhesive should be used. Low bond strength indicates that contamination may be present or that voiding is reducing bond area. Voiding can have many sources, and help from the adhesive supplier usually can solve the problem. Plastic BGA packages, however, can experience a number of moisture-related problems that relate to the die-attach adhesive. Popcorning, the explosive evolution of steam from trapped moisture during solder heating, can be related to the adhesive. However, this is a complex issue where substrate, encapsulant, adhesive, and design must be examined as a system.

5.7 REWORK

5.7.1 Thermosets

Thermosets are not considered to be reworkable in the common sense of the term. However, thermosets have been designed that can undergo depolymerization reactions. Some are made from epoxy resin with a thermally unstable link that will break down at 225 to 250°C. Other polymer systems have been designed that are attacked by specific agents such as acids. However, in most regards, only the thermoplastics are reworkable in the practical sense.

5.7.2 Thermoplastics

Removal. Rework can be essential in many multichip package (MCP) applications. Large, complex dies can have low wafer yields, perhaps under 20 percent. Test-at-speed is required but difficult in the bare die state, although several known good die (KGD) schemes are in use and there is continual progress. The fixed softening point and predictable temperature-viscosity relationship of thermoplastic adhesives ensure easy rework. The adhesive is brought to a temperature where the reduced viscosity allows the die to be removed by low-force shearing or torquing with a vacuum collet or simply by lifting a corner. The same time-temperature-pressure relationship used for bonding can be applied to rework because no property change occurred during the attachment process. The temperature-viscosity curve previously shown in Fig. 5.1 also can be used to estimate removal force because it is related directly to viscosity.

Although the thermoplastic adhesive is chemically unchanged after die repair, shape change may have occurred, and new material is recommended for rebonding unless reuse of the same adhesive pad is verified by testing. After die removal, the adhesive pad may remain on the die or substrate. The old adhesive pad can be removed by peeling it away with tweezers during heating. Not only does a clean surface result, the first adhesive actually may remove contamination so that the reworked bond is stronger.

Special temporary thermoplastic adhesives can be used when the intent is to bond, test, and remove the die. These materials can have the same electrical and thermal performance as the products intended for permanent die attach. Specific solvents affect the temporary adhesives. The bonding process is the same as that described earlier. After test and burn-in, the die is readily removed by adding solvent to the bond site. The adhesive swells and loses strength so that the die is removed with very little force. The die can now be bonded to the final assembly with standard thermoplastic die-attach adhesive. The temporary adhesives are being used in certain KGD applications. Note that the intent is not to dissolve the adhesive but to cause debonding by swelling. This results in a clean, noncontaminating process.

Rebonding. Since a thermoplastic die-attach film can be removed cleanly, no special cleaning step or preparation is needed. Rebonding simply involves starting over with a new die and fresh adhesive.

5.8 SUMMARY AND CONCLUSIONS

Die attachment is an important step in manufacturing area array packages. Nearly all use adhesives. The exception is the flip chip (Chap. 1, 2), where the die is connected from the face. While thermoset die-attach adhesives are the common standard, thermoplastics offer two advantages for area array packaging. The first is higher performance because of lower ionics and lower modulus where beneficial. The second is easy die removability and the ability to rework. This feature is especially valuable where an expensive package can be salvaged or multiple dies are used where the odds of a bad die upsetting the economics are significant.

5.9 REFERENCES

1. Martin, S., "A Practical Approach to Producing Known-Good Die," in *ICEMM Proceedings, '93*, IMAPS, Washington, DC, April 1993, pp. 139–151.
2. Trigwell, S., "Failure Mechanisms of Wire and Die Bonding," *Solid State Technology,* May 1993, pp. 45–49.
3. Corbett, S., and Ongley, P. E., "New Innovations in Thermoplastic Die Attach Adhesives for Microelectronic Packaging," *Proceedings—ISHM-UK European Multichip Module Conference,* London, England, 1996.
4. Gilleo, K., Cinque, T., Corbett, S., et al., "Die Attachment for Today's Packaging Challenges," *Electronic Packaging and Production,* Feb. 1994, pp. 109, 110, 112.

CHAPTER 6

LIQUID ENCAPSULATION EQUIPMENT AND PROCESSES

C. Lazinsky
Speedline Technologies

K. Gilleo
Cookson Electronics

6.0 INTRODUCTION

The area array revolution has brought on a plethora of new packaging designs, new and modified materials, innovative processes, and advanced encapsulation equipment. This chapter briefly describes the package types, provides the basics of the materials, and highlights the most common processes before going into the details of equipment for liquid encapsulation. Major advances in materials, processes, and equipment have occurred during the 1990s, and the pace of change may even be accelerating today. Dispensers are now available that apply material with vacuum injection to eliminate air. Very high capacity machines are also available that employ up to four dispensing heads that can run the same or different liquid encapsulants. All these developments have occurred while the equipment has become more robust, increasingly precise, and easier to set up.

6.1 MODERN TRENDS IN PACKAGING

The advent of area array packaging has greatly increased the use of liquid encapsulants and has resulted in new processes. Although transfer molding using solid epoxy molding compound (EMC) is still dominant, liquid encapsulation has become popular for many chip scale packages (CSPs) and larger plastic ball grid arrays (PBGAs). Liquids offer certain advantages that offset their slightly higher cost. Direct chip attach (DCA) assembly, better known as *flip chip* (FC), uses liquid dispensing almost exclusively for underfill. The next subsection will describe the basic packaging concepts and processes.

6.1.1 COB

The chip-on-board (COB) process involves first attaching the bare die or semiconductor chip to a circuit board or to a chip carrier packaging substrate. A polymeric die-attach adhesive is used commonly, as described in Chap. 5. Next, first-level electrical connections are made between the chip pads and substrate using wire bonding. The system is now ready to function, but it is unprotected from the environment and could not survive very long in service. A polymeric material must now be added around and over the die and all the wire bond structure. This material is called *encapsulant,* and the process is referred to as *encapsulation.* The encapsulant must be applied in liquid form to surround and cover the chip and wire structure. Both solid and liquid encapsulants can be used here, but this chapter is focused on liquid. (See Chap. 7 for encapsulation using solid molding.)

COB has been used for many decades for ceramic hybrid and later for building modules on organic circuit boards. However, these products were circuits, not packages, and they typically contained an assortment of packaged components along with the one or more bare dies that were wire bonded. However, when IBM and Motorola commercialized the ball grid array (BGA) package, the COB concept was applied to packaging. Motorola's Ompack used COB as the first-level connection method to the organic chip carrier. The COB structure was then overmolded using the common transfer molding process. However, transfer molding requires significant and costly tooling. The molds can take months to build, costing tens of thousands of dollars. Since the new BGA packaging was undergoing continual design changes in the early 1990s, an encapsulation process with low tooling, simple setup, and good economics for short runs was sought. The answer was to dispense liquid encapsulant onto the COB area.

6.1.2 BGAs

There are two basic BGA designs: *die-up* and *die-down* (also called *cavity-up* and *cavity-down,* but this terminology is less accurate). Both constructions can use liquid encapsulant. In fact, the die-down design cannot be encapsulated easily without using liquid dispensing because the package actually has a recessed cavity. BGAs made from organic, ceramic, and metal-ceramic platforms can all be encapsulated with liquids using needle dispensing. BGAs with the die-down configuration are simpler to fill because the BGA substrate often can contain the liquid if there is a deep enough cavity. However, the die-up BGA built on a planar carrier requires an additional step of first adding a retaining structure to hold the liquid encapsulant. The retaining methods will be covered in Sec. 6.3. Flux also may be needle dispensed onto BGA substrate for the ball attachment step, although it is more common to apply flux to the solder balls while they are held in a vacuum pickup tool. Camalot Matrixx applies flux with pin transfer and then uses positive pin placement to place solder spheres accurately. This avoids contaminating the spheres and eliminates potential clogging of any vacuum mechanism.

6.1.3 Flip Chips (FCs)

Nearly all modern FC packages require underfill, and most underfill is applied by liquid dispensing, although molded underfill methods have been commercialized in the past several years. Underfill materials and processes have been covered in several other chapters.

6.1.4 Chip Scale Packages (CSPs)

CSPs come in a wide variety of designs and therefore can require different encapsulation methods. The simplest constructions use the die-up design, similar to the larger BGAs. The only difference is one of dimensions, where the CSP must be kept within a limit where the package area does not exceed chip area by more than 10 percent. This type of CSP can be encapsulated by the same methods used for the larger BGA kin. A retaining structure is added, and the area around and over the COB structure is filled with liquid encapsulant.

More complex structures such as the Tessera μBGA require special processes. The encapsulant is applied to the inside of the package through an opening. Special dispensing machines with vacuum assistance have been designed and are in use for this type of package.

6.2 ENCAPSULATION MATERIALS

Although a large number of liquid encapsulants are on the market, most are based on epoxies, which have been the workhorse of electronics for many decades. However, the encapsulants contain enough ingredients to make the number of combinations endless. Furthermore, different classes of liquids are available for specific purposes, such as forming a retaining dam. Other classes of polymers are being offered as liquid encapsulants with very good properties. If brominated epoxy resin is banned, nonepoxy systems with inherent flame retardancy could come into use. Brominated compounds impart flame retardancy and are considered safe. However, there seems to be an initiative to ban bromine compounds without any scientific support or even a rationale. Bromine and other halogens are "mined" from the ocean and are absolutely required for life.

6.2.1 Basic Chemistry

Briefly, the liquid encapsulants are a solventless blend of polymer precursors and filler. In the case of the common epoxy system, hardener and epoxy resin react together to produce long polymer chains that are solids that cannot melt. The polymerized structure is a very good barrier to most but not all external agents. A small amount of moisture and gases penetrate the polymer enclosure. The barrier properties are good enough to protect all but a few types of semiconductors. The liquid starting point of these encapsulants means that they can be pumped and dispensed by machines and then hardened to a permanent solid at a later stage. The liquid epoxies have a composition very similar to the EMCs that are so common. The filler in both is simple silica (SiO_2) that is used primarily to reduce the coefficient of thermal expansion. Wetting and flow agents also may be added.

6.2.2 Functionality

There are several types of encapsulants, and each has a different function. Retaining wall or damming compounds are used to form the wall around the die that is needed to hold back the liquid encapsulant that is often called the *fill.* The precured properties are very important because the damming compound must stay in place. The viscosity is high and must remain so even during the curing stage, which can be as high as 175°C. The uncured retaining dam also must hold the low-viscosity liquid fill encapsulant so that both materials can be cured together in a single step. The cured properties of the retaining dam usually are similar to those of the fill, although this is not a requirement because the dam does not contact the die. The ideal system of dam and fill will fuse together during the thermal cure to provide a strong, resistant, and aesthetically appealing structure. The appearance is important because the package will be seen and handled by assemblers and others. Also, a smooth finish is need so that nomenclature can be printed on the package and an automatic assembly machine can handle the package.

Packages that do not need an added retaining wall can be encapsulated with fill material directly. Low viscosity allows the material to flow around the fine-wire structure and also helps reduce air entrapment. However, too low a viscosity can cause the filler to settle before the cure step is complete. Heating nearly always causes the prepolymer system to drop in viscosity and thus accelerates settling.

6.2.3 Performance Characteristics

First, all polymer-enclosed packages are considered nonhermetic. The die is exposed to small amounts of gases that diffuse through the polymer structure. The most significant gases in any measurable quantity are oxygen and water vapor. Oxygen generally is not a problem for common silicon chips that are already protected from further oxidation by silicon oxide. The aluminum (or copper) pads on the chip readily oxidize, but the resulting oxide is self-limiting. Moisture is a larger problem, however, because it serves as a medium for many chemical reactions, especially corrosion. Ions such as chloride (Cl^-) are especially damaging in the presence of water and can degrade chip pads and the wire bond. Chloride is present as a by-product of nearly all epoxies, and too high a level will cause reliability problems. Liquid encapsulants are made with epoxies that are highly purified to remove chloride and other ions so that the finished product will have an ion level of 10 ppm or less. The moisture-barrier properties of a good encapsulant will only allow a fraction of a percent of water absorption, even in a high-humidity environment. Therefore, the epoxies provide a highly reliable enclosure, provided that chloride is kept low and moisture resistance is high. Encapsulants have continued to improve in both regards, and these products are considered to be highly reliable.

There is another reliability concern for BGAs made with organic substrates that involves performance during the heating in the solder assembly process. The small amount of moisture within the encapsulant and the chip carrier can outgas at an explosive rate during the soldering step. Enough pressure can be generated to cause delamination or even cracking of the encapsulant in a phenomenon called *popcorning.* While the source of moisture entry can be the package, any reduction in absorption by the encapsulant should improve the situation. The industry has continued to improve moisture resistance of epoxies, often by adding water repellants. However, some epoxy encapsulants contain copolymers such as cyanate esters that improve moisture resistance and other properties. The package may require predrying prior to soldering depending on moisture-absorption properties and storage conditions.

6.2.4 Handling and Storage

Liquid encapsulants are commonly supplied as a premix of hardener and resin that is reactive even at room temperature. The reaction rate depends directly on temperature, however. Raising the temperature to around 150°C will cause the polymerization to complete in 30 min or even less. Conversely, cooling the encapsulant to −40 to −50°C will retard the reaction and thus provide a storage lifetime of up to 12 months. The vast majority of liquid encapsulants are shipped with ice packs or dry ice and stored in freezers by users. The material must be brought to room temperature before dispensing, and the time required depends on the size and construction of the container. Many products are sealed in plastic bags that should not be opened until the product has warmed up. This keeps any water that condenses out of the atmosphere away from the primary container, which typically is a syringe designed to fit a needle dispenser.

Epoxies are not very hazardous, but they should be handled with the respect afforded any chemical. The liquids can irritate skin and cause dermatitis in some people, and gloves are recommended if there is any chance of skin contact. The materials are not very volatile but should still be handled with good ventilation. Any pretesting should be done in a fume hood, especially if the material will be heated. Many liquid epoxy encapsulants use acid anhydride hardener that can be released during open heating. The anhydride reacts with moisture in the air or in the nasal passages to form weak but irritating organic acids. The uncured epoxies are soluble in alcohols, such as isopropyl alcohol (IPA), and most organic solvents. They are not soluble in water but can still be washed off with detergent, which will emulsify them. As with all chemicals, they should be disposed of in keeping with state and federal laws. Since solid plastics are safer and easier to deal with than reactive liquid precursors, it is a common practice to harden any of the waste liquid.

6.3 ENCAPSULATING METHODS

6.3.1 Overmolding

Overmolding, or *transfer molding,* is the more common method for encapsulating standard packages. This is covered in Chap. 7 and is discussed briefly here for reference as the alternative to liquid dispensing. The method is capital-intensive, and a good machine can cost over $1 million. However, the method is very well suited for high volume because it is a parallel process that can produce hundreds of parts per cycle.

6.3.2 Overmolding Equipment

The modern molding machine is large and costly but has very high output that justifies this method for high-volume production. Figure 6.1 presents a simple diagram of a molding press. Solid epoxy molding compound preform is preheated and then physically transferred to the mold, where it is liquefied and injected under pressure. The liquid epoxy enters the mold cavities that contain the electronic packages to be encapsulated. The heated mold is held at a temperature that causes the liquid epoxy resin-hardener compound to polymerize into a permanent solid. The mold is opened, and the parts are ejected. The parts may be exposed to more heat in a postcuring operation. The process is efficient for high-volume production but not well suited to short runs.

6.3.3 Needle Dispensing

Needle dispensing must deal with liquid materials, but the viscosity can range from water-thin (alcohol and flux) to high-viscosity pastes. The process can be as simple as using a syringe that relies on human force or as complex as an automatic dispensing machine that even calibrates the volume or weight of material and adjusts for changes in its properties. The key feature of dispensing is that no tooling is required. The programming is the virtual tooling.

Modern equipment can check fiducials on the substrate, calibrate material, position the needle, set the height, and dispense the correct volume of material. Needle dispensing is the most versatile of all the encapsulation methods, but its serial nature means that it is not inherently as fast as mass methods such as pin transfer. Viscosities can be very high as seen with solder paste, conductive adhesives, and damming compounds or very low, as found in fluxes and underfills. Another valuable feature is that material can be applied in the presence of components or even on three-dimensional (3D) structures. In fact, many applications involve placing material next to, into, or onto components already in place. While this data-driven "no tooling" method has superb versatility and precision, the most significant limitation has been modest throughput. Fortunately, the productivity issue has been solved, and this will be the theme in the remaining sections.

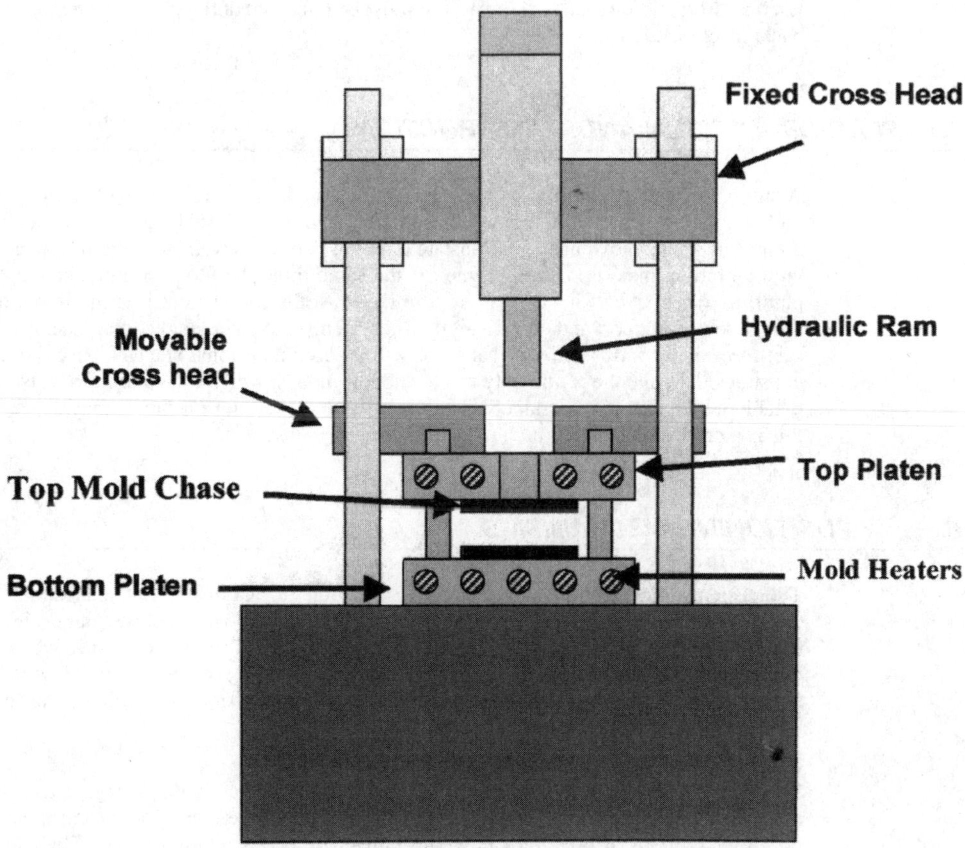

FIGURE 6.1 Transfer molding press.

6.3.4 Needle Dispensing Equipment

Basic requirements are accurate positioning of the needle over the target zone, automatic height adjustment, and precise control of the dispensing process. The amount of material deposited and its shape are determined by a number of factors, including needle height, pump rate, pressure, temperature, and rate of needle withdrawal (*z* height). In many ways, needle dispensing is a complex process requiring more machine control than other methods.

The dispensing machine consists of a frame that holds the subsystems, a platform that produces *xy* positioning movement, a moving head that contains the needle and dispensing control, a pump with fluid delivery, and the master control system. Some machines move the base that holds the substrate instead of the needle. Fully automated equipment has substrate handling to move parts in and out of the dispensing zone. The dispenser may have a vision system, heaters, and even an atmosphere-control enclosure.

Components of the subsystems include

1. Machine platform
2. *xyz* positioner
3. Control head
4. Fluid handling system including the pump
5. Total system control unit

Each section is critical, and their design must be balanced and well integrated. Figure 6.2 shows the basic machine elements.

6.4 PLATFORM DESIGN AND CONSTRUCTION

A sturdy and robust platform is at the top of the list. Just as a reliable building or vehicle must begin with a sound foundation or frame, the same holds true for durable and reliable assembly-line equipment. Any frame movement can translate to motion errors as well as maintenance problems associated with vibration, distortion, and fatigue. At the same time, the footprint must kept at a minimum. The platform, often taken for granted, is more than just the foundation; it is the frame that must be well thought out to support and contain all the subsystems that make up the total machine. Some frames are cast from metal and "aged" so that there will be no dimensional changes once the machine has been constructed. While the frame only has to support modest weight, the ability to resist movement during machine operation is important. Thus, while some dispensing machines may seem to be "overbuilt," this is a prudent design feature and a sign of true quality.

6.5 XY POSITIONING MECHANISMS

The dispensing needle must move in an *xy* direction in relationship to the substrate, and this requires that either the needle assembly or the piece parts be moved. The most popular design concept is a stationary piece under a moving head. The design challenge is to provide fast movement with high precision and repeatability. However, the system also must be capable of very slow movement required for dispensing damming compound in a rectangular pattern or underfill along the edge of a tiny chip.

6.5.1 Conventional Positioning Assemblies

Standard dispensing machines typically are comprised of a single gantry, an *xyz* motion table, that accommodates one or two valves and a single- or dual-lane conveyor system to shuttle parts in and out of the equipment. Accuracy down to 2 mils (50 μm) is required today. There are many architectures, but most involve separate *x* and *y* drives that work in unison to move the dispensing head to a precise location. Each directional unit has a motor and a mechanical movement with a controller. The platform also includes a z-axis movement to lift and drop the dispensing head (Fig. 6.2).

A common motion system may use ball screws and servo motors with closed loop encoders to provide smooth point-to-point movement with input-output zones to reduce substrate transport time. Lower-cost systems still use pulleys and cables, but the penalty is reduced accuracy and repeatability. More than one needle dispenser may be configured onto the gantry.

6.5.2 z Positioning (Vertical Control)

The tip of the dispensing needle must be positioned at the right height above the substrate, and the distance may be different for various applications and materials. Height control is also important if large boards or long strips of packaging substrate are to be managed and the deposit is to have optimal accuracy. The z height, the space between the needle and the substrate, is more important than many people realize. Too great a distance and a small drop may not even deposit, whereas too small a distance will deform the shape and even interfere with flow. Improper z spacing also can cause tailing as the needle moves away. Effective designs would employ an active touch probe with a transducer to accurately access pressure and thereby control height to the desired preset values. Alternatively, a noncontact laser sensor may be employed to measure z height on densely populated boards or on delicate substrates. Since the substrate can have warp, it is desirable to have a means of adjusting for the height variation. Height-sensing technology can be used, but a mechanical "foot" has proven to solve the problem much more simply.

6.5.3 Multigantry or Modular-Style Systems

Since needle dispensing is a serial process applying only one "dot" at a time, efforts have been made to boost the rate to better compete with screen printing, which is a parallel deposition method. Note

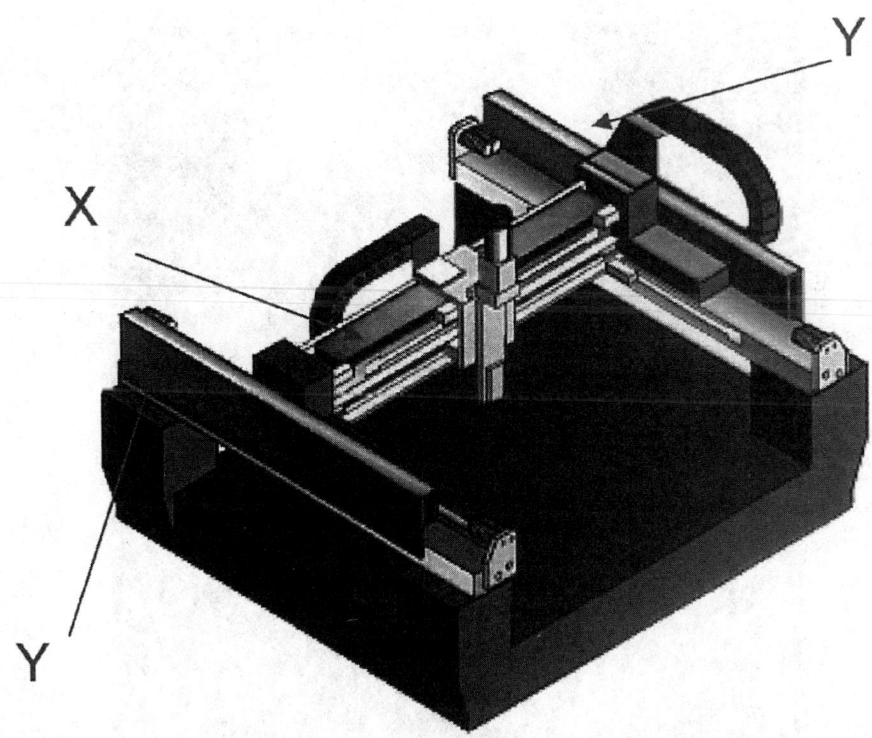

FIGURE 6.2 Typical single-gantry platform of a dispensing system.

that screen printing can be used for surface mount device (SMD) attachment adhesive dots, but the method is not practical for underfilling and the encapsulation process in most situations. One alternative to increase the overall production capacity is to link multiple machines. Each one of them requires a handling system and an area to create a work cell. Clean-room floor space is at a premium, and any solution to avoid redundancy in the line (such as reducing the handling system and conveyor) can save real estate. Camalot XyflexPro is a compact, modular dispenser that can be linked to one, two, three, or four units in-line to provide required throughput. All machines are controlled from a single host computer and user interface. They can be converted easily from SMD assembly to semiconductor packaging applications in the field. Dual-lane single-head dispensers are another possibility, but they are used very rarely because of additional cost for the handling system for a minimal gain in units per hour. The last alternative is to design a multiple-gantry machine that minimizes the footprint but still achieves greater units per hour (uph). Traditional gantry designs where the y axis carries the x and z axes, determine the overall machine footprint. Innovative designs using two y-axis gantries allow a four-head system to occupy the floor space of only one and a half times the size of a standard dispenser yet deliver three to four times more units per hour (Fig. 6.3).

Since the gantry occupies the greatest area in the system, it dictates the amount of floor space required by the dispenser. Incorporating traditional xy gantries into a multihead system leads to an extremely large platform, which would occupy the same floor space as linking multiple dispensers side by side. Using this innovative gantry design, the Camalot Xyflex is a four-head dispensing system that occupies about 1½ times the floor space of a single-head dispenser and dispenses at rates up to three to four times faster (application-dependent). This is accomplished with a gantry design that is driven by a single computer that helps the system to achieve its speed with four independently operating heads, each with a dispense unit installed on a y-axis gantry only (Fig. 6.4).

FIGURE 6.3 Multiple gantries (Speedline Camalot Xyflex).

High throughput is attained without the higher purchase cost and system complexity of linking two or more dispensers in series. Gantry robustness and accuracy are also critical to achieving high-volume, high-yield production. Lead-screw-driven gantry systems historically have proven more accurate at high speed and more robust than alternative designs. With the demands for higher units per hour, the gantry must now keep up with the pump dispense rate. Most dispensing valves are capable of dispense rates greater than the ones use in production, but dispensers with superior gantry systems are going to achieve higher yields and greater units per hour. Dispensing patterns can be run at greater speed without overshoot or misalignment and without causing mechanical defects (i.e., wire-bond sweep). Innovative dispensing patterns are also an important part of void-free encapsulation. Advanced software allows the end user to optimize and balance the dispenser's multiple heads. Two modes are used frequently: parallel mode, in which every gantry executes the same pattern, or serial mode, in which each head executes a section of the program. Typically, for a BGA encapsulation, one or two heads dispense the dam, and two to three heads deposit the fill pattern.

6.6 DISPENSING HEAD

The head contains the vision camera, the z probe, and the all-important dispensing unit (DU). This is where the action is—where the material meets the product. The dispensing unit must respond quickly and accurately regardless of the rheology of the material. A needle calibration station can be used where the highest accuracy and consistency are important. Quick change is also required for high throughput. The lowest mass would facilitate fast, accurate movement. At the same time, a rugged structure is a fundamental requisite.

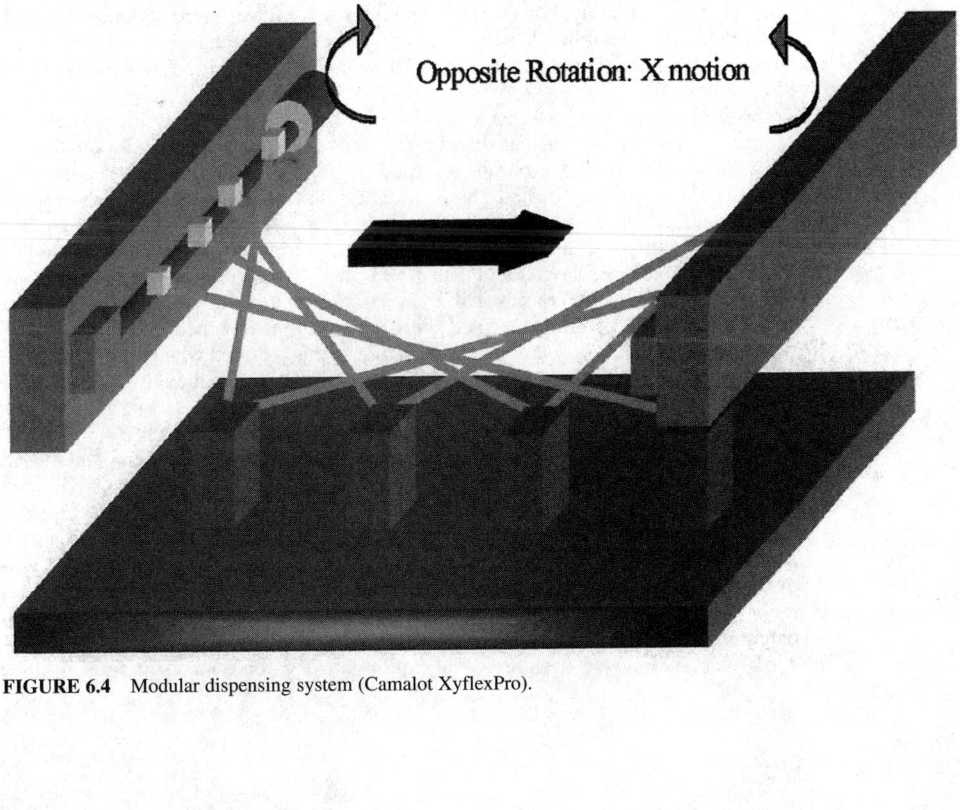

FIGURE 6.4 Modular dispensing system (Camalot XyflexPro).

6.7 NEEDLES

The industry is well supplied with manufacturers of precise needles covering a wide range of diameters.

6.8 PUMPS

The pump is the heart of the dispensing machine, and much engineering has gone into improving designs and constructions. The simplest mechanism is to apply manual pressure to a plunger, and this can be automated with a piston pump. Air pressure also can be used, and such systems are known as *pneumatic* pumps. Auger pumps add more precision and control and are popular in automatic dispensing machines. Desire for maximum control, however, has led to several pump concepts that fall under the heading of *positive-displacement.* Each pump style will be covered in the following subsections.

6.8.1 Pneumatic Pumps

The simplest method for rudimentary automation is the pneumatic approach, where air pressure pushes liquid out of a syringe. Precision is only moderate, but the mechanism is quite simple and is used as an adjunct to manual assembly. The manual plunger is removed from a disposable plastic syringe, and the end is fitted with an adapter that connects to an air controller. A simple pressure controller, often with a settable timer, is used to regulate air pressure and duration.

6.8.2 Piston-Type Pumps

Piston pumps apply force directly to the liquid material and are considered to be positive-displacement pumps. They can be single- or multiple-piston types and work well with thin and high-viscosity materials. This type of pump, however, has found favor for lower-viscosity underfills and fill-type encapsulants. Volume repeatability is better than ± 1 percent, and recharge time is about 1 s. Figure 6.5 shows a cross section of a single-piston pump. These pumps can be used with disposable plastic syringes for lower cost of use.

Multipiston pumps have been developed more recently to provide a faster recharging time and more accuracy at higher dispensing rates. In fact, the flow rate is continuous, so recharge time essentially drops to zero. These pumps are used for high-rate dispensers of encapsulants and underfills. Advanced designs have heated pump chambers so that the viscosity of the encapsulant or underfill can be reduced and kept constant. Figure 6.6 shows the multipiston design and how it works.

Why have dispensing machines moved to new pumps such as the multipiston type? A single-piston pump is capable of achieving less than 1 percent volume repeatability, and the smaller micropackages can require increased tolerances. Different designs of single-piston pumps are available, but all use the same principle. A chamber is filled with material that is expelled with a piston. Even though the repeatability and accuracy are good, a recharge step is required when the chamber is empty. It typically takes 1 to 10 s to recharge a 2-cc chamber, which results in productivity loss. Using an Auer boat (parts carrier) containing five BGA packages requiring 900 mg of material, a recharge step is required to finish the entire boat, thus interrupting production for a minimum of 10 s.

Technology has evolved to the multiple-piston pump design to eliminate the recharge step inherent in the single-piston pump design. While one chamber is being emptied, one or two others are being recharged simultaneously. In a dual-chamber piston pump, the recharge rate is equal to or faster than the dispense rate. One commercial multipiston pump incorporates three pistons driven by a cam–gate valve assembly. This continuous-flow design allows the pump to dispense material through one chamber onto the parts and simultaneously recharge the other two chambers. The slow recharge rate allows the material to feed the chamber uniformly without creating a vacuum. Its patented design prevents dripping from occurring. On the other hand, commercially available standard rotary-gate-valve designs usually allow air and material to migrate into the assembly when

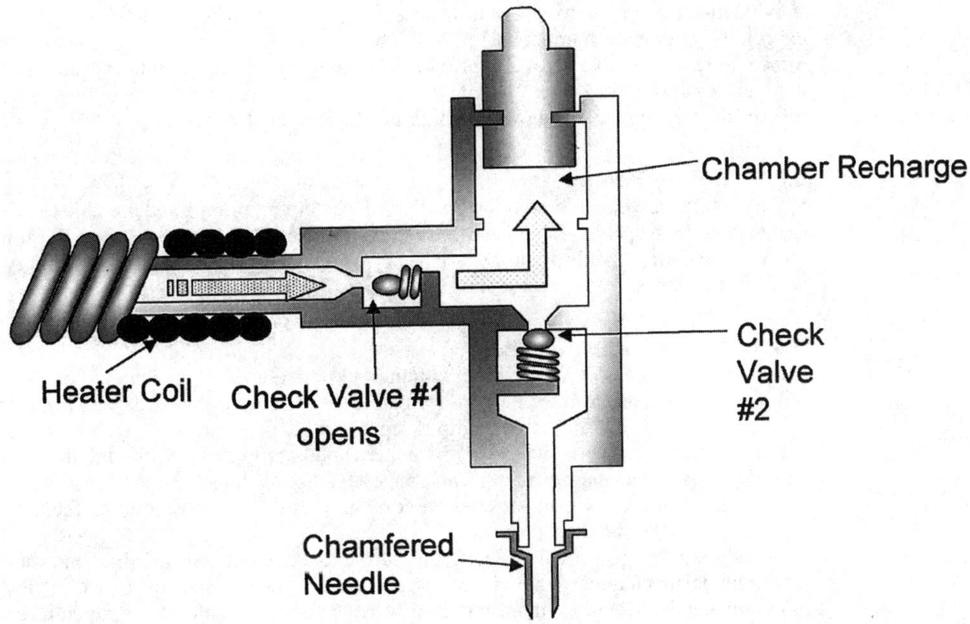

FIGURE 6.5 Single-piston pump.

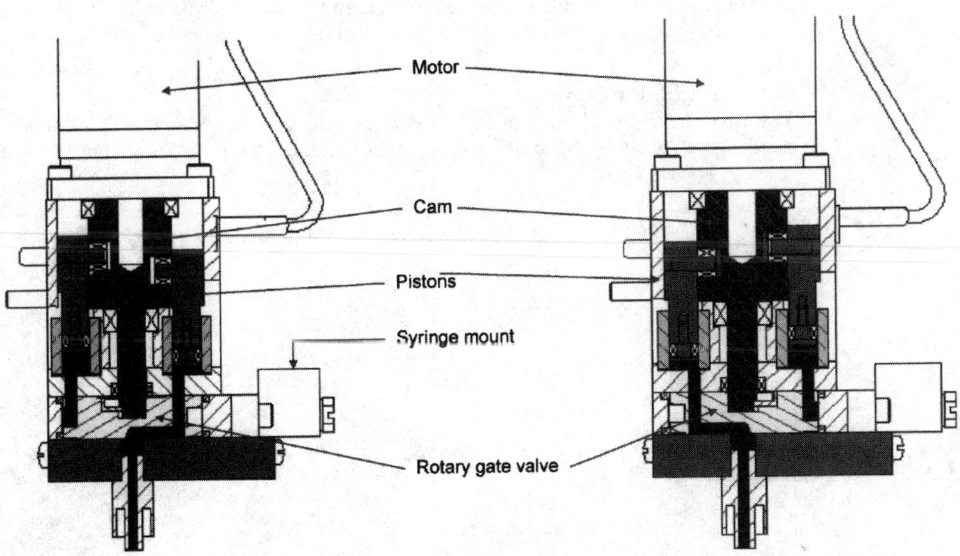

FIGURE 6.6 Multipiston pump.

moving from chamber to chamber. The entrapped air causes dripping or inconsistent dispensing, especially after the chamber is refilled. The multipiston pump design maximizes units per hour (since no recharge is needed) and provides less than ±1 percent weight repeatability. Combined with a weight-calibration system, customers are able to achieve process capability index (Cpk) numbers greater than 2. Figure 6.7 shows the high level of repeatability.

6.8.3 Rotary Pumps

Rotary pumps are considered to be positive-displacement pumps and are able to meet high precision demands at high speeds. Repeatability is excellent even with a very fast rate. Cleaning is simple, and changeover is fast. Multiple nozzles can be configured. Figure 6.8 shows a rotary pump assembly with two syringes.

6.8.4 Auger Pumps

The traditional auger design for semiconductor applications is perceived to be outdated technology, but it is still the preferred technology for dispensing the retaining dam. Very high viscosity materials perform marginally with piston pump designs, whereas augers are capable of transferring viscous material with ease. Typically, dual-helix augers and components exposed to the material are made of carbide to enhance their strength and longevity. Again, these pumps must be fast, accurate, and repeatable to produce a dam of consistent height, width, and cosmetic perfection (i.e., no bulge at the start or end of the dam).

The auger design provides steady pressure and excellent repeatability, and the mechanism does not produce mechanical friction heating that will change viscosity. However, the lower cartridge assembly can be heated for materials that should be applied at elevated temperatures. Pressure is very constant and can be held at a low value for very fluid materials. One design keeps the motor at a constant speed, and a clutch is used to transmit motion to the pump when required. The dripping tendency is very low with augers. Figure 6.9 shows an auger pump and head configuration.

Modern auger pumps have a double-helix thread for higher precision, and the lead screw and insert are made of carbide to reduce wear caused by mineral fillers in most encapsulants. Advanced dispensing machines use a closed-loop rmp controller.

6.8.5 Fluid Jet Pumps

The fluid jet approach to deposition involves a noncontact method where liquid is "shot" from the head at the substrate. Strictly speaking, this is not a needle dispenser but is included under pumps

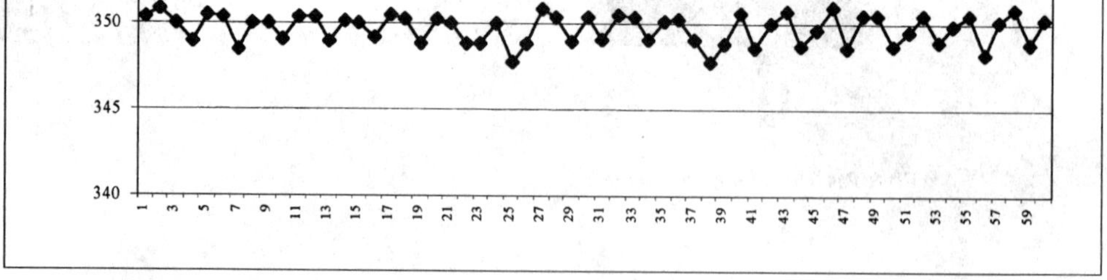

FIGURE 6.7 Repeatability.

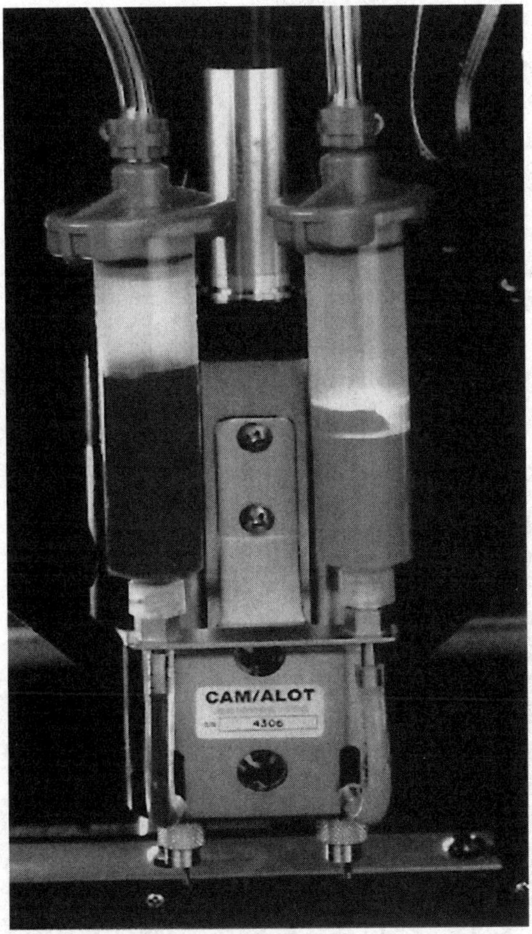

FIGURE 6.8 Rotary pump assembly.

for thoroughness. Fluid jet dispensing is much less versatile than needle dispensing and is used primarily to apply dots of adhesive for SMD assembly. The method has not found significant use in the encapsulation area.

6.8.6 Overall Fluid-Handling System

Earlier dispensing designs consisted of a fluid reservoir, a pump connected to the reservoir, valving, and a needle. Today's systems are much more integrated into a dispensing head. The basic fluid-handling system consists of a reservoir or container in a holder, at least one pump, and a valving mechanism. Advanced equipment should detect low material levels and even report changes in material properties, especially a viscosity drift. While a number of pump types are used, rotary-auger or positive-displacement pump designs are best. Air pressure mechanisms are considered inferior. Multipiston and rotary pump designs were considered, and both versions were implemented. The pumping system must respond quickly and repeatably, dealing with very small or large volumes. A constant-pressure design was considered ideal. The wide variety of materials, from peanut butter-like pastes to watery fluids, requires an extremely versatile pump system.

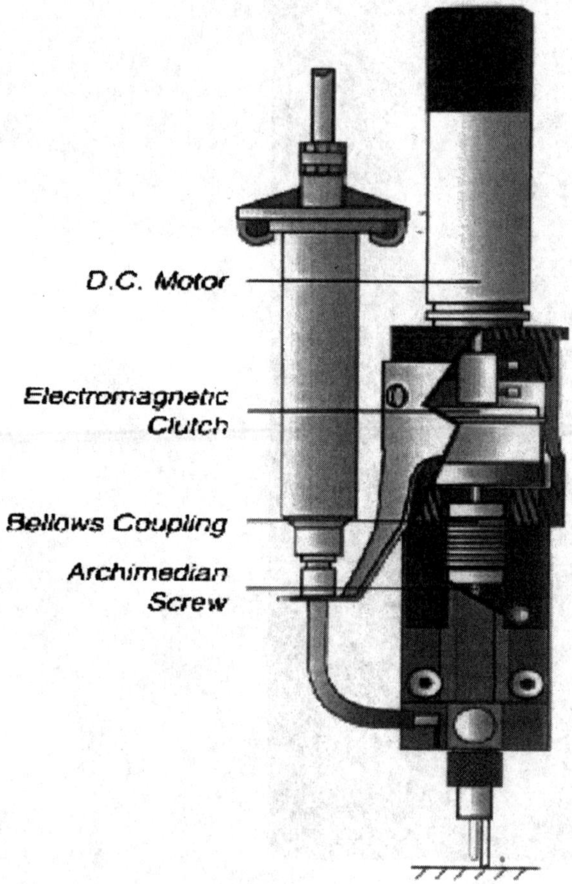

FIGURE 6.9 Auger pump assembly.

6.9 SYSTEM CONTROL

6.9.1 Advanced Process/Software Controls

The alignment system controls the *xy* positioner motion so that material will be deposited precisely where desired even if the substrate is dimensionally unstable. The best system uses a digital camera, advanced illumination, and sophisticated vision-recognition software along with the appropriate computer hardware. Such an intelligent vision system is able to detect inconsistencies in the substrate and "no component" areas, as well as do a quality check on the deposited material.

All the subsystems must work in concert with good balance. The central system is the nerve center and mission control for the entire machine. High-speed data access and analysis are essential. The system must deal with simultaneous input, make a sound decision, and initiate the correct action. The most modern, state-of-the-art technology is employed. Fiberoptics, found in advanced aircraft such as the Boeing 777, is used for maximum rate, extreme reliability, and immunity from interference.

The technical advancements in materials are forcing the dispensing equipment industry not only to be faster and more accurate but also to provide new software advancement to help the end user improve productivity and reduce down time. The new high-flow or single-pass materials require equipment that can handle higher dispensing rates and can still maintain superior positioning accuracy. Most underfill materials have a pot life of 8 h, which basically means that after 8 h of exposure at room temperature, the material doubles in viscosity. The viscosity being a function of time and temperature, it is critical to maintain the temperature of the material in the syringe constant at room temperature. The material can then be heated just prior dispensing to improve its flow characteristics. Several software features are available to help optimize pot life. Timers can be set to warn the operator that the material is reaching its limits. Alarms are also available to ensure that the substrate has reached the optimal flow temperature of the material. The fast cure materials have a tendency to thicken in the needle tip if the flow is interrupted for too long, causing poor underfill during the next dispense cycle. An autopurge timer ensures that the material is moved for a settable amount of time at regular intervals.

Also, when the size of the die requires several successive passes, it becomes critical to provide material in a timely manner so that there is no material flow interruption and limited air entrapment risk. Timer commands can be inserted into the pattern to ensure that the proper interval between successive passes is respected. Additionally, advanced routines allow the dispenser to process as many parts as possible in a given period of time before returning to the first part to complete the process. Similarly, when dealing with multiple array panels or partially filled trays, new software routines are now capable of optimizing the process regardless of the number of dies or parts. Most of the cycle time is spent dispensing material and not waiting for it to flow. The new features offer the end user an automatic optimization and process-balancing system that maximizes the material characteristics but minimizes cycle time and down time.

Along with the linear positive-displacement pump, volumetric and weight-measurement systems were developed to target and monitor the process by either weight or volume. Weight-calibration tools integrated into the dispenser are more common. They monitor either the flow rate or the weight of the pattern to be dispensed. Typically, when the integration of the scale is well designed, settling time of the tool is fast. The finest resolution can then be used (to a tenth of a milligram), increasing the accuracy of the system. Vibrations, air drafts, and static can generate false readings. Even a half-milligram error caused by an air draft or vibrations can offset the setup of the machine and result in scrap parts. On the more advanced systems, a calculator is integrated to help the user set up the pattern for the right volume. Target weight, tolerance, and weight-calibration frequency are also programmable parameters. The weight adjustment is achieved by automatically adjusting either the travel speed of the dispense head or the speed (rpms) of the pump. On the more advanced systems, averaging can be used when the amount dispensed is low, thus minimizing the effect of weight-scale limitation (to a tenth of milligram). For example, at 10 mg, a variation of 0.1 mg is 10 percent.

6.9.2 Human Interface

While dispensing machines have become more complex, the human interface has been designed to be friendlier. Multiple screens are used with high-resolution graphics that show just what is happening and what function is being queried or altered. Today's machines are easy to program and have a graphic interface with real-time process monitoring.

6.9.3 Computers

Today's fast and sophisticated dispensing machines use a relatively powerful computer to control functions, handle the operator interface, monitor the process data, and store both programs and data. These computers are industrial hardened and can have uninterrupted power supplies so that no data are lost in the event of factory power disruptions. Redundancy of critical systems is found in the best machines. Quick change-out and "hot swapping" of computer boards and controllers are valuable, especially if there is easy access.

6.9.4 Software

Today's software has a high-level graphic interface to make the system friendlier and less prone to operator errors. The entire process can be monitored in real time. The setup screen takes the operator through the process step by step and "flags" unlikely parameter inputs. Figure 6.10 shows a setup screen. Powerful software allows the end user to create complex patterns with multiple passes without any idle time. Self-diagnosis and preemptive troubleshooting are valuable in the software.

6.9.5 System Calibration

Advanced dispensing machines have calibration-integrated hardware and software that can be set for periodic calibration of material weight, volume, or both (Fig. 6.10). When used in a multiple-gantry system, greater units per hour are achieved with minimal machine footprint. Along with the linear positive-displacement pump, volumetric and weight-measurement systems were developed to target and monitor the process either by weight or by volume.

6.10 THE DISPENSING PROCESS

6.10.1 Material Constraints and Considerations

A large number of materials are used, and dispensers have been designed to handle products ranging from viscous damming compounds to very fluid fast-flow underfills. The industry also has specialized in and built machines for BGA encapsulation, others for underfill, and still others with vacuum assist for filling internal regions of CSPs.

6.10.2 Substrates

Substrates vary in both material composition and shape. The most common packaging substrates are ceramic, rigid organic, and flexible circuitry. While single packages are used, productivity demands have led to the use of strips and even reels of substrate in the case of flexible carriers.

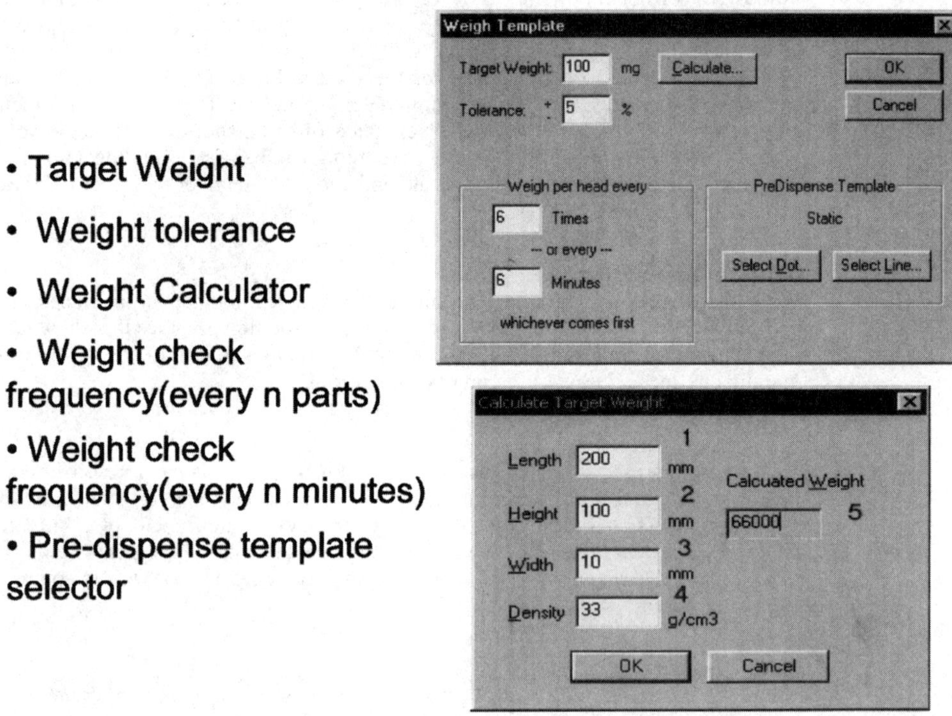

FIGURE 6.10 Calibration screen.

6.10.3 Applications and Processes

The two main applications areas for liquid dispensing in packaging are underfill for the flip-chip-in-package (FCIP) and encapsulation of the many types of BGAs. While many of the CSPs are little more than small BGAs, some have unique structures that require different dispensing methods. The μBGA, for example, requires a vacuum-assisted filling process because an internal cavity is involved, and air must be displaced. Table 6.1 lists a few of the products where liquid dispensing is employed. However, the use of area array packaging has become so pervasive that it includes nearly all electronics except very low end products that still use low-density feed-through devices.

6.10.4 Chip-on-Board (COB)

From a packaging perspective, COB is a common method for building most BGAs and some CSPs. The chip can be electrically interconnected to the package by three basic methods, and each represents a different challenge. The most common, wire bond connections, can present the problem of covering a large number of very fine and closely spaced wires. Air can be trapped, and filler in the encapsulant can aggravate the problem by blocking space between closely spaced wires. The dispensing head can be programmed to move around the wire structure in a spiral pattern or one that has been found to reduce any propensity for air entrapment. In addition, a dam may be formed first around the perimeter of the package if there is no recess to hold the filling encapsulant.

Tape carriers, really flexible circuitry, are becoming more common as area array packages. In a sense, the wires for bonding are built into the circuit or carrier. An array of fine metal beam leads is formed to correspond to the chip pad layout. The beam leads are bonded to the chip pads by a tape automated bonding (TAB) bonder that applies heat and force. All the bonds may be made simultaneously (gang bonding) or only one side at a time. For very fine pitch, the wires can be bonded one at a time like conventional wire bonding. The resulting chip and wire type of connection is easier to encapsulate that the traditional wire-bonded chip. The array is much more orderly and planar, making it unlikely that any two wires will be too close to permit encapsulant flow. Since there is only a single set of bonds, those at the chip, there is less impedance to flow.

The densest connection means is direct chip attach (DCA) or FC, where the chip is flipped over and interconnected directly to the packaging substrate. The bumped chip usually is soldered, but some processes use conductive adhesives. The assembly will now have a gap between the chip and substrate that must be filled with encapsulant called *underfill*. This is covered in several other chapters, and the reader should refer to the index for the particular underfill process of interest. The underfilled package may now be ready to ship. However, the back of the die is exposed, and some packages are further encapsulated to hide the chip. However, this seals in heat and is not recommended for any integrated circuits (ICs) that generate substantial heat. Since the back of the die is not very sensitive to handling, leaving it exposed allows heat to be removed. For example, a heat spreader or heat sink can be connected to the die either as part of the package or after assembly.

TABLE 6.1 Typical Large-Scale Electronics Assemblers' Applications

Industry	Typical end product
Communications	TV cable boxes
	Telephone modems
	Satellite applications
Automotive	Air bag sensors
	Car radios
	Brake systems
Consumer electronics	Personal digital assistants (PDAs)
	Mobile phones
	Pagers
	Personal computers
Medical	Analyzers/diagnosis equipment
	Hearing aids

6.10.5 Dam/Fill

The retaining dam application step involves first applying a viscous paste called *damming compound.* A square or rectangular pattern is applied close to the edge of the package to retain the more fluid fill encapsulant in the second step. The dispensing machine is programmed to automatically form the dam and then fill the area. Two separate machines can be used, but a multihead type is ideal. Large BGAs with high-density wire bonds require several dispensing steps. A first layer of material is dispensed on top of the wires of every package, typically at a moderate speed, to avoid any sweep. This allows the epoxy to adequately flow around the wires and prevent any air entrapment. A second pass dispensed at a higher rate to finish the encapsulation then follows. Figure 6.11 shows the dam and fill steps.

(a)

(b)

FIGURE 6.11 (*a*) Applying the dam. (*b*) Fill step.

6.10.6 Underfill

The aim of the preceding steps is to be able to bring the needle of the valve as close to the die edge as possible in order to avoid contamination around the die perimeter. It is also necessary to position the needle tip just below the lower surface of the die to maximize the flow underneath the die. As long as the position of the needle is close enough to the edge of the die and at the correct height, the material dispensed along the edge of the die will flow underneath rapidly and uniformly.

Figure 6.12 shows the results of insufficient, excessive, and correct dispensing. However, it is better to err on the side of excessive rather than insufficient dispense, but with the equipment available today, it is possible to control dispense volume to within ±2 percent.

Another issue that has been debated recently is the need to carry out a fillet dispense around the other three sides of the die. Until recently, this was thought to be purely cosmetic and a waste of valuable dispensing time and material. However, recent studies have shown that if the three-side fillet dispense is not carried out, the dispense side has a much larger fillet. During curing of the underfill material, the die may be subjected to uneven stresses that in some cases have lifted the die up and fractured the interconnections on one side. The needle height and position are important, and Fig. 6.13 shows the correct relationship to the chip.

It may be necessary with larger dies to make more than one pass along the edge of the die, and in other cases, the pattern of the pass also may vary. The way in which the needle is moved along the edge of the die and in which sequence can prevent air entrapment. For dies smaller than 3 mm^2, one pass may be sufficient. For dies larger than 3 mm^2 and smaller than 6 mm^2, a two-sided pass may be preferred in an L shape. These special L-shaped passes may be done in different ways, as shown in Fig. 6.14. Figure 6.15 shows the flow-out process. Also see the various FC and underfill chapters.

Temperature Control. When dispensing underfill materials onto any substrate, the temperature of the surface and the die must be 70 to 90°C. One reason for this is that it will create an upward-moving

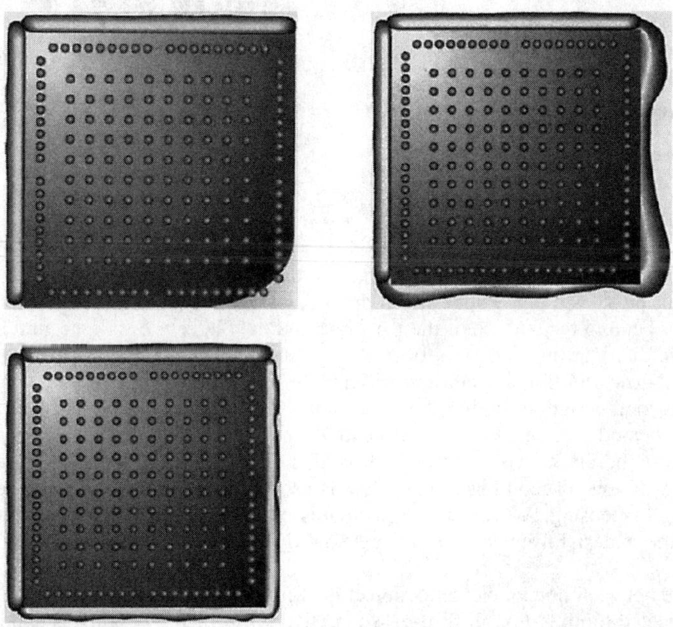

FIGURE 6.12 (*a*) Insufficient underfill. (*b*) Excessive underfill. (*c*) Correct amount of underfill.

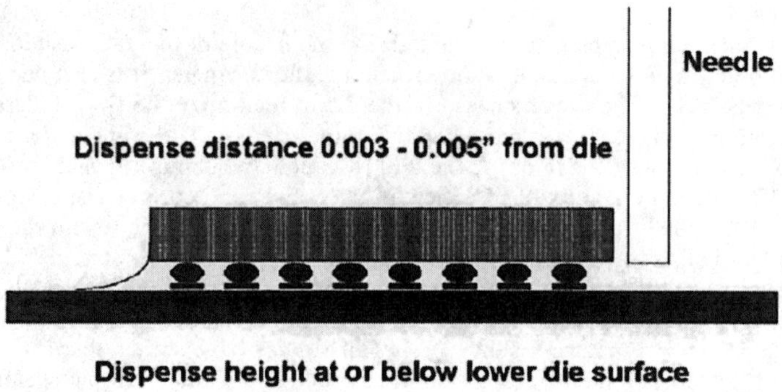

FIGURE 6.13 Needle position.

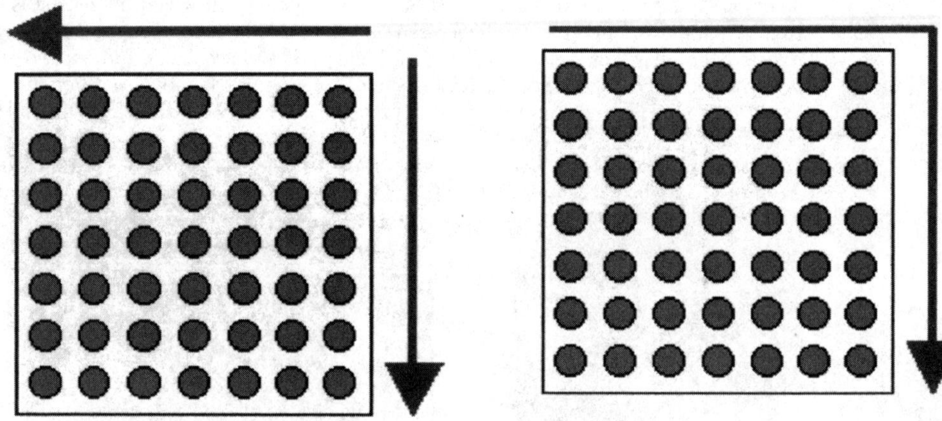

FIGURE 6.14 Dispensing patterns.

airflow from underneath the die and the surface of the substrate, degassing the surface and helping to eliminate air bubbles trapped underneath the die, which will result in long-term failure. The second reason for heating the surface of the substrate and the die is to ensure accurate flow of material underneath the die. Varying or nonuniform temperatures of either the substrate or the die can result in a nonlinear flow and thus air entrapment. The temperature of the substrate must be controlled using a closed-loop system to within ±5°C for noncontact systems or better than ±2°C for contact systems using heated vacuum chucks underneath the substrate. The substrate must be reheated to the ideal temperature before it arrives in the dispensing zone; a preheat station therefore is necessary. At the dispensing station, a second heating system is used to maintain the temperature of the substrate and die during dispensing. A third station, which is optional depending on the application, may be used to help the material flow correctly or gel slightly before being transported to a batch oven for curing.

The choice between noncontact and contact heating systems is a fairly simple one between flexibility or dedicated tooling. For all film-based applications, the only choice is contact heating using vacuum hot chucks under the film at the preheat, maintain heat, and flow stations. For all those applications using Auer boats or any dedicated transport tooling, as well as lead frame applications, the

FIGURE 6.15 Flow-out process.

contact heating system is the preferred method. Contact heating has the advantages of a more unified temperature distribution and the ability to heat only part of the substrate. In FC applications, for example, there may exist more than one size of die, perhaps ranging from 3 up to 12 or even 20 mm^2. By manufacturing dedicated tooling, it is possible for one set of tooling to handle the whole range of components. The only drawback to using contact heating systems is their inherent inflexibility, but often the advantages, such as speed of temperature ramp-up and temperature uniformity, outweigh the disadvantage of a 10-min tooling changeover.

In applications using standard FR-4 printed wiring boards (PWBs) as the substrate, such as FC underfill for contract houses where board sizes may vary considerably, noncontact heating can be used successfully. Although there is a tradeoff between the speed of temperature ramp-up and thermal uniformity, the possibility of changing the dimension of the substrate as required, without stopping the production, is an advantage for subcontractors. With the use of FC technologies becoming more and more widespread in mainstream electronics, manufacturers will be required to offer flexible noncontact heating systems with the advantages of contact heating where possible.

6.11 ENCAPSULATING WITH VACUUM ASSIST FOR CSPs AND OTHER PACKAGES

As semiconductor packaging continues to shrink in size and grow in number of inputs-outputs (I/Os) and complexity, more demands are placed on the equipment supplier to develop process solutions to aid the back-end semiconductor packaging industry to increase encapsulation yields and throughput. As the package sizes of CSPs get smaller, they become more difficult to encapsulate with standard dispensing processes. MCPs also are improved by vacuum encapsulation because their complexity tends to trap air. Two methods of increasing encapsulation yields and throughput have emerged recently. Vacuum-assisted dispensing and vacuum-assisted injection are now used as a result of new down-sizing assembly techniques. Encapsulating materials use highly refined mixing and packaging methods to avoid even the slightest introduction of air into encapsulation materials loaded into the final shipping media (e.g., syringes and cartridges). Thus the need for vacuum is not a result of problems with materials but rather is due to the ever-increasing complexity of the microelectronic assemblies they are used in or on.

One of the other problems with trapped air can show up just after the encapsulation process while curing the dispensed encapsulation. After encapsulant is dispensed into or onto the package, it must be placed in a cure oven to raise the temperature in the curing or hardening process. During this curing process, most encapsulation materials tend to decrease slightly in viscosity with the initial onset of the heated cure cycle. Then during the cure cycle, the encapsulation material's viscosity begins to increase rapidly to the hard or cured state. During this process, as the initial drop in viscosity occurs, trapped air can rise slowly to the surface of the component, leaving a pinhole that extends from point of origin to the surface. This pinhole will allow moisture and air to reach the components and cause premature failure of the device.

Vacuum applied during the encapsulation process can reduce void-type defects significantly. The encapsulation process for some CSPs was unsuccessful until vacuum was used. The extra care taken to slowly cover the wire bonds and capacitor was, however, time-consuming. The vacuum-modified dispensing process was an effort to reduce dispense time (and increase throughput). One process involves simple layering—initial layer, degas, and then the finish layer. Throughput is increased significantly, and the quality of the finished surface remains unblemished.

The Tessera μBGA package offers the size, weight, and performance advantages of an FC in a mechanically robust package that is compatible with existing SMD placement equipment. The μBGA is a popular CSP licensed by Tessera to over two dozen companies worldwide. This package uses a special low-modulus encapsulant (silicon-based). This material flows around the nubbin spacer columns that separate the silicon die and the flexible circuit tape. This material has a high viscosity, and although it is easy to dispense, the flow of this material is hampered by entrapped air underneath the die. Tessera and Speedline Technologies engineers have developed a method of dispensing material under vacuum that aids in the flow and allows complete coverage underneath the die.

The main advantage of a CSP, such as the Tessera μBGA package, is its inherent flexibility or compliance between the interconnect solder bumps/spheres on the bottom of the package and the die that overcomes issues related to CTE mismatch. This compliant interlayer or buffer (low-modulus silicone encapsulant) gives this package its compliant properties and makes underfill used in FC manufacturing redundant, which allows faster throughput times in circuit board assembly. Other than the fact that there is no longer a requirement for underfilling, the main advantage is that the packages are smaller than standard BGA packages but only 25 percent larger than an FC or bare die. Add to this the fact that the packages can be tested more easily than bare dies and can be placed with conventional pick-and-place production equipment, and we have a winning combination.

The way in which the compliant interlayer is achieved differs from package to package; however, one of the most popular at this moment is the μBGA package. This package uses a special low-modulus encapsulant, developed by Dow Corning, around the nubbin spacers that separates the silicon die and the flexible circuit tape. Although this material is easy to dispense, its flow underneath the die is hampered due to the viscosity of the material (50,000 to 60,000 cP) and to entrapped air between the silicon die and the flexible interposer. We cannot rely on capillary action as we do with FC underfill encapsulation, which has a viscosity typically in the less than 10,000-cP range.

Because of the high viscosity of this low-modulus encapsulant, approximately 50,000 to 60,000 cP, the material tends to flow very slowly. Unlike FC underfill, heating the part or the material inhibits (rather than enhances) the flow because of the curing characteristics of the encapsulant. The process thus required the dispensing platform to dispense a bead of encapsulant along three sides of the die, which are arranged on the circuit tape in a matrix format.

Vacuum dispensing equipment was designed by Speedline Technologies, with support from CSP maker Tessera, for dispensing in a vacuum environment and the associated manufacturing process. This system is not a single-chamber system, which has proven to be very slow. The system discussed in this section is equipped with three independent vacuum chambers, with the dispensing head actually working inside the main chamber at levels of vacuum equaling or exceeding 28 in Hg (710 torr).

The Speedline System 3900 has a small predispense vacuum chamber at the entrance end of the machine in which the boat, strip, or individual part is brought to vacuum at high speed. Once the required level of vacuum is obtained in the predispense chamber, an airtight interlock door is opened so that the part can move into the dispensing chamber, which is kept under constant vacuum while the machine is in production. The door separating the predispense chamber from the main dispense chamber is then closed, and the predispense chamber is brought back to atmospheric pressure so that it can receive another strip or boat. Inside the dispense chamber, an *xyz* movement system and dispensing pump with specifically engineered vacuum-capable components and lubricants dispenses the required pattern. Air is absent from the chamber, so no air can be trapped under the die. However, this alone does not ensure complete flow of the material under the die. Once the dispensing cycle is completed, the airtight interlock door between the main dispense chamber and the postdispense chamber opens, and the boat, strip, or individual part moves into the postdispense vacuum chamber. The door between the main dispensing chamber and the postdispense chamber then closes. In this third chamber, the environment is programmed to ramp slowly back to normal atmospheric pressure. As the pressure changes from vacuum back to normal atmosphere, the differential in chamber pressure forces the dispensed encapsulation material under the die into the void created while the device was under vacuum.

6.12 PERFORMANCE

The subcontracting business has increased dramatically as original equipment manufacturers (OEMs) continue to outsource more work. Subcontractors are now required to demonstrate their capability to produce better-quality components because they now compete against each other. Statistical tools such as CpK are now being used to qualify processes. In a BGA encapsulation, the critical dimensions are height and flatness of the package after curing. Because the wire bonds need to be covered and the surface of the encapsulation must be flat, the amount dispensed throughout the shift needs to be tightly controlled and monitored. In the past, 1 to 2 percent weight repeatability was required without CpK or accuracy requirements. Today, statistical process control and parameters are being used to qualify a process. CpK is a statistical number used to qualify a process. It indicates the number of defects that may be expected from a process. In the past, a CpK of 1.33 (or 31.5 defects per million) was the standard. Today, a CpK of 1.66 to 2 (100 defects per billion) is more commonly demanded from leading subcontractors. CpK relates the measurement of equipment capability as a ratio to the tolerance range established by the customer for the particular process. A higher CpK ratio is better because fewer devices need to be scrapped. To minimize the number of defects per million given a set of tolerances (USL, LSL), established by the customer, the process variations must be minimal. For example, to get a CpK of 2 given a USL and LSL of 2 percent, the 3SD (standard deviation) variation of the process has to be less than or equal to 1 percent (when mean is equal to the target).

6.12.1 Positioning Accuracy

An *xy* position accuracy of 2 mils (50 μm) is required today, and this can be met readily with the present equipment. Several gantry systems are even more precise. Vision systems make sure that positioning is accurate even when substrate is not precisely dimensioned. High precision is most critical for FC underfilling, where the needle must be located at a specific distance from the edge.

The *z*-height accuracy is also very important. Excessive distance from the substrate can produce too small a drop and may not even permit deposition. Insufficient clearance can deform the shape and even interfere with flow. Improper *z* spacing also can cause tailing as the needle moves away. Effective designs employ an active touch probe with a transducer to accurately access pressure and thereby control height to the desired preset values.

6.12.2 Shot Accuracy

Advanced dispensing machines have calibration-integrated hardware and software to ensure a high level of absolute shot-size accuracy. Weight-calibration tools integrated into the dispenser are becoming more common. Resolution of 0.1 mg can be achieved with an advanced system, and this level is becoming essential. Even a half-milligram error caused by an air draft or vibrations can offset the setup of the machine and result in scrap parts.

6.12.3 Application Rate

Multigantry systems with up to four dispensers and high-volume pumps now provide high application rates with a reasonable footprint.

6.12.4 Selecting the Right Equipment

The large variety of basic dispensing machine types and models within the categories makes the selection process more complicated than it was even a few years ago. The following subsections represent important areas for consideration.

Flexibility. How many types of materials will be run? Must a single dispenser run everything from flux to underfill? High machine flexibility is required when one dispenser must do many tasks. Although flexible equipment is available, there is always a compromise. As has been shown in earlier sections, some pumps are better for low-viscosity materials, whereas others are more suitable for thick pastes. One must balance flexibility against quality and productivity. Production may be served best by setting up dedicated lines.

Footprint. Factory or even laboratory floor space has considerable cost associated with it. Clean rooms add even more cost from both an initial construction figure and the annual upkeep. Fortunately, the industry has worked hard to decrease footprint while adding capability. Multigantry machines appear to provide the most efficient use of floor space.

Equipment Cost. While most salespeople will suggest that capital cost is the least important factor in determining annual running costs, productivity analysis bears this out. Productivity differences soon overwhelm initial investment dollars. Throughput is a major multiplier, and even a 10 percent increase offsets many thousands of dollars of investment. Another value to factor in is uptime. A more costly but more robust production machine quickly gives a return on investment if there is low downtime. Downtime should include actual time for repairs as well as maintenance. Modern features, such as self-diagnosis and hot swapping of control cards, reduce both factors.

6.13 SUMMARY AND CONCLUSIONS

Needle dispensing has become a key process to produce several types of area array products. Many BGAs use the dam and fill method of encapsulation that can only be done with needle dispensing. The die-down (cavity-down) packages also require this type of dispensing method. FCIP designs also use needle dispensing extensively.

Modern needle dispensers are highly automated and are self-calibrating. Software provides an easy interface for fast setup. Programs are stored for recall by part number, and many systems come with a library of common dispensing patterns. The latest equipment can provide very fast rates. The highest-productivity machines have up to four gantries and heads that can supply different materials that may be required in one package.

6.14 REFERENCES

1. Lee, C., "Ultra-High Speed Dispensing of Surface Mount Materials to Enhance Production Throughput," paper presented at APEX 2000, Long Beach, CA, 2000.
2. Miquel, B., and Lee, C., "Advanced Process Control for BGA Encapsulation and Flip Chip Underfill," paper presented at Semicon Taiwan, September 2000.
3. Lawing, C., "Preventing Voids in BGA Packages," *Chip Scale Review,* March 1998.
4. Lawing, C., "Vacuum Dispensing of Encapsulants for μBGA Manufacturing," *Electronics Packaging and Production,* February 1998.

CHAPTER 7
MOLDING FOR AREA ARRAY PACKAGES

K. Gilleo
Cookson Electronics

B. Cotterman
Cookson Semiconductor Packaging Materials

T. Chen
Intel Corp.

7.0 INTRODUCTION

Electronic packaging must provide physical protection during component storage, handling, and assembly to the circuit board and long-term operation within the finished product. Environmental protection is also required, especially resistance to moisture and ionic materials. The end user typically will expose the package to extremes of temperature, mechanical shock, and harsh environments. The finished package must have proper dimensions for machine handling while also allowing identification nomenclature to be applied. All these requirements can be achieved using polymers in the transfer-molding process. Plastic packages make up a large segment of area array packages. Both larger plastic ball grid arrays (PBGAs) and smaller chip scale packages (CSPs) use molding processes. Transfer molding is by far the predominant process employed because it has a long and successful history in the production of all types of packages. Transfer-molding processes and materials will be described in detail with this chapter. For now, it will suffice to describe the method that uses solid preforms of polymer precursors that are melted, injected into molds, and permanently hardened by thermally induced polymerization.

The most common materials in encapsulants are epoxy mold compounds (EMCs). The mixture contains reactive epoxy resins, a hardener, at least one reaction accelerator, flame retardants, a filler, a lubricant, a coupling agent, a flow promoter, a stress modifier, an ion getter (chemical scavenger), and a colorant. The basic epoxy-hardener prepolymer mixture consists of low-melting solids that will liquefy, flow, and polymerize to a permanent solid in the heated mold. Table 7.1 lists the effects produced by the components.

The solid materials are mixed, melted, and formed into preforms. A large EMC plant may make a powder or blocks of material that are stored and later formed into the desired preforms, often at other facilities. The in-process material must be kept at low temperature to avoid slowly polymerizing.

7.1 MOLDING MATERIALS

Epoxies are the predominate polymers used for transfer molding for a number of reasons. First, the cured epoxy provides very good mechanical and electrical characteristics. Perhaps the most important

TABLE 7.1 Molding Compound Composition

Component	Percent	Function(s)
Polymer bulk		
Coupling agent	<1	Bond between hardener and filler
Resin	5–20	Provide cross-linking reaction
Hardener (curing agent)	3–10	Provide cross-linking reaction
Accelerators		
Catalyst	<2	Accelerate cross-linking reaction
Fillers		
Silica	70–92	Improve physical properties
Additives		
Flame retardant	<3	To meet UL-94 flammability requirement
Pigment	<1	Color for marking purposes
Lubricant	<1	Improve mold release
Stress modifier	<3	Reduce internal package stress
Flow promoter	<3	Improve compound flow length
Adhesion promoter	<2	Improve compound adhesion to different surfaces
Ion getter	<1	Improve corrosion resistances

feature of epoxies is that starting materials can be solids. In fact, the range of resins and hardeners is so broad that the melting point can be formulated precisely. This is important for EMCs because the materials must be melted and transferred to the mold. The cure rate also can be tailored by the judicious choice of resins, hardeners, catalysts, and accelerators. This is important because material must remain in a liquid state long enough to be transferred to the molds. Also, the rheology of the molding material under heating must change in a way that meets the requirements of the molding equipment. Heating reduces viscosity but also triggers the polymerization reactions that increase viscosity. These two viscosity change forces must be balanced by skillful formulation.

The EMC contains significant mineral filler, typically silica, to reduce thermomechanical expansion. While standard EMCs have a filler level in the 70 to 75 percent range, advanced BGA materials have levels as high as 85 percent by weight. The filler size, shape, and quantity are critical for good performance during the molding process and for finished properties of the package. Once again, there must be skillful balancing of the formulation. Increasing the filler level decreases the coefficient of thermal expansion (CTE) but increases the viscosity. If the viscosity of the melted EMC is too high, flow will be impeded, and the molds will not fill properly. However, smoother and more spherical filler will reduce viscosity at the same loading. This allows molding compounds to be produce with good flow properties but low CTE values. High filler loadings are especially beneficial for area array packaging because stress is reduced and a more planar package is produced. Higher filler means that the total polymer content is lower, and this reduces shrinkage. Epoxies and most other polymers shrink on curing. Shrinking, as would be expected, tends to warp the packaging substrate, especially when it is a thin organic material. High filler loadings using smooth spherical filler offset shrinking to give the flattest package. Although some warpage is acceptable, a warped area array package is more difficult to assemble because not all the solder balls may make contact with the circuit board. Low flammability that is required for some packages is achieved by adding brominated hardeners and resins at levels up to 5 percent by weight of total composition.

Molding compounds are provided in a variety of solid forms designed to fit available equipment and mold sizes. Products for area array packaging, while similar to standard molding compounds, tend to be optimized for this type of package. One of the most important optimizations is low stress to minimize warpage that can be improved with biphenyl epoxy resins that have less shrinkage and the addition of elastomerics, such as silicone rubbers. The particles are extremely small and may be less than 0.25 μm in diameter. Specialty EMCs are also offered with such properties as low popcorning and higher thermal conductivity, which is achieved with different fillers with greater thermal conductivity that include aluminum nitride, boron nitride, silicon carbide, and even diamond dust.

The preforms are made by mixing the solid ingredients (dry mix), melting them, further mixing (melt mix) in the liquid phase, and then cooling back to a solid before pulverizing or granulizing into a powder. The powder may be stored at low temperatures as an intermediate raw material. The mix must be very thorough for best results. Powder is shaped into preforms by applying heat and pressure before shipping to customers. Preforms typically are stored at 5°C. Figure 7.1 shows typical EMC preforms.

7.2 TRANSFER-MOLDING EQUIPMENT

Molding machines vary from manual types to fully automatic. Manual molding has obvious limitations, such as high labor requirements, longer cycle times, and lower yield. Automatic gang pot molding is 100 percent automatic, requiring very little attention. Yield is virtual 100 percent, cycle time is short, and waste is very small. This process is used for high-value-added packages and high volumes. Figure 7.2 shows a simple molding machine diagram, whereas Figure 7.3 represents a production transfer-molding machine with a loader/unloader for automatic handling. Figure 7.4 is a diagram of a mold.

7.3 MOLDING PROCESSES

The chip carrier is first prepared, and this involves attaching a die and making the electrical connections, usually by wire bonding. Although single-package structures can be molded, it is more

FIGURE 7.1 EMC preforms.

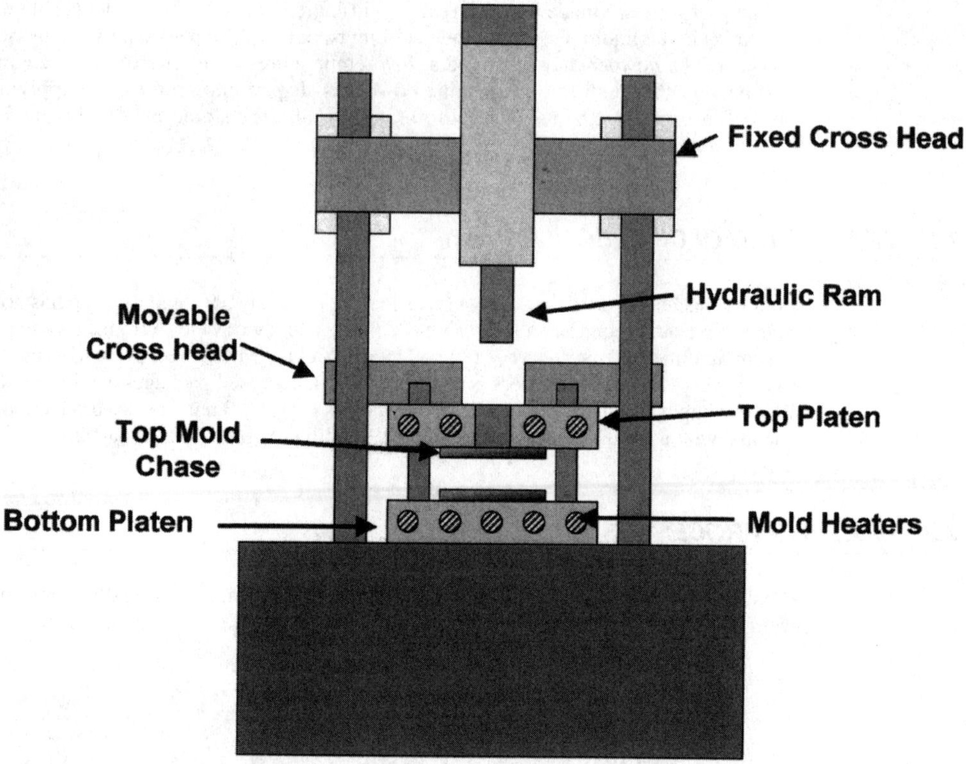

FIGURE 7.2 Transfer molding press diagram.

common to use strips or even full sheets of product for manufacturing efficiency. Strips loaded with dies are placed in the open mold, which is preheated. Solid preforms of EMC are placed into the mold. The mold is closed, and the cycle begins. The EMC melts, and a plunger pushes or transfers the liquid to the mold cavities. The material must be fluid enough so that the dies and fine-connection structure are surrounded by the encapsulant. Pressure and flow rate are regulated so that no damage occurs to the thin wire bonds (wire sweep damage) but so that the material fills the mold before thickening and hardening occur. The heat from the mold causes the polymerization reaction to accelerate to polymerize the EMC into a permanent solid that cannot be melted again. Cycle times range from 60 to 120 s for conventional molds and from 60 to 90 s for smaller multiplunger gang pot systems for high-value parts. The conventional molding process yields from 50 to 1000 packages per cycle, whereas the gang pot multiplunger method produces only 4 to 40 parts.

Equipment used in the molding process is continually changing, so when a problem is encountered, the solution is not always readily apparent. Any of the following variables can affect the encapsulation process:

- Mold temperature profile
- Transfer pressure
- Transfer ram speed
- Clamp pressure

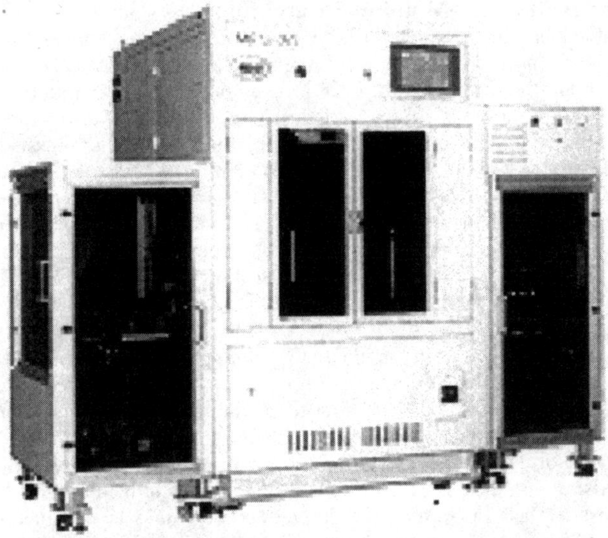

FIGURE 7.3 Transfer molding machine.

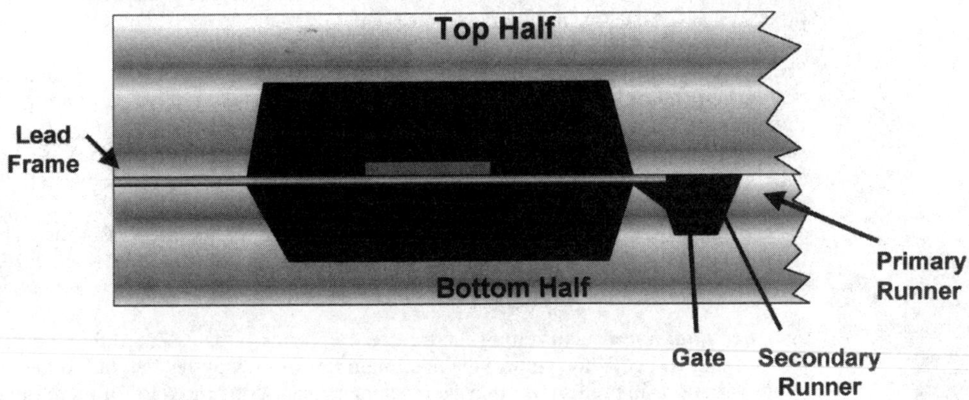

FIGURE 7.4 Mold.

- Preheat time and temperature
- Cure time and temperature
- Operator technique
- Material variations
- Material handling variations
- Tooling designs
- Preform quality

The next subsection will go into detail describing the parameters and the rationale for the changes.

7.3.1 Recommended Transfer-Molding Parameters

A daily log is invaluable in setting a standard for optimal molding parameters and will assist in problem solving. EMCs can be molded over a range of temperatures from 165 to 188°C (300 to 370°F). Molding parameters should be optimized for each process depending on mold design, device type, package type, and molding compound characteristics. Generally, low transfer pressures and slow transfer times are recommended for most encapsulation materials. Molding pressures for encapsulation materials typically are within the range of 500 to 1500 psia (pounds per square inch, actual).

The use of dielectically preheated preforms is important for minimizing cure cycles and improving moldability. Preheat temperatures of 80 to 100°C (175 to 210°F) are recommended. Uniform preheat temperature can best be achieved by using roller preheaters. Uniform-quality preforms, exhibiting a density in the range of 80 to 90 percent of molded density (0.8 to 0.9 preform bulk density), also improve preheat consistency.

Mold Temperature. It is very important that the temperature profile, across both the top and bottom of the mold, be within ±3°C (5°F). To obtain a temperature profile, measure the temperature of at least six points across the mold surface. This tight temperature differential is essential to maintain uniform molding conditions for all cavities of the mold.

For example, if one cavity block is 5.5°C (10°F) hotter than another, significantly different molding conditions will exist within the same cycle. This difference might result in insufficient fill in one cavity block and excessive flash in another. Similarly, good parts may be obtained from one side of the mold and parts exhibiting cosmetic defects may be produced on the other.

The temperature differential between top and bottom dies should not be more than 5.5°C (10°F). This will ensure uniform flow and cure and will reduce bowing of large parts.

The temperature profile should be measured periodically throughout the day and recorded on a process sheet. Adjustment of temperature should be made as needed.

Transfer Pressure. The actual transfer pressure, measured at the bottom of the transfer ram, should be checked and recorded routinely. A commonly used technique is to place a force gauge under the transfer ram to measure the total applied force directly. This will serve to verify the hydraulic pressure gauges on the press and will indicate that the pot and plunger frictions are constant. If low or erratic readings are obtained, this could indicate friction or improper fit between the transfer ram and the pot.

The molding press transfer pressure gauge should be read periodically throughout the day and recorded on a process sheet. These readings should be validated routinely (e.g., weekly) per the force gauge method described above.

Transfer Ram Speed. In setting the transfer ram speed, it is essential that the ram speed control, and not some other factor, control the maximum rate of flow of the material. If the ram is set too fast, slight variations in preheat or other parameters can cause the mold to fill more rapidly and result in voids in the molded parts. The transfer ram speed should be fast enough to ensure that all cavities fill before gelling takes place but not so fast as to cause air entrapment.

Transfer ram speed should be checked periodically throughout the day and recorded on the process sheet.

Preheat Time and Temperature. To obtain maximum preheat efficiency, the preforms should be heated to 80 to 100°C (175 to 210°F). Standard high-frequency preheating equipment is recommended. Preheaters operating at low frequencies require longer heating cycles and are not recommended for use with Plaskon epoxy molding compounds. When the preheater's electrodes are adjusted properly, Plaskon epoxy molding compound preforms should heat in approximately 20 to 30 s.

If heating is too rapid, a runaway exotherm could cause premature curing of the preforms. Conversely, if the preheat cycle is too long, the preforms may begin to cure slightly while still in the preheater, resulting in marginal cavity filling and/or damage to the device being encapsulated. The preform should be soft and pliable but not so soft as to stick to the operator's glove. The transfer time between completion of preheating and activation of the transfer ram press should be minimal.

The preheat time and temperature should be checked periodically throughout the day and recorded on the process sheet. If any changes in these parameters are made during operation, the changes also should be noted on the process sheet.

Operator Technique. The press operator can be a major source of variations in the molding process. The operator's performance should be monitored from time to time. The supervisor should check preheat time and temperature, transfer time, molding pressure on the hydraulic gauge, cure time, and other variables periodically to see that the proper molding conditions are being used.

Material Variation. Variations in the molding compound that would result in molding difficulties are rare. If a problem develops, all molding parameters should be checked. In most cases, simple adjustments in the transfer pressure or speed and/or cure time will solve the problem. If the difficulty persists, contact your local Plaskon sales representative.

Once a systematic pattern of in-process checks has been established, these historical data are invaluable in tracing molding problems. After adequate information is established, the frequency of calibration can be reduced for those parameters which are shown to remain constant. For additional information on solving molding problems, refer to the troubleshooting guide (Sec. 7.5).

Postcure of Molded Parts. Postcure is recommended to obtain optimal properties of molded parts. A typical postcure schedule of 4 h at 175°C (347°F) usually will maximize the physical properties of the encapsulant. However, each semiconductor device–encapsulant assembly should be evaluated at several different postcure schedules to determine the best postcure time and temperature for individual applications.

Plasticity or Flow. Materials for encapsulation are supplied with various melt viscosities, as indicated by the flow range listed on the individual data sheets. The most common method of measuring plasticity or flow characteristics of Plaskon epoxy molding compounds is ASTM-D3123-72, "Spiral Flow of Low Pressure Thermosetting Molding Compounds." For most semiconductor encapsulants, spiral flow is run at 177°C (350°F) with 1000 psi transfer pressure.

Calculation of Molding Pressure. Use the following formula to calculate actual hydraulic pressure (psia) being exerted on the molding compound:

Hydraulic pressure ÷ transfer cylinder diameter^2 × pressure = molding compound transfer plunger diameter^2 (psig) in the pot (psia)

Example

Diameter of transfer cylinder = 4 in

Diameter of transfer plunger = 1.75 in

Gauge pressure of hydraulic press = 150 psig

Therefore:

$$\frac{4^2}{1.75^2} \times 150 = 784 \text{ psia}$$

Testing Methods. The following test methods are used to determine the properties and specifications of epoxy molding compound encapsulants.

Property	Procedure
Physical	
Specific gravity	ASTM-D792*
Spiral flow and ram follower gel time	S-724-TO†
Flammability rating per UL-94-VO	PD-1007†
Ash content	U-815-TO†
Hydrolyzable chloride	U-814-TO
Water extract conductance and pH	PD-1026
Nitrogen content	U-819-TO
Thermal	
Expansion coefficients and glass transition temperatures	U-858-TO
Mechanical	
Flexural strength and modulus	PD-1025
Tensile strength	PD-1027
Hot hardness and mold appearance	PD-1008
Electrical	
Dielectric strength	ASTM-D149
Dielectric constant	ASTM-D150
Dissipation factor	ASTM-D150

*ASTM = American Society for Testing Materials.

†Cookson Semiconductor Packaging Materials Test Methods. Copies available on request.

Shipping, Handling, and Storage. To maximize retention of the spiral flow of the material, Plaskon epoxy molding compounds should be shipped by refrigerated truck or packed in dry ice if shipped by air. Temperature recorders, using 30-, 60-, or 90-day charts, are available, and it is suggested that one be included in each shipment to monitor temperature during transit. Plaskon can furnish disposable temperature recorders on request.

Shipping and storage should be at or below 5°C (40°F). A cold storage room can be designed so that new material enters one door and material to be used is withdrawn from a door at the opposite end. This ensures use of the oldest material first. Allow 2 ft^3 for every 50-lb pail of Plaskon epoxy molding compound.

Plaskon Electronic Materials' semiconductor-grade epoxy molding compounds are supplied in moisture-resistant plastic pails designed for overseas shipment. These heavy-duty, self-locking polyethylene pails provide excellent resistance to moisture ingress and physical damage during shipment or storage.

After removal from cold storage, the pails should be kept closed and allowed to equilibrate to ambient conditions before opening. Recommended time is 24 h. This ensures that the material will preheat uniformly and prevents moisture from condensing on the molding compound. Care should be exercised in opening pails to see that no moisture drops onto the molding compound.

Shelf Life. If the molding compound is to be kept for extended periods, it should be stored at 5°C (40°F) or below. Materials stored at room temperature (21°C, 70°F) will retain at least 70 percent of their original flow characteristics for no more than 2 weeks. Material always should be stored in tightly closed containers.

Mold Design. The mold design for epoxy molding compounds, with few exceptions, is similar to the mold design used for other thermosetting resins. Epoxy molding compounds have low-melt viscosities, which means that molds must be designed with close tolerances to minimize flashing around knockout pins, core pins, and the pot and transfer ram. Shallow, wide gates and small runner systems are recommended to minimize spurting and "jetting."

Adequate venting will ensure uniform density and minimize molding problems. Venting should be provided for each cavity.

When "breaking in" new molds, carnauba wax or some other release agent should be applied to the mold surface. If molds previously have used materials containing a different lubricating system (or other molding compounds, e.g., phenolic), general-purpose melamine cleaning compounds should be used to purge the molds before using epoxy molding systems.

Mold Cavities. Relatively low shrinkage is one of the characteristics of Plaskon epoxy molding compounds, and it must be taken into account when designing a mold. Mold cavities should be designed with adequate draft around all areas perpendicular to the parting line. This is especially true when the part to be ejected is very delicate, as are most semiconductor devices. Adequate draft minimizes the ejection pressure required to remove parts from molding cavities.

Gating. Gates approximately 1/8 in wide can be as thin as 0.010 to 0.015 in. Gate width is limited by lead-frame and package designs. Gates should be located in the thickest sections, if possible.

Venting. To ensure adequate venting of entrapped air, vents should be added to each cavity of the mold. For epoxy molding compounds, vents about 0.0005 to 0.001 in deep and about 1/4 in wide should prove satisfactory.

Molding Surface. The mold should be made of high-quality, heat-treated tool steel having a Rockwell hardness of at least 60C. The molding surface should be highly polished, with a finish of about 10 to 15 μin on all molding surfaces, including areas of the parting line that may be exposed to flash. All the molding surfaces should then be plated with a hard, highly polished chrome deposit.

Lead Slots. Lead slots should conform as closely as possible to the lead dimensions. Tolerances in this area should be approximately 0.0005 in or less. Minimizing flashing around the leads facilitates molding and ejection and alleviates lead deflashing after molding. Where practical, an interference fit should be employed.

Knockout System. All production molds should have some type of mechanical knockout system. If possible, the knockouts should be located in the cavities with the maximum surface area exposed for ejection from the mold. If more than one knockout pin is used per cavity, it is essential that the pins remove the part in a level manner. Knockout pins should be placed against solid, thick, or inflexible sections. Whenever possible, knockout pins also should be placed in runner and cull areas.

Mold Cleaning. To produce well-molded parts, molds must be maintained in top condition at all times. Specific mold cleaning compounds are described in separate brochures (26AI05 and 26AI06).

Preforming. EMCs can be formed into various size and weight preforms. This provides several advantages in handling the molding compound. Preforms

- Minimize entrapped air in transfer molding
- Reduce dusting
- Improve material handling
- Optimize dielectric preheating

If preforms are made in the molder's plant, the pressures listed in Table 7.2 are recommended to achieve the noted hardness.

Mold Maintenance. To produce well-molded parts, molds must be maintained in top condition at all times. Regularly scheduled in-process mold cleaning is the best way to maintain productivity. Molds should be cleaned periodically with either compression or transfer molding of a mold-cleaning compound. Mold-cleaning compounds designed specifically for this job are available.

In-Process Cleaning. EMCs are formulated to release easily from all areas of the mold. Before starting a new molding cycle, the upper and lower mold halves should be inspected for any loose

TABLE 7.2 Preform Pressures and Hardness

Degree of hardness	Pressure, t/in^2 (approx.)	Shore D hardness
Minimum	2.5	62–75
Average	5	69–80
Maximum	7.5	74–82

plastic. Brushing the surfaces of the mold with a brass or high-temperature-nylon brush should remove excess plastic. Stubborn pieces can be removed with a brass pick. After brushing, any remaining loose plastic can be blown off the mold surface with clean, dry compressed air. It is very important that no excess plastic be left on the mold from the previous shot, particularly in the vent area. Such residue in the vents can result in incomplete cavity fill on successive shots.

Caution: Never use any material other than brass, beryllium, copper, or high-temperature nylon (melting point $> 400°F$) for mold cleaning. Harder materials can scratch the mold surface, and the scratches will appear on the molded epoxy surface.

Note: Do not remove any plastic that accumulates in the grooves of the transfer plunger. These grooves are designed to fill with plastic and act as a seal to prevent plastic from flowing upward past the plunger.

Periodic Cleaning. The most effective cleaning method for removing mold stains is with the compression molding of Plaskon Melamine Mold Cleaning Compound. This melamine molding compound does not harm mold surfaces but will attract deposits causing surface staining. For those operations in which transfer-mold-cleaning compounds are preferred, materials such as Plaskon's Nikalet ECR are recommended.

As an alternative to compression cleaning, the industry offers a transfer-grade melamine cleaning compound for cleaning. Typical transfer-molding pressures for 14-lead DIP molds range from 3000 to 4500 psi. This procedure may require more shots to obtain as clean a mold as when a compression mold cleaning compound is used.

Whenever possible, it is better not to use lead frames in cleaning. When lead frames are not used, the compound will flow into vents and land areas, cleaning these parts of the mold as well as the cavities and runners. It should be noted, however, that some molds can be damaged if lead frames are not in place during molding. Whether using preforms or powder, the cleaning compound should be preheated. The transfer time should be approximately 15 s.

7.3.2 Area or Flood Molding

Flood molding is a newer process that provides several advantages over the conventional transfer-molding approach. A general-purpose single-cavity mold is used that can hold the entire sheet of substrate that is loaded with dies. This means that different-sized packages can be molded with one general-purpose tool. The EMC is melted as with the conventional process and injected into the single large cavity. The hot mold cures the epoxy. The mold opens, and the molded sheet is ejected. The dies must now be singulated into separate packages. The sheet containing hundreds of dies is sawn in a process similar to wafer sawing. The PBGAs appear similar to those molded in separate cavity tools except that the edges are flat. Although flood molding adds the extra step of sawing, the economics are very good because molding is faster and tooling cost and time are reduced to near zero. The flood-molding process has become increasingly popular because of faster molding and then elimination of custom tooling.

7.3.3 Injection Molding

While thermoset polymers such as epoxies have dominated molding, high-temperature thermoplastics can and are being used. Thermosets dominate because they can be formulated as low-melting solids that harden to nonmelting solids. However, since heating can always remelt thermoplastics, they are a greater challenge. The main problem is selecting a plastic with a high enough melting point for package performance but not so high that it cannot be melted for the injection-molding process. Several thermoplastics have been found acceptable and are used to mold some packages today.

Injection molding is a very common process that is used to make the thousands of plastic parts used everywhere from toys to automobile parts. The process has some similarities to transfer molding but is really quite different. Granular plastic pellets are fed into the molding machine through a hopper or even by means of a piping system. The pellets are melted and then moved toward the mold by a screw that mixes the hot mass while generating force. The mold is closed, and a nozzle is inserted into the mold that will inject hot molten plastic. The plastic is injected under high pressure while the two halves of the mold are clamped together. The liquid flows through runners into the cavity. Unlike the transfer mold, the injection mold is usually not heated, although it may be kept at some optimal temperature that allows the liquefied plastic to flow completely before it solidifies. Fluid typically is circulated through the mold or through heat exchangers in contact with the mold to remove heat from the hot plastic. The mold is opened, and the finished parts are ejected.

Injection molding is a simpler process in that the thermoplastic will not cure like the thermoset. This means that the plastic can remain hot for long periods of time. In fact, the barrel that contains the screw serves as a reservoir for material. Injection molding is a fast and efficient process that is well supported by a large machine and tool industry. However, because the finished product can still soften and distort at higher temperatures, the method has not gained widespread use. One concern is that poorly controlled solder ovens could damage the package.

7.3.4 Molding Underfill (MUF)

Materials. Some suppliers have started to offer molded underfill (MUF) materials that are based on special EMCs used in molds designed for this special process. The EMC must convert to a low-viscosity fluid on heating that will flow between the small gap between a flip chip (FC) and a packaging substrate. The final properties must meet those achieved by the common liquid underfills. In some cases, the mold is designed to allow product to flow under and over the chip so that it is an encapsulant as well as an underfill. The MUF material is similar to EMCs but must have filler particles that are small enough that they will fit in the small gap. The material must mold into a spacing down to a minimum of 2.5 mils (0.06 mm) and fill completely without voiding or other defects.

Tooling. The mold is designed to hold the FC assembled to the packaging substrate, but with gating that allows input to the edge of the chip. A vent may be required to allow air to escape from the chip gap. Some processes require a small vent hole in the substrate underneath the chip. One alternative is to use a vacuum.

The MUF Process. This is a one-step process that simultaneously underfills and overmolds to provide a robust, one-step total encapsulation solution, as shown in Fig. 7.5.

Optimal molding conditions for the MUF process have been identified as

Preform temperature: 75–80°C

Transfer time: 15 s

Material pressure: 350–400 psi

Mold temperature: 165°C

In-mold cure time: 105 s

MUF Performance. While transfer-molded underfill is relatively new, the process appears to run well, and good performance has been measured. Tests by Cookson Semiconductor Packaging Materials with large FCs did not show any problems with voiding or incomplete filling. Figures 7.6 and 7.7 show results after molding.

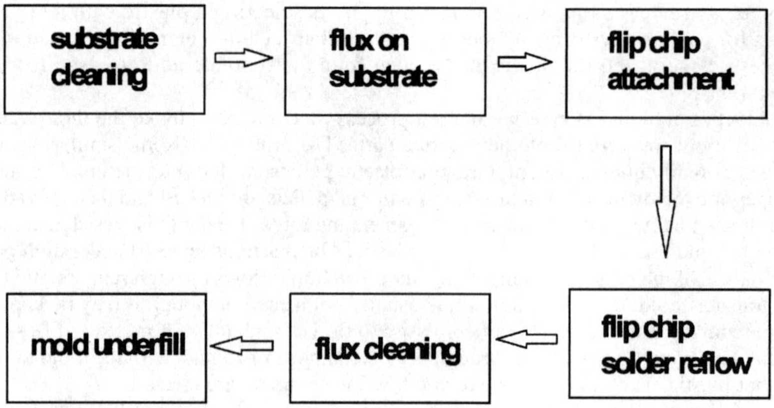

FIGURE 7.5 Molded underfill process steps.

Chips with 2.5-mil bumped spacing (400 × 600-mil bumped die, bump pitch 8 mils, bump diameter 5 mils, gap between bumps 3 mils) are used with bumps intact, and all the gaps and spacing were filled completely with no voids (microscope). Further tests showed that the gap filling limitation for the current formulation (153.405) was 2.1 mils.

The MUF approach is a fast process with no liquid-related issues (particle settling, nonhomogeneous). MUF is a traditional process (both material and equipment) that is a more automated, simple, robust process and less expensive relative to present underfill and overmold processes and materials. The method is ideal for FC BGAs, where strips of loaded substrate can be fed into a transfer-molding machine. Higher filler levels can be used than possible with liquids because force is used to push the material under the chip.

7.3.5 Preventing Molding Process Problems

The first step in preventing any molding problem is the use of proper molding techniques and maintenance. Sometimes, however, even with recommended molding techniques and mold maintenance, problems still arise. This section presents recommendations for molding techniques and mold maintenance, as well as troubleshooting for the prevention and solution of these molding problems.

Important Transfer-Molding Parameters

Effect of molding parameters

- Mold temperature
- Transfer pressure
- Transfer ram speed
- Preheat time and temperature
- Operator technique
- Material variation

Postcure of molded parts

Plasticity or flow

Pressure

Testing methods to ensure quality

- Physical
- Thermal

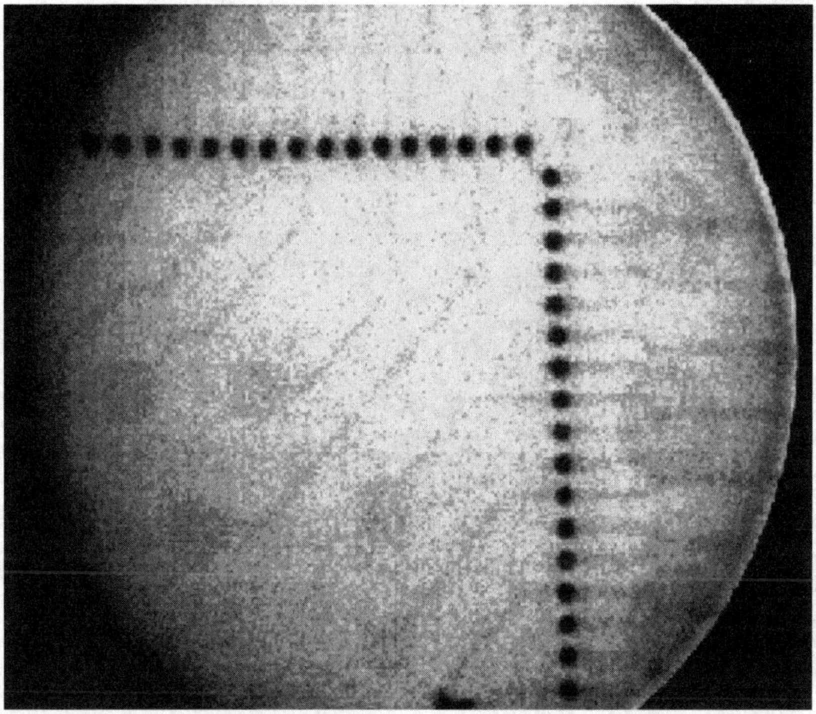

FIGURE 7.6 X-ray of molded underfill part.

- Mechanical
- Electrical

Shipping and handling of material

Storage of material

Shelf life, age of material

Mold design

- Mold cavities
- Grating
- Venting
- Molding surface
- Lead slots
- Knockout system
- Mold cleaning
- Preforming

7.4 TROUBLESHOOTING GUIDE

The following troubleshooting section is presented to address commonly encountered molding problems with a **cause** → remedy format.

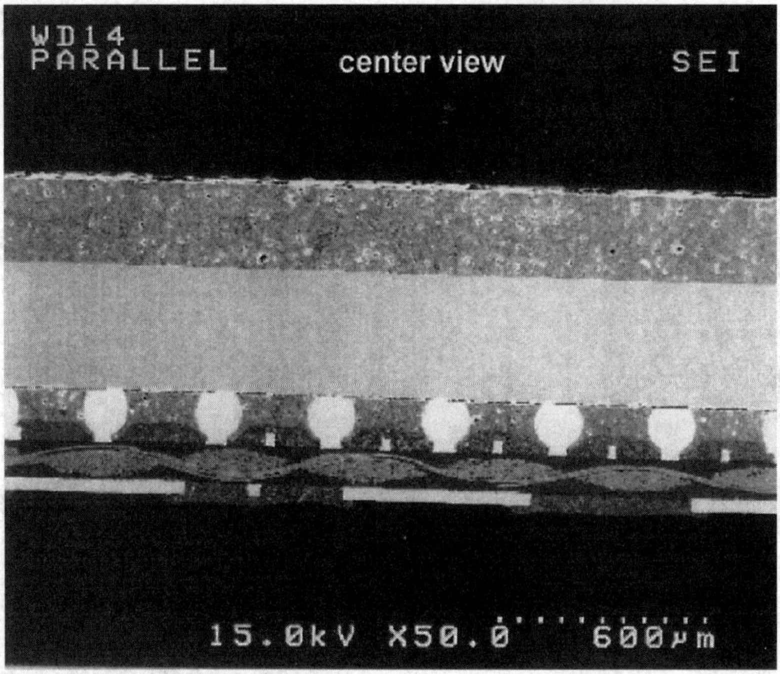

FIGURE 7.7 SEM of MUF CSP.

7.4.1 Incomplete Fill and Knit Lines

Short or Overcharge. Check cull thickness to ensure proper charge weight. Check preform weights. An excessive charge weight can cause filling problems when the material does not attain optimal viscosities during transfer.

Incorrect Transfer Pressure. Check pressure with force gauge to verify pressure gauge, and check wear on plunger and pot. May need to adjust pressure or align plunger.

Improper Mold Temperature. Verify that heating elements are operating effectively and that mold is heating evenly by obtaining accurate pyrometer or temperature salt readings over the mold surface. Low temperatures can cause higher viscosities, whereas high temperatures can cause premature gelling. Temperatures that are either too low or too high can result in incomplete filling.

Incorrect Transfer Speed. Adjust the transfer speed. A fast transfer speed can cause turbulent flow, resulting in air entrapment. A transfer speed that is too low can permit material to gel before the mold is filled completely.

Improper Preheat Temperature. Closely control preheat temperature to avoid filling problems associated with high melt viscosities (low temperature) or premature gelling (high temperature).

Poor Venting. Check to verify that the vents are clean. Adequate venting is required to ensure that air in the runners and cavities is not entrapped. Vents normally are located opposite the gate area, and dimensions should be optimized for each mold. Some molds may require multiple vents.

Short Flow Material. Optimize preheat, mold temperature, and/or transfer pressure. Consult vendor to optimize spiral flow, cure rate, and melt viscosity. Check in-house storage conditions.

7.4.2 Voids and Blisters

Fast Transfer Time. Reduce the transfer time to allow the trapped air to escape from the runners and cavities through the vents.

Incorrect Preheat Temperature. Ensure that preheat temperature is optimized and that inserts, if used, are not cold.

Low Melt Viscosity. Consult vendor to obtain higher melt viscosity or longer gel time materials. Check transfer speed control of press. Material with a very low melt viscosity or a high plunger speed setting can cause turbulent filling and entrapped air in the runners and cavities.

Preform Density. Ensure that preforms are sufficiently dense (at 0.83 of molded density) by measuring dimensions and weight and calculating actual preform density. Consult vendor and/or preformer if required.

Molding Temperature. You may need to adjust molding temperature if blisters appear in molded parts. Excessive temperature and/or insufficient cure can cause blisters.

Poor Venting. Adequate venting is required to prevent air entrapment inside runners and cavities. Clean vents if needed.

High Volatiles. Proper handling procedures must be used to ensure that material is not contaminated with high moisture levels. To prevent condensation, material coming from cold storage should remain at room temperature for 24 h prior to opening pails.

7.4.3 Flashing and Resin Bleed

Transfer Pressure and Time. Reduce the time in which the transfer pressure is acting on the cavities. Lower the transfer pressure.

Clamp Pressure. Check the clamp pressure, and increase it if needed.

Worn Molds. Replace worn parts.

Out-of-Spec Inserts. Ensure that insert tolerances are maintained properly.

External Release Agent. Minimize the use of external release agents that can act as a plasticizer for the molding compound.

Low-Viscosity Molding Compound. Materials with very low viscosities can generate more flash and bleed. Higher-viscosity products may be necessary. Reducing the preheat temperature and reducing the transfer pressure can help to reduce flash.

7.4.4 Mold Staining and High Molding Temperature

Lower molding temperature but do *not* sacrifice cure. Use increased curing time if needed. Implement periodic thorough cleaning.

External Release Agent. Minimize or avoid using an external release agent. Use molding compound and longer cure time (5 min) for conditioning of the mold after cleaning.

Molding Compound. Consult vendor for products with better staining resistance.

7.4.5 Sticking

Low Mold Temperature. Check temperature profile of the mold.

High Mold Temperature. Reduce mold temperature. Check for hot spots. Excessive mold temperature accelerates degradation of mold-release agents that can form a buildup on the mold surface and interfere with release.

Short Cycle Time. Increase cure time to ensure effective cure.

Changing Materials. When changing molding compounds, thorough cleaning and conditioning of the mold are required.

Mold Design. Materials vary in shrinkage, and the draft angles may need to be modified.

Molding Compound. Inadequate release agent levels in the molding compound can cause sticking.

Moisture Contamination. Anhydride materials will stick if contaminated with moisture. Check handling procedures and aging time.

7.5 TOOLING

The tooling consists of a precision metal mold that can have many hundreds of mold cavities. Tooling can be expensive, especially for high-volume molds with several hundred cavities. The lead time can extend for several months. However, newer methods, such as flood or area molding, have reduced tooling substantially. Flood molding involves using a generic mold that is loaded with package carriers. Molding compound is injected into the mold to encapsulate all the packages as if they were a single giant package. The entire array is removed, and sawing or singulating forms the individual packages. This adds a new process, but savings in mold costs and a fast process offset the cost of mechanical singulation.

7.6 PRODUCTIVITY AND COST

7.6.1 Volume Dependency

Transfer molding is intended to be a relatively high-volume process. While capital equipment cost and tooling cost are considered high, transfer molding provides substantial savings over liquid dispensing once the volumes reach several million units per year. When volumes are low and many different package sizes are being made, liquid dispensing is the better process.

7.6.2 Advantages of Area Molding

As mentioned earlier, area (flood) molding can offer very good economics for reducing tooling costs. A few generic molds can be used for a variety of different-sized packages. This also reduces time to market because there is no wait for tooling that can take several months for large molds with individual cavities. The disadvantage of area molding is that mechanical singulation is required. The array, which is made up of hundreds of packages joined by the hardened molding compound, must be singulated mechanically. Sawing, the same process used to singulate semiconductor chips, is used here. Laser, water jet cutting, and other processes have not been able to displace the mechanical saw method. The sawing process leaves a slightly rougher edge than molding, thus reducing the cosmetic appearance slightly. Overall, the area molding process has been successful and probably will become the standard package method for BGAs and CSPs.

7.6.3 Liquid versus Solid Encapsulants

Transfer molding provides the best cosmetics and the most mechanical precision for packaging. While liquid dispensing is more convenient for small runs, cosmetics are not nearly as good for die-up packaging, where a retaining dam must be applied. EMCs also provide slightly higher performance than liquids that must flow by gravity. The simple rule for choice is to use liquid encapsulants for shorter runs or where molding is not suitable, such as with die-down cavity packages. Transfer molding is the de facto standard for high volume, and its use will increase because of the ability to achieve good, low-volume economics with area molding.

7.7 RELIABILITY

Transfer molding is a tried and true method for producing packages. The method has been used successfully for many decades and should be considered the de facto standard for producing plastic packages. Transfer molding was used on the earliest PBGA packages starting with Motorola's OMPACK. In fact, the name stands for *o*ver*m*olded *pack*age. The molding materials, equipment, tools, and processes have been optimized, and substantial test data support the reliability of this method. While reliability of the molded BGA is high, the very nature of the package design creates some minor problems that are addressed by proper handling and storage. Since the bottom of the substrate is often not encapsulated, the packaging barrier properties are influenced by the substrate properties. The most significant problem has been moisture entry through the substrate. Absorbed moisture can produce the undesirable phenomenon known as *popcorning*. During heating from the solder assembly process, instantaneous conversion of this moisture to steam can delaminate the molding compound from the substrate or even cause encapsulant fracture.

While the development of very low moisture absorption materials eventually may solve this problem, the process solution is to keep the package dry or to dry it just prior to assembly. Not all packages are affected by popcorning, but those which are known to be sensitive can be stored in a dry box. The alternative is to predry in an oven at 125°C for 2 h.

7.8 SUMMARY AND CONCLUSIONS

Molding is the most common method for encapsulating area array packages such as the PBGA and some of the CSPs. Transfer molding, the same process used for nearly all packages, is the dominant one here, although needle dispensing of liquid encapsulants is also used. The large, thin area array package has offered more challenges than the common perimeter-lead packages, however. The very concept of area array requires a relatively flat package for successful board assembly. The construction of the PBGA package, where molding compound typically is applied to just the top, sets up the mechanism for warpage if materials are not balanced. Mold compound shrinkage, while less important for fully encapsulated designs, will promote warpage with top or overmolded BGAs.

The encapsulant producers have responded to area array packages with low-stress, reduced-warpage EMCs. Today's EMCs designed for area array produce flat packages. The warpage has been so reduced that full sheets of die-loaded substrate can be overmolded using the newer flood-molding process. The ability to mold hundreds of packages in a single cycle with one mold has substantially boosted productivity. Area or flood molding is now a popular process for high-volume BGAs and μBGAs.

This segment of the packaging industry is one of the most innovative, since area array packages are leading-edge enabling technology. The previously mentioned flood-molding process is one example of innovation, but there are others. Molded underfill for FC packages is the newest and will grow in importance as increasing lead counts move us from wire bonding to FC in-package (FCIP). The response of the materials, equipment, and tooling companies will continue as part of the activities that make molding and area array packaging a rapidly growing business.

7.9 FUTURE EXPECTATIONS

We should expect molding compounds to continue to improve, especially as regards a reduction in moisture absorption. Improvements in low-warp materials also will continue because this feature will be more in demand as packages grow larger because of larger dies and multiple components. This feature is also of value for the flood-molding process, although it is less critical because the sheets of packages are sawn.

Depending on legislative activities, bromine may be eliminated as a flame retardant, although no negative environmental case has been made for the elimination. However, should the use of bromine be curtailed, the industry could move to nonepoxy chemistry.

Improvements in molding processes will continue. The most notable new processes are flood molding and molded underfill. We can expect larger arrays to be molded in a cycle with parts per shot exceeding 1000. Singulating will improve, and perhaps methods other than sawing will be adopted. Work will continue aimed at simplifying the MUF process and incorporating flood molding to boost productivity.

INDEX

AATC (*see Air to air thermal cycling*)
Activity-based costing (ABC), of FC assembly process, 7–9
Adhesives, die-attach, 205–207
Air to air thermal cycling (AATC), 28–31, 33–35, 37–39
Alpha particles, 139
Area array packages:
  FC design, 104, 107–110
  rework of (*see Rework*)
Area molding, 250–251
Assembly, IC, and flip chip processing, 5–7
Assembly process(es):
  for FC test vehicle assemblies, 17, 24, 25
  of FCOBs, 2
  for flip chips (*see Flip chip processing*)
Auger pumps (needle dispensing), 220, 222
Autoclave, 138

Balancing of profile settings, 156–158
Ball grid arrays (BGAs), 171, 172
  basic design of, 210
  flip chips vs., 172–173
  QFPs vs., 172–174
  reliability testing of, 174–193
    CBGA failure mechanisms, 180–182
    CBGA thermal cycling results, 185–188
    failure criteria in, 178, 180, 181
    PBGA failure mechanisms, 181, 183–184
    PBGA thermal cycling results, 184–185
    shock and thermal cycling synergism effects on, 188–192
    thermal cycling in, 177–180
  rework of (*see Rework*)
  (*See also specific types, e.g.: Plastic ball grid arrays*)
BGAs (*see Ball grid arrays*)
Board thickness analysis, 15, 16
Board-level testing, 133–140
Boeing, 180
Bond strength (die attach), 203
Bonding temperature/curing, 204
Brushing, mechanical, 122–124, 150

C4 (*see Controlled collapse chip connection*)
Camalot Xyflex, 216, 217
Capillary flow underfills:
  cost analysis of processes using, 7, 8
  curing of, 120, 121
  dispensing of, 116–121
  in flip chip assembly process, 98, 101, 109–124
  flip chip substrate design with, 109–124
Capture voids, 63–65
Ceramic ball grid array (CBGA) packages:
  failure mechanisms of, 180–182
  shock and thermal cycling synergism effects on, 188–192
  thermal cycling testing results for, 185–188
Chemical resistance (die-attach adhesives), 205
Chip floating modeling and analysis, 23, 79–86
Chip parameters main effects analysis, 16
Chip scale packages (CSPs), 172
  encapsulation of, 210
  μBGA (*see μBGA CSP*)
  rework of (*see Rework*)
Chip-and-wire DCA, 171
Chip-on-board (COB), 171, 172, 209–210, 225
Cleaning:
  in FC assembly, 111, 113, 114
  mechanical brushing, 122–124
  of molds, 243, 244
  pad, 149, 152
C-M fatigue-life prediction (*see Coffin-Manson fatigue-life prediction*)
Coating(s):
  experimental wafer, 129–130
  extrusion, 129
  spin coating on wafer, 199
  wafer-level FC processing, 47–54
    desired characteristics for, 47–49
    methods/materials, 49–54
  (*See also Surface finishes*)
COB (*see Chip-on-board*)
Coefficient of thermal expansion (CTE):
  in FCOB assemblies, 2
  matching, 98, 204
  of silicon die, 173

Coffin-Manson (C-M) fatigue-life prediction, 186–187
Collapse time (chips), 91–94
Collapsible solder balls, 144, 145
Component-level testing, 131–134
Components:
  placement of, during rework, 154–155
  prebaking, 148
  removal of, 148–151
    placement, 154–155
    reballing, 161–164
    reflow profiling, 155, 156
Compression flow underfill processing (flip chips), 65–77
  placement force characterization of, 65–67
  placement model and analysis, 77–79
  simulation analysis of, 67–77
    discussion, 73–76
    finite element mesh and boundary conditions, 70
    finite element simulations, 67–69
    parametric studies, 68
    results, 70–75
Compression voids, 62–64
Computers and peripherals, for needle dispensing systems, 223
Continuity testing (FCs), 26
Controlled collapse chip connection (C4), 97
Cooling stage (reflow), 155, 156
Copper, for flip chip substrate design, 107–109
Cost analysis (FC processing), 7–12
  conventional processing using capillary flow underfills, 7, 8
  low-cost, high-throughput processing using no-flow underfills, 7–10
  throughput, chip size, and number of chips per panel in, 10–12
Cost(s), and equipment selection, 232–233
CpK, 231–232
Cracking, 135, 137, 138
Creep (solder), 133, 134, 144, 145
C-SAM analysis, 26, 39, 41, 42, 45
CSPs (*see Chip scale packages*)
CTE (*see Coefficient of thermal expansion*)
Curing, of capillary flow underfill, 120, 121

Damming compounds, 226
Die attach, 195, 197–208
  bond strength, 203
  film application, 199–202
  fluid jetting, 199
  materials used in, 197
  needle dispensing, 198
  properties of, 203–206
  reliability of, 206
  rework of, 207
  screen printing, 199
  spin coating on wafer, 199
  stenciling, 198, 199
Die-down configuration, 210
Die-up configuration, 210
Dip transfer process, 154
Direct flip chip, 171
Dispensing, 150
  of capillary flow underfill, 116–121
  die attach, 198
  for liquid encapsulation, 217, 224–230
  needle, 211–233
    applications/processes for, 225
    conventional positioning assemblies for, 214–215
    of damming compounds, 226
    die attach, 198
    dispensing head for, 217
    equipment used in, 213, 215
    fluid-handling system for, 221
    for liquid encapsulation, 213–217
    material constraints in, 224
    multigantry/modular-style systems for, 215–217
    platform/design/construction for, 214
    pumps for, 218–222
    and substrates, 224
    system controls for, 222–224
    of underfill, 226–230
    with vacuum assist, 230–231
    z positioning (vertical control), 215
Dow Corning, 231

Electrical conductivity, for die grounding, 204
Electromigration, 139, 140
Electronic packages:
  current trends in, 171
  miniaturization in, 171, 172, 174
  (*See also specific headings*)
EMCs (*see Epoxy molding compounds*)
Encapsulants:
  basic chemistry of, 211
  functionality of, 211
    handling and storage of, 211
    solid vs., 251
Encapsulation, 209–210
  liquid, 209–233
    dispensing head for, 217
    dispensing process in, 224–230
    materials used in, 210–212
    needle dispensing, 213–215
    overmolding, 212–213
    performance of, 231–233
    platform design/construction for, 214
    positioning mechanisms for, 214–217
    pumps for, 218–222
    system controls for, 222–224
    with vacuum assist, 230–231
Environmental testing, 133–140
Epoxies, handling and storage of, 211

Epoxy molding compounds (EMCs), 235–236, 238
mold design for, 242–243
Extrusion coating, 129

Fabrication, of printed wiring boards, 102–106
Failure analysis, of FCs, 26
Failure mechanisms:
of attached dies, 207
of CBGAs, 180–182
in PBGAs, 181, 183–184
Fatigue, 133
FCIPs (*see Flip-chips-in-package*)
FCOB (*see Flip-chip-on-board*)
FCOC (*see Flip-chip-on-ceramic*)
FCs (*see Flip chips*)
Fill/filters:
definition of, 211
incomplete, 248
Filleting (underfill), 118–120
Fine pitch (FP) surface-mount components, 172
Flashing (molding), 249
Flex testing, 137
Flip chip processing, 1–96, 98–102, 109–130
with capillary flow underfills, 98, 101
chip floating modeling and analysis, 79–86
compression flow placement model and analysis, 77–79
compression flow underfill processing, 65–77
placement force characterization of, 65–67
simulation analysis of, 67–77
concept of, 1–7
conventional processing using capillary flow underfills, 7, 8
cost analysis of, 7–12
conventional processing using capillary flow underfills, 7, 8
low-cost, high-throughput processing using no-flow underfills, 7–10
throughput, chip size, and number of chips per panel in, 10–12
design guidelines for, 64, 65
with fluxing underfills, 98, 101
high-throughput, using no-flow underfills, 4–5, 11–42
failure mode analysis of, 35–46
placement voids, 22, 23
process yield analysis of, 14–23
reliability analysis of, 23–35
and IC assembly requirements, 5–7
interconnection yield analysis during reflow process, 85–95
bumps in contact due to placement, 93, 95
chip collapse time, 91
interconnection density effect, 93–94
processing windows for complete chip collapse, 92–94
solder force, 88–90
Flip chip processing, interconnection yield analysis during reflow process (*Cont.*):
steady state forces, 86–89
underfill force, 90–93
underfill viscosity effect, 91–93
limits of conventional, 2
low cost:
based on fast-flow, snap cure underfills, 2–4
based on wafer-applied underfills, 5
high-throughput processing using no-flow underfills, 7–10
next-generation technology for, 2–5
potential void formation in, 62–65
with wafer-applied underfills, 101, 102
wafer-level, 42, 46–62
analysis of underfill coated chip placement accuracy, 59, 61
automated vision analysis of underfill coated chip, 55–60
chip imaging and placement, 53, 55
coating methods/materials, 49–54
desired coating characteristics for, 47–49
Flip chips (FCs), 97–140, 142
advantages of, 97–98
assembly processes for (*see Flip chip processing*)
BGAs vs., 172–173
liquid dispensing of underfill for, 210
reliability of, 130–140
component-level testing, 131–134
environmental and board-level testing, 133–140
substrates in, 101–109
with capillary flow underfill, 109–124
copper and solder mask requirements, 107–109
with fluxing underfills, 123–127
layout, 101–110
with wafer-applied underfills, 128–131
wire bonding vs., 1
Flip-chip-on-board (FCOB), 2
Flip-chip-on-ceramic (FCOC), 97
Flip-chips-in-package (FCIPs), 1
Flood molding, 250–251
Flow, underfill, 119–122
Fluid jetting:
die attach, 199
pumps for, 220, 221
Flux(es):
in FC assembly, 110–114
solder paste vs., for rework, 146–148
Fluxing underfills:
in flip chip assembly process, 98, 101
flip chip substrate design with, 123–127
FP surface-mount components (*see Fine pitch surface-mount components*)

Gates (molding), 243
Glass transition temperature, 204

Handling, of molding compounds, 242
HAST (highly accelerated stress test), 138
HDI substrates (*see High-density interconnect substrates*)
Heat aging, 206
High-density interconnect (HDI) substrates, 104, 107–109
Highly accelerated stress test (HAST), 138
High-temperature storage, 138–139
High-throughput FC processing, using no-flow underfills, 4–5, 11–42
  failure mode analysis of, 35–46
  placement voids, 22, 23
  process yield analysis of, 14–23
  reliability analysis of, 23–35
Humidity, and die-attach adhesives, 206
Hydrostatic squeeze film theory, 90

IBM, FC/SLT technology introduced by, 1, 97
ICs (*see Integrated circuits*)
Injection molding, 244–245
Inspection:
  of pads, 169
  X-ray, 164–168
Integrated circuit (IC) chips:
  compliant packaging for (*see μBGA (microBGA) CSP*)
  smaller feature sizes on, 171
Integrated circuits (ICs), assembly requirements for, 5–7
Interconnections:
  collapse time and density of, 93, 94
  density effect, interconnection, 93–94
  IC-to-package/substrate, 5–7
  and rework, 144–145
  yield analysis during FC processing reflow, 85–95
    bumps in contact due to placement, 93, 95
    chip collapse time, 91
    interconnection density effect, 93–94
    processing windows for complete chip collapse, 92–94
    solder force, 88–90
    steady state forces, 86–89
    underfill force, 90–93
    underfill viscosity effect, 91–93

JEDEC (*see Joint Electron Device Engineering Council*)
Jet Propulsion Laboratory (JPL), 174, 178, 179
Joint Electron Device Engineering Council (JEDEC), Solid State Technology Association, 132, 133
Joints, (*see Interconnections*)
JPL (*see Jet Propulsion Laboratory*)
JPL BGA consortium, 174

Knit lines (molding), 248
Knockout system (molds), 243

Liquid encapsulants, 211, 251
Liquid encapsulation, 209–233
  dispensing head for, 217
  dispensing process in, 224–230
  materials used in, 210–212
  needle dispensing of, 213–215
  overmolding, 212–213
  performance of, 231–233
  platform design/construction for, 214
  positioning mechanisms for, 214–217
  pumps for, 218–222
  system controls for, 222–224
  with vacuum assist, 230–231
Liquid-to-liquid thermal shock (LLTS), 26, 30, 33–35, 37–39

Material deposition techniques (rework), 150, 153–154
Materials, used in die attach, 197
μBGA (microBGA) CSP, 230, 231
  encapsulant application in, 210
Mechanical brushing, 122–124, 150
MicroBGA CSP (*see μBGA CSP*)
Miniaturization, 171, 174
Modeling:
  chip floating modeling and analysis, 79–86
  compression flow placement model and analysis, 77–79
Modulus, in die-attach adhesives, 205
Moisture:
  as medium for chemical reactions, 211, 212
  preconditioning, moisture, 132, 133
Moisture sensitivity, test methods/definition of, 132, 133
Molding, 235–252
  equipment used for, 237–239
  flashing and resin bleed in, 249
  incomplete fill and knit lines in, 248
  materials used for, 235–237
  mold staining and high molding temperature in, 249–250
  processes of, 237–248
    flood molding, 244
    injection molding, 244–245
    molding underfill (MUF), 245–246
    preventing problems in, 246–247
    recommended transfer-molding parameters, 240–244
  productivity and cost of, 250–251
  reliability of, 251
  sticking in, 250
  tooling for, 250
  troubleshooting guide for, 247–250
  voids and blisters in, 249
Motorola, 110

Natural frequencies, 189, 190
Needle dispensing, 211, 213
  applications/processes for, 225

Needle dispensing (*Cont.*):
conventional positioning assemblies for, 214–215
of damming compounds, 226
die attach, 198
dispensing head for, 217
equipment used in, 213, 215
fluid-handling system for, 221
for liquid encapsulation, 214–217
material constraints in, 224
multigantry/modular-style systems for, 215–217
platform/design/construction for, 214
pumps for, 218–222
and substrates, 224
system controls for, 222–224
of underfill, 226–230
with vacuum assist, 230–231
z positioning (vertical control), 215
No-clean fluxes, 111, 113, 114
No-flow underfills, curing of, 85, 86
Nonhermetic packages, 211
Nonmelt solder balls/columns, 144–146
No-residue fluxes, 113

Open circuits, 166, 167
Organic solderability preservative (OSP) finishes, 107, 108
OSP finishes (*see Organic solderability preservative finishes*)
Outgassing, 62, 204, 212
Overmolding, 212–213
Oxygen, 211

Packages (*see Electronic packages*)
Packaging, trends in, 209–210
Pads:
cleaning of, 149, 152
inspection of, 169
wafer-applied underfill printing of, 128, 129
Paste, solder (*see Solder paste*)
PBGAs (*see Plastic ball grid arrays*)
PCBs (printed circuit boards) (*see Printed wiring boards*)
Performance, of liquid encapsulation, 231–233
Pick and place, for FC dies, 113–115
Piston-type pumps (needle dispensing), 218–220
Pitch, in BGAs vs. QFPs, 173, 174, 176
Placement:
accuracy of, underfill coated chips, 59, 61
bumps in contract due to, 93, 95
of coated chips, 53, 55
component, during rework, 154–155
of FC dies, 113–115, 125–126, 130
force/velocity interaction of, 16, 18, 20
split-field prism system, 155
void analysis, 16–17, 20, 22
Plaskon Electronic Materials, 242
Plaskon epoxy molding compounds, 242, 243
Plaskon Melamine Mold Cleaning Compound, 244
Plastic ball grid arrays (PBGAs):
failure mechanisms in, 181, 183–184
thermal cycling testing results in, 184–185
Plastic leaded chip carrier, 97
Platform design/construction, 214
Pneumatic pumps (needle dispensing), 218
POLYFLOW software, 68
Polymers, 195–196
thermoplastics, 196
rework of, 207
thermosets, 196
rework of, 207
Popcorning, 132, 148, 212, 251
Positioning, for liquid encapsulation, 214–217
Power cycling, 137
Prebaking, 148
Preforms:
EMC, 237, 243, 244
reballing using, 161–163
Preheat stage (reflow), 155
Printed circuit boards (PCBs) (*see Printed wiring boards*)
Printed wiring boards (PWBs):
fabrication of, 102–106
prebaking, 148
rework of, 167, 169
warpage of, 159
Printing, 150, 164
screen printing:
die attach, 199
for flux application, 111
with wafer-applied underfills, 128
of solder paste, 163
stenciling:
die attach, 198, 199
for flux application, 111
with wafer-applied underfills, 128
Process variable main effects analysis, 18
Process(es) (*see Assembly process[es]; Flip chip processing*)
Productivity, and cost of molding, 250–251
Pumps:
for liquid encapsulation, 218–222
needle dispensing, 218–222
PWBs (*see Printed wiring boards*)

QFPs (*see Quad flat packs*)
Quad flat packs (QFPs), 97, 172
BGAs vs., 172–174
rework of, 143

Ramp rate, 135, 177
Rapid ramp reflow profile, 16, 18
Reballing, 161–164
using fixtures, 163–164
using solder preforms, 161–163
Rebonding (thermoplastics), 207

Reflow:
with capillary flow underfills, 114, 116, 117
with fluxing underfills, 126–127
interconnection yield analysis during FC reflow, 85–95
bumps in contact due to placement, 93, 95
chip collapse time, 91
interconnection density effect, 93–94
processing windows for complete chip collapse, 92–94
solder force, 88–90
steady state forces, 86–89
underfill force, 90–93
underfill viscosity effect, 91–93
nozzle clearance during, 167
profiling, 16, 155–158
rapid ramp, 16, 18
simultaneous, 159–161
solder ball, 163, 164
step, 16
volcano, 126, 127
with wafer-applied underfills, 130, 131
Reliability:
of die attach, 206
of flip chips, 130–140
component-level testing, 131–134
environmental and board-level testing, 133–140
FCOCs, 6
of molding, 251
testing, of BGAs, 174–193
CBGA failure mechanisms, 180–182
CBGA thermal cycling results, 185–188
PBGA failure mechanisms, 181, 183–184
PBGA thermal cycling results, 184–185
shock and thermal cycling synergism effects on, 188–192
thermal cycling in, 177–180
underfill to increase, 98
Removability, 207
Resin bleed (molding), 249
Rework, 143–169
component placement, 154–155
component reballing considerations, 161–164
component reflow profiling, 155, 156
component removal, 148–151
of die attachments, 207
and interconnection method, 144–145
material deposition techniques, 150, 153–154
pad cleaning and preparation methods, 149, 152
prebaking components and PWBs, 148
process control in, 143
process yield problems, 156–161
balancing profile settings, 156–158
caused by warping, 158–159
simultaneous reflow issues, 159–161
PWB design guidelines for, 167, 169
solder paste vs. flux for, 146–148
Rework (*Cont.*):
x-ray inspection in, 164–168
Rotary pumps (needle dispensing), 220, 221

SAM (*see Scanning acoustic microscopy*)
Sandia National Laboratories, 98, 121
Scanning acoustic microscopy (SAM), 120, 121
Screen printing:
die attach, 199
for flux application, 111
with wafer-applied underfills, 128
Shelf life, of molding compounds, 242
Shipping, of molding compounds, 242
Shock, BGAs and effects of, 188–192
Shorts, solder, 159, 166, 168
Simulation, compression flow underfill processing, 67–77
Simultaneous reflow, 159–161
Single-row perimeter FC design, 102–105
SLC technology (*see Surface laminar circuitry technology*)
Slump/slumping, 111
Small outline (SO) packages, 97
SMT (*see Surface mount technology*)
Snap cure underfills, FC processing, 2–4
SO packages (*see Small outline packages*)
Soak stage (reflow), 155
Soft errors, 139
Software:
for needle dispensing systems, 224
POLYFLOW, 68
Solder balls:
application of loose, 163, 164
collapsible, 144, 145
nonmelt, 144–146
Solder columns, nonmelt, 144–146
Solder masks, 107–109
Solder paste:
flux vs., for rework, 146–148
printing process for, 163
typical process window, 156
Soldering, solder-joints, X-ray inspection of, 165, 166
SolderQuik, 162
Solder(s):
paste, solder (*see Solder paste*)
removal methods for, 149
and surface energy forces, 88–90
warping and shorts in, 159
X-ray inspection of shorts in, 166, 168
Solubility, of die-attach adhesives, 205
Speedline Camalot Xyflex, 216, 217
Speedline System 3900, 231
Speedline Technologies, 231
Spin coating on wafer, 199
Split-field prism system (placement), 155
Squeeze flow process (underfill), 125
Steady state forces, 86–89

Stenciling:
  die attach, 198, 199
  for flux applications, 111
  with wafer-applied underfills, 128
Stencils, clearances for, 167
Step reflow profile, 16
Sticking (molding), 250
Storage, of molding compounds, 242
Substrates:
  flip chips, 101–109
    with capillary flow underfill, 109–124
    copper and solder mask requirements, 107–109
    with fluxing underfills, 123–127
    layout, 101–110
    with wafer-applied underfills, 128–131
  high-density interconnect, 104, 107–109
  mounting on organic vs. ceramic, 171
  and needle dispensing process, 224
Surface finishes, for flip chips, 107, 108
Surface laminar circuitry (SLC) technology, 104, 107
Surface mount technology (SMT), 97
System controls (liquid encapsulation), 222–224

TAB (*see Tape automated bonding*)
Tape automated bonding (TAB), 171
  costs of, 6
Tape carrier package (TCP) DCA, 171
Tape carrier packages (TCPs) (*See also Tape automated bonding*)
TCP DCA (tape carrier package DCA), 171
TCPs (*see Tape carrier packages*)
Temperature humidity (TH) testing, 32–34
Temperature-humidity-bias aging, 137, 138
Temperature(s):
  and die-attach adhesives, 206
  glass transition temperature, 204
  and underfill dispensing, 227–230
Tessera, Inc, 230, 231
Tessera-compliant chips (TCCs) (*see μBGA CSP*)
Testing:
  of die-attach adhesives, 206
  of FC assembly, 133–140
    alpha particles, 139
    autoclave and HAST, 138
    component-level, 131–134
    electromigration, 139, 140
    environmental/board-level, 133–140
    flex testing, 137
    high-temperature storage, 138–139
    power cycling, 137
    temperature-humidity-bias aging, 137, 138
    thermal cycling/thermal shock, 133–138
Thermal conductivity, for die-generated heat, 205
Thermal cycling:
  in BGA testing, 177–180
  BGAs and effects of, 188–192
  CBGA damage induced by, 189, 190
  CBGA testing results, 185–188
Thermal cycling (*Cont.*):
  failure of FC solder joints in, 134–137
  PBGA testing results, 184–185
  thermal shock vs., 135
  vibration and, 191–192
Thermal shock:
  failure of FC solder joints in, 134–137
  thermal cycling vs., 135
Thermal stability, of polymers/polyimides, 205
Thermally reworkable underfills, 122, 123
Thermocycling, 206
Thermoplastics, 196
  rework of, 207
Thermosets, 196
  rework of, 207
Transfer molding, 212–213, 240–244
  machines, 238, 239
  parameters for, 246, 247
  recommended parameters for, 240–244
Two-row perimeter FC design, 103–106

Ultra fine pitch (UFP) surface-mount components, 172
Underfill(s):
  application of:
    capillary flow, 110–114
    fluxing, 125
    wafer-applied, 128–130
  capillary flow, 7, 8, 98, 101
    cost analysis of processes using, 7, 8
    curing of, 120, 121
    die presentation, 109, 110
    dispensing of, 116–121
    in flip chip assembly process, 98, 101, 109–124
    flip chip substrate design with, 109–124
    flux and flux application, 110–114
    pick and place, 113–115
    reflow, 114, 116, 117
    rework, 121–124
    substrate dehydration, 116, 118
  in flip chip assembly process, 98–102
    capillary flow underfills, 98, 101, 109–124
    fluxing underfills, 98, 101, 123–127
    wafer-applied underfills, 101, 102, 128–131
  fluxing, 98, 101
    application of, 125
    die placement, 125–126
    reflow, 126–127
    rework, 127
    substrate dehydration, 124, 125
  force exerted by, 90–93
  needle dispensing of, 226–230
    curing of, 85, 86
  no-flow, high-throughput processing using, 4–5, 7–10
  potential void formation in, 62–65
  reflow:
    capillary flow, 114, 116, 117

Underfill(s), reflow (*Cont.*):
fluxing, 126–127
wafer-applied, 130, 131
rework of:
capillary flow, 121–124
fluxing, 127
wafer-applied, 130
snap cure, 2–4
substrate dehydration in
capillary flow, 116, 118
fluxing, 124, 125
viscosity effect, 91, 92
wafer-applied, 5, 101, 102
application of, 128–130
placement, 130
reflow, 130, 131
rework, 130

Vacuum assist (liquid encapsulation), 230–231
Venting (molding), 243
Vibration, 189–192
Viscosity, 117
Vision analysis (FC processing), 55–60
Voids:
formation of (FC processing), 62–65
in molding, 249
placement, analysis of, 16–17, 20, 22
X-ray inspection of, 165–167

Wafer-applied underfills, 5, 101, 102
application of, 128–130
in flip chip assembly process, 101, 102
flip chip substrate design with, 128–131
placement, 130
Wafer-applied underfills (*Cont.*):
reflow, 130, 131
rework, 130
Wafer-level FC processing, 2, 5, 42, 46–62
analysis of underfill coated chip placement accuracy, 59, 61
automated vision analysis of underfill coated chip, 55–60
chip imaging and placement, 53, 55
coating methods/materials, 49–54
desired coating characteristics for, 47–49
Warping, 158–159
WB (*see Wire bonding)*
Weibull parameters, 173, 185
Wetting:
of flux, 111–113
wafer-level underfills, 49
Wire bonding (WB), flip chips vs., 1

X-ray inspection, 113, 114
of molded underfill, 247
in rework, 164–168

Yield, rework problems with, 156–161
Yield analysis (FC processing reflow), 85–95
bumps in contact due to placement, 93, 95
chip collapse time, 91
interconnection density effect, 93–94
processing windows for complete chip collapse, 92–94
solder force, 88–90
steady state forces, 86–89
underfill force, 90–93
underfill viscosity effect, 91–93

ABOUT THE EDITOR

KEN GILLEO, PH.D., is a well-known author/editor/columnist in the fields of circuitry, MEMS, and microelectronics packaging, and CEO of ET-Trends LLC, a consultancy and intellectual property company focused on emerging technology. His *Area Array Packaging Handbook: Manufacturing and Assembly*, also published by McGraw-Hill, is the premier professional guide in the field. Dr. Gilleo also has edited the *Handbook of Flexible Circuits* and *Polymer Thick Film*, and has written over 400 articles on packaging, circuitry, MEMS, and materials. He holds 35 U.S. patents, currently serves on the Board of Directors of the Surface Mount Technology Association, and is active in IEEE, IMAPS, and other professional organizations. He lives in Warwick, Rhode Island.